D1462220

CYCLES OF SOIL

CYCLES OF SOIL
CARBON, NITROGEN, PHOSPHORUS, SULFUR, MICRONUTRIENTS

F. J. STEVENSON
Department of Agronomy
University of Illinois

A WILEY-INTERSCIENCE PUBLICATION

JOHN WILEY & SONS

New York   Chichester   Brisbane   Toronto   Singapore

Library of Congress Cataloging in Publication Data:

Stevenson, F. J.
    Cycles of soil.

    "A Wiley-Interscience publication."
    Includes bibliographies and index.
    1. Soil biochemistry. 2. Biogeochemical
cycles. 3. Soil ecology. I. Title.
S592.7.S73 1985   631.4'1   85-12042
ISBN 0-471-82218-3

Printed in the United States of America

10 9 8 7 6 5 4 3 2

TO MY CHILDREN:
MARK, DIANA, FRANK

PREFACE

Living organisms and the transformations they bring about have a profound effect on the ability of soils to provide food and fiber for an expanding world population. Of paramount importance is the cycling of carbon (C), nitrogen (N), phosphorus (P), sulfur (S), and the micronutrient cations (B, Cu, Fe, Mn, Mo, and Zn). An understanding of the various cycles and their interactions is essential for the intelligent use of soil as a medium for plant growth and for the rational use of natural and synthetic fertilizers. Since the biochemical cycles constitute the lifeline of planet earth, information about their functioning in terrestrial soils has a direct application to other ecosystems.

A common feature of the biochemical cycles is that they are strongly mediated by microorganisms. As plant residues undergo decay in soil, N, P, S, and the micronutrient cations appear in plant-available forms; portions are assimilated into microbial tissue (the soil biomass). With time, N, P, and S are stabilized by conversion into recalcitrant organic forms and by interactions with inorganic substances. Nutrient losses occur through erosion, leaching, and, for N and S, volatilization. Although separate pathways are operative for specific cycles (e.g., denitrification in the case of N; chelation for micronutrients), most cycles have some processes in common, notably mineralization and immobilization of N, P, and S. Interest in the various cycles is at an all time high, as attested by the large number of books that have been published in recent years. Unfortunately, most volumes represent collections of symposium papers and are of interest only to the research specialist.

A unique feature of this book is that it is devoted exclusively to the biochemical cycles in soils. Many facets of the C, N, P, S, and micronutrient cycles are covered, including fluxes with other ecosystems, biochemical pathways and chemical transformations, gains, losses, and recycling, plant availability, and environmental pollution. Considerable latitude was taken in developing the various chapters, with the result that both panoramic and specific views have been presented for each major cycle. Because of the voluminous literature that has accumulated, an exhaustive coverage of the literature was not possible, and both selection of references and depth of documentation has been arbitrary. One function of the book is to present a critical account of our general knowledge of each cycle. In doing so, some of the research discussed may be misrepresented inadvertently. I apologize for any omissions of important work.

In a sense, the C cycle acts as a driving force for the other cycles. Accordingly, this topic is covered first (Chapters 1 to 3), followed in order by the N cycle (Chapters 4 to 6), the P cycle (Chapter 7), the S cycle (Chapter 8), and the micronutrient cycle (Chapter 9).

The book is intended for the soil scientist but will also interest researchers in microbiology, forestry, horticulture, organic geochemistry, environmental science, and a host of other disciplines concerned with global cycling of C, N, P, S, and micronutrients. As a companion volume to *Humus Chemistry: Genesis, Composition, Reactions* (Wiley-Interscience, 1982), the book is well-suited as a graduate or advanced undergraduate reference text for courses in soil microbiology and soil biochemistry.

Appreciation is expressed to graduate students and staff members at the University of Illinois for encouragement in the book's preparation. Special thanks go to Cynthia L. Stubbs, Frank J. Sikora, and George F. Vance for editorial assistance.

F. J. STEVENSON

Urbana, Illinois
July 1985

CONTENTS

CYCLES OF SOIL

ONE

THE CARBON CYCLE

The decomposition of plant and animal remains in soil constitutes a basic biological process in that carbon (C) is recirculated to the atmosphere as carbon dioxide (CO_2), nitrogen (N) is made available as ammonium (NH_4^+) and nitrate (NO_3^-), and other associated elements (phosphorus, sulfur, and various micronutrients) appear in forms required by higher plants.[1-4] In the process, part of the nutrients is assimilated by microorganisms and incorporated into microbial tissues (the soil biomass). The conversion of C, N, P, and S to mineral forms is called *mineralization;* the reverse process is called *immobilization.*

The photosynthetic process is of primary importance in providing raw material for humus synthesis. By using solar energy plus nutrients derived from the soil, higher plants produce lignin, cellulose, protein, and other organic substances that make up their structures. During decomposition by microorganisms, some of the C is released to the atmosphere as CO_2 and the remainder becomes part of the soil organic matter. Part of the native humus is mineralized concurrently. Results of studies with ^{14}C-labeled plant residues applied to temperate zone soils have shown that approximately one-third of the applied C remains behind in the soil after the first year. Mean residence times of the residual C range from weeks to years for the biomass components to centuries for the humic material.

In this volume the C, N, P, and S cycles will be examined separately. It should be noted, however, that the four cycles do not operate independently but are linked by biological processes common to each, notably mineralization and immobilization. A schematic illustration of interrelationships of C, N, P, and S cycling within the soil–plant system is shown in Fig. 1.1.

The decay process is emphasized in this chapter and results obtained using ^{14}C-labeled substrates will be stressed. A brief account is also given on the extraction, fractionation, and general chemical properties of soil organic matter. The disposal of organic wastes in soil is discussed separately in Chapter 3. Supplemental information regarding the C cycle can be found in several excellent reviews.[1-8]

GLOBAL ASPECTS

The global C cycle is a two-compartment system that involves a cycle on land and one on the sea. The original source of the C was the fundamental

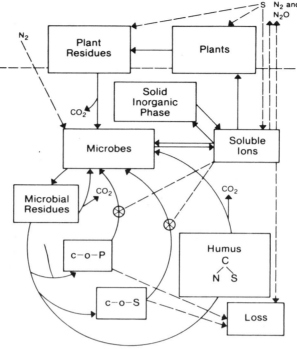

Fig. 1.1

Schematic illustration of C, N, S, and P cycling within the soil–plant system. (From McGill and Cole,[4] reproduced by permission of Elsevier Scientific Publishing Co.)

rocks, from which CO_2 was evolved by outgassing during periods of intense volcanic activity. A secondary source was the primitive atmosphere, which may have contained appreciable amounts of methane gas.

Sizes of the reservoirs and fluxes for the portion of the soil C cycle that interfaces with the global cycle are illustrated in Fig. 1.2; the distribution of C in some of the main compartments is given in Table 1.1. A unique feature of the overall C cycle is that much of the C that has passed through the biosphere is now preserved in sediments.

The amount of C contained in the organic matter of terrestrial soils (about $2,500 \times 10^{12}$ kg) is three to four times the C content of the atmosphere (700 $\times 10^{12}$ kg) and five to six times the land biomass (480×10^{12} kg). In contrast, the amount contained in sediments (6×10^{19} kg) is many orders higher. The annual input of C to the soil is about 110×10^{12} kg/year, or about 15% of the atmospheric CO_2. However, an equivalent amount of C is returned to the atmosphere by decay. A small quantity of C finds its way to the sea, a portion of which is deposited in sediments. The CO_2 content of the atmosphere has been increasing at a steady rate over the past century because of burning of fossil fuel and deforestation, as discussed later in Chapter 3.

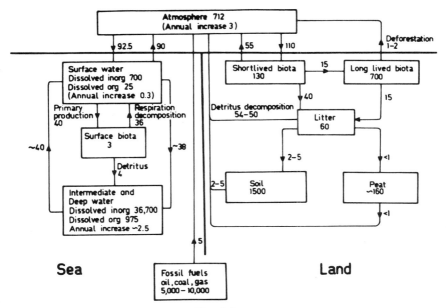

Fig. 1.2

Sizes of reservoirs (10^{12} kg) and fluxes (10^{12} kg/year) for the part of the C cycle that is in a state of comparatively rapid turnover (<1000 years). Other estimates for C in the main compartments are given in Table 1.1. (From B. Bolin, *The Carbon Cycle in SCOPE 21, the Major Biogeochemical Cycles and Their Interactions,* edited by B. Bolin and R. B. Cook, John Wiley & Sons Ltd., 1983.)

Table 1.1

Distribution of C in Some of the Main Compartments Involved in the Soil C Cycle[a]

Compartment	Amount of C, $\times$ 10^{12} kg
Atmosphere	700
Soil organic matter	2,500
Marine humus	3,000
Land life forms	480
Ocean life forms	50
Dissolved carbonate–bicarbonate in oceans	3,840
Coal and petroleum	1×10^4
Sediments	6×10^7

[a] From Delwiche.[3] For other estimates see Fig. 1.2 and references 5–8.

Table 1.2

Organic C in Soil Associations of North America, South America, and Other Areas[a,b]

Association	Area, $\times 10^5$ km	Total Organic C, $\times 10^{12}$ kg	C Content, kg/m^2
North America			
Gelic Histosols	9.1	182	200
Dystric Histosols	4.2	84	200
Podzols	31.5	66	21
Cambisols	9.4	56	60
Haplic Krastanozems	32.0	51	16
All others	116.8	226	
	203.0	665	
South America			
Ferralsols	89.8	108	12
Dystric Histosols	3.8	76	200
Cambisol–Andisols	6.2	35	57
Cambisols	4.2	26	60
All others	78.0	56	
	182	301	
Asia, Africa, Europe, Oceania			
Histosols	26	520	200
Cambisols	47	280	60
Podzols	130	270	21
Krastanozems	131	210	16
Chernozems	51	200	40
Ferralsols	141	170	12
Cambisol–Vertisols	28	110	40
All others	282	220	40
	836	1,980	

[a] As reported by Bohn.[5]

[b] The C values represent amounts to a soil depth of 1 m.

Geographical Distribution of Soil C

Data for the distribution of C in soil associations of North America, South America, and four other continents (combined soil depth of 1 m) are given in Table 1.2. Histosols (organic soils), because of their very high C content, are major contributors to the total soil C, amounting to 762×10^{12} kg, or slightly more than one-third of the total soil C. The contribution from mineral soils of South America is relatively low, primarily because most of them are tropical soils. Factors affecting the C content of terrestrial soils are discussed at some length in Chapter 2.

Isotopes of C

Carbon has two important stable isotopes (^{12}C and ^{13}C) and one radioactive isotope (^{14}C). The radioactive species is formed in the earth's upper atmosphere through the bombardment of N with cosmic neutrons. The resulting ^{14}C finds its way into the biosphere through photosynthesis, and hence into the pedosphere and hydrosphere. As will be shown in Chapter 2, ^{14}C provides a means of determining the absolute ages of buried soils and the mean residence time of organic matter in modern soils.

The ratio $^{12}C/^{13}C$ varies widely in nature, depending upon the source and origin of the C and the conditions under which the C was assimilated. Natural variations in the relative abundance of the stable isotopes have been examined almost exclusively from a geochemical standpoint and will not be considered further.

THE DECAY PROCESS

Because of the complex nature of organic remains, numerous species of microorganisms are involved in the decay process.[9,10] Some of the C is converted to CO_2, some is incorporated into microbial tissue, and some is converted into stable humus. Native humus is mineralized concurrently; thus although prodigious quantities of organic residues may be returned to the soil following harvest, decay does not necessarily lead to an increase in organic matter content. Maintenance of steady-state levels of organic matter, as mentioned in Chapter 2, requires a return of CO_2 to the atmosphere through soil respiration in an amount equal to gains by photosynthesis. Thus since the content of organic matter in any given soil remains essentially constant from year to year, an estimate of the quantity of C returned to the atmosphere as CO_2 can be obtained from the amount of residues added, which for any given soil will vary from one season to another, depending on cropping sequence, climate, and other environmental factors.

Several stages can be delineated in the decay of organic remains in soil. Earthworms and other soil animals play an important role in reducing the size of fresh plant material; further transformations are carried out by enzymes produced by microorganisms. The initial phase of microbial attack is characterized by rapid loss of readily decomposable organic substances. Depending on the nature of the soil microflora and quantity of synthesized microbial cells, the amount of substrate C utilized for cell synthesis will vary from 10 to 70%. Molds and spore-forming bacteria are especially active in consuming proteins, starches, and cellulose. Byproducts include NH_3, hydrogen sulfide, CO_2, organic acids, and other incompletely oxidized substances. In subsequent phases, organic intermediates and newly formed biomass tissues are attacked by a wide variety of microorganisms, with production of new biomass and further C loss as CO_2. The final stage of

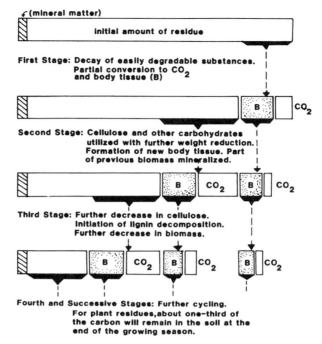

Fig. 1.3

Stages in the microbial decomposition of organic residues in soil. The letter B refers to the biomass. Each stage involves partial conversion of C to CO_2 and synthesis of microbial tissue. (Adapted from a drawing by Burges.[9])

decomposition is characterized by gradual decomposition of the more resistant plant parts, such as lignin, for which the actinomycetes and fungi play a major role.

Stages in the microbial decomposition of organic residues are depicted in Fig. 1.3. Each stage involves partial conversion of the C to CO_2 and synthesis of body tissues.[9] For many soils of the temperate region, about one-third of the C will remain in the soil at the end of the growing season (discussed later). The quantity of C in microbial tissue is usually insignificant in comparison to the amount tied up as stable humus and, for this reason, is often ignored in considering the C balance of the soil.

As suggested, different constituents of organic residues decompose at different rates. Simple sugars, amino acids, most proteins, and certain polysaccharides decompose very quickly and can be completely utilized in a matter of hours or days. Large macromolecules, which make up the bulk of plant residues, must first be broken down into simpler units before they can be utilized further for energy and cell synthesis. This is accomplished by enzymes excreted by microorganisms. Polysaccharides, such as cellulose, are cleaved into oligosaccharides (Greek *oligo,* meaning "few"), and finally simple sugars; proteins are broken down into peptides and amino acids.

Polysaccharides → oligosaccharides → simple sugars

Proteins → peptides → amino acids

The decomposition of lignin by microscopic fungi leads to the release of the phenylpropane (C_6–C_3) units of which lignin is composed, as follows:

Coniferyl alcohol   p-Hydroxycinnamyl   Sinapyl
                          alcohol   alcohol

The C_6–C_3 units, in turn, are broken down into simple phenolic compounds.[11,12] A pathway for the breakdown of lignin is shown in Fig. 1.4.

As noted earlier (see also Fig. 1.3), utilization of residue components and their breakdown products (sugars, amino acids, phenolic compounds, and others) leads to the production of microbial cells, which are further degraded following death of the organisms. As was the case with plant and animal residues, different components of microbial cells decompose at different rates. Proteins and simple biochemical compounds (e.g., sugars, amino acids) decompose quickly, while cell walls and certain microbial melanins resist decomposition.[13,14]

The Soil Biomass

A vast array of microorganisms live in the soil (Table 1.3). Estimates given for bacteria, actinomycetes, fungi, and yeasts are based on plate counts and must be regarded as highly conservative. Numbers of bacteria are particularly high, often reaching 500 million per gram and higher. Actinomycetes are the next most numerous, usually of the order of 1 to 20 million per gram. Numbers of fungi vary widely and can reach up to a million per gram. Algae numbers are somewhat lower, with up to 500,000 per gram being typical. Protozoa are also found in numbers up to 500,000 per gram; nematode numbers are 50 or more per gram. Despite their high numbers, bacteria probably occupy no more than 0.4 to 1.0% of the soil volume not occupied by solids or liquids and considered to be theoretically available spaces for growth.[15]

Considerable attention has been shown in recent years to the estimation

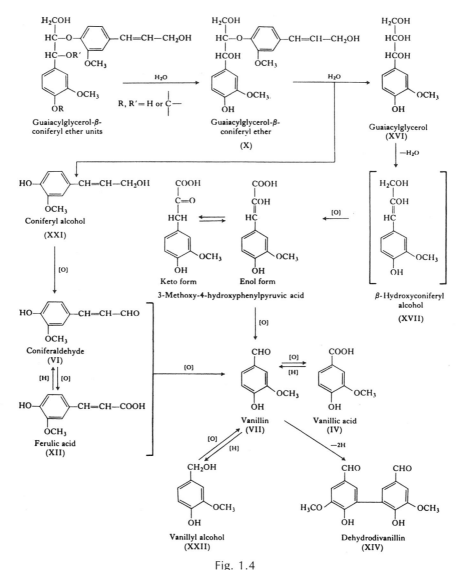

Fig. 1.4

Pathway for the breakdown of lignin by microorganisms. (From Schubert,[12] reproduced by permission of Academic Press, Inc.)

Table 1.3
Approximate Numbers of Organisms Commonly
Found in Soils[a]

Organism[b]	Estimated Numbers/g
Bacteria	3,000,000–500,000,000
Actinomycetes	1,000,000–20,000,000
Fungi	5,000–900,000
Yeasts	1,000–100,000
Algae	1,000–500,000
Protozoa	1,000–500,000
Nematodes	50–200

[a] From Martin and Focht.[10]
[b] Numbers for bacteria, actinomycetes, fungi, and yeasts are based on plate counts. Other organisms found in soil include viruses, arthropods, and earthworms.

of the microbial biomass in soil. Interest in this subject has arisen from the realization that microorganisms play a major role in the retention and release of energy and nutrients. Reviews concerning the microbial biomass in soil include those of Jenkinson and Ladd[16] and Paul and Voroney.[17]

Several approaches have been used in attempts to estimate the quantity of biomass tissues in soils and sediments, including counts for total microorganisms and measurements for metabolic activities. Included with the latter are O_2 uptake, CO_2 production, enzymatic activity, heat production, and adenosine triphosphate (ATP) content. Babiuk and Paul[18] concluded that microscopic count gives the most accurate estimate. However, this method is rather laborious; furthermore, dead microbial cells are included with the biomass estimates. The plate count method would provide a better estimate of metabolizing cells, provided a nonselective growth medium could be found.

Biomass estimates based on microbial counts require a knowledge of cell weights, which can be obtained from microscopic observations for cell dimensions. Thus

$$\text{Biomass} = \text{number of cells} \times \text{volume} \times \text{density}$$

From data for numbers and shapes of bacteria, along with assumed densities for microbial cells (~ 1.5 g/cm^3), it can be demonstrated that the total live weight of bacteria in soil is of the order of 0.15 to 1.5 g per kg of soil, or about 336 kg/ha to plow depth (300 lb/acre) at the lower value. Using a similar approach, the weight of fungal tissue in a kilogram of soil can be shown to range from 0.24 to 2.4 g, or 540 kg/ha (480 lb) to plow depth of soil at the lower value. Thus although numbers of bacteria greatly exceed those for fungi in most soils, the total mass of the latter is larger due to the larger cell weights. Numbers of microorganisms in soil, and thus the biomass, are subject to daily and seasonal variations, as dictated by such factors as moisture, temperature, pH, and energy supply.

Measurements of O_2 uptake and CO_2 production often have been used as a measure of microbial activity, but the values obtained cannot easily be interpreted in terms of the microbial biomass. A popular approach currently being used to estimate the magnitude of the biomass is to measure the increase (flush) in CO_2 upon incubation of fumigated soil.[19-23] The flush in CO_2 evolution is believed to result from the decomposition of microbial cells killed by fumigation, thereby providing an indication of the size of the biomass. Assumptions underlying the fumigation method include the following:

1 Carbon in dead organisms is more rapidly mineralized than that in living organisms.
2 All microorganisms are killed during fumigation.
3 Death of organisms in the unfumigated (control) soil is negligible as compared to the fumigated soil.
4 The only effect of fumigation is to kill the biomass.
5 The fraction of dead biomass C mineralized over a given time period is similar for all soils.

Attempts have been made to use chemical methods for biomass estimates, such as muramic acid for bacteria and chitin for fungi.[24,25] Assays for nucleic acid (DNA and RNA)[26,27] and adenosine triphosphate (ATP) have also been suggested.[28-29] A universal chemical technique for estimating the active microbial biomass should meet the following requirements.[30]

1 The component to be measured must be present in all living cells and absent from all dead cells.
2 It must be absent from all nonliving material.
3 It must exist in fairly uniform concentration in all cells, regardless of environmental stresses.
4 The technique employed must be relatively rapid and easy to use for large-scale programs involving numerous samples.

Many scientists feel that these conditions are met reasonably well with the bioluminescence technique for determining adenosine triphosphate (ATP). The method, which is based on measurement of the light emitted from the interaction of ATP with luciferin, luciferase, and atmospheric O_2, is described in greater detail in Chapter 7.

Table 1.4
Partial List of Enzymes in Soil

Enzyme or Enzyme System	Reaction Catalyzed
α- and β-Amylase	Hydrolysis of 1,4-glucosidic bonds
Arylsulfatases	$R\text{-}SO_3^- + H_2O \rightarrow ROH + H^+ + SO_4^{2-}$
Asparaginase	Asparagine $+ H_2O \rightarrow$ aspartate $+ NH_3$
Cellulase	Hydrolysis of β-1,4-glucan bonds
Deamidase	Carboxylic acid amide $+ H_2O \rightarrow$ carboxylic acid $+$ NH_3
Dehydrogenases	$XH_2 +$ acceptor $\rightarrow X +$ acceptor·H_2
α- and β-Galactosidase	Galactoside $+ H_2O \rightarrow ROH +$ galactose
α- and β-Glucosidase	Glucoside $+ H_2O \rightarrow ROH +$ glucose
Lichenase	Hydrolysis of β-1,3-cellotriose bonds
Lipase	Triglyceride $+ 3H_2O \rightarrow$ glycerol $+$ 3 fatty acids
Nucleotidases	Dephosphorylation of nucleotides
Phenoloxidases	Diphenol $+ \frac{1}{2}O_2 \rightarrow$ quinone $+ H_2O$
Phosphatase	Phosphate ester $+ H_2O \rightarrow ROH + PO_4^{3-}$
Phytases	Inositol hexaphosphate $+ 6H_2O \rightarrow$ inositol $+$ $6PO_4^{3-}$
Proteases	Proteins $\rightarrow$ peptides and amino acids
Pyrophosphatase	Pyrophosphate $+ H_2O \rightarrow 2PO_4^{3-}$
Urease	Urea $\rightarrow 2NH_3 + CO_2$

Soil Enzymes

All biochemical action is dependent on, or related to, the presence of enzymes. Because of the complex and variable substrates that serve as energy sources for microorganisms, the soil would be expected to contain a wide array of enzymes.[28-30] Each soil may have its own characteristic pattern of specific enzymes, as stated by Kuprevich and Shcherbakova:[29]

> The differences in level of enzymatic activity are caused primarily by the fact that every soil type, depending on its origin and developmental conditions, is distinct from every other in its content of organic matter, in the composition and activity of living organisms inhabiting it, and consequently, in the intensity of biological processes. Obviously, it is probable that each type of soil has its own inherent level of enzymatic activity.

A partial list of enzymes found in soils is given in Table 1.4. It should be noted that enzymes in soil are determined not by direct analysis but indirectly through their ability to transform a given organic substrate into known products. For example, urease activity is estimated from the conversion of added urea to NH_3 and CO_2 (urea $\rightarrow$ $2NH_3$ + CO_2). Methods for the determination of soil enzymes have been described by Tabatabai.[30]

Certain enzymes are ubiquitous; that is, they are found in virtually all soils. Typical examples are urease, catalase, phosphatase, and peptide hydrolysases of various types. Other enzymes may be produced in soil only under special circumstances. Dehydrogenase activity appears to be related to the quantity of decomposable organic matter and is closely related to the soil biomass.

Like microbial numbers in soil, enzymatic activity is not static but fluctuates with biotic and abiotic conditions. Such factors as moisture, temperature, aeration, soil structure, organic matter content, seasonal variations, and soil treatment all have an influence on the presence and abundance of enzymes. A marked change occurs in the kinds and amounts of soil enzymes when virgin lands are first placed under cultivation. The possible use of enzymatic data as a means of characterizing soil for forensic purposes has been investigated by Thornton et al.[31]

Soil enzymes are not readily recovered from soil; only rarely have they been isolated in a pure form. McLaren[32] concluded that most enzymes are complexed, or immobilized, with clay minerals and/or humus.

Attempts have been made to correlate enzyme activities to the overall metabolic activity of the soil, but they have been only partially successful. Best results have been obtained with dehydrogenase.[33,34] This enzyme is involved in the transfer of hydrogen to molecular O_2, such as through the nicotinamide dinucleotide (NAD)–flavin adenine dinucleotide (FAD)–Fe^{3+} cytochrome system.

BIOCHEMISTRY OF HUMUS FORMATION

Soil organic matter, or humus, consists of two major types of compounds, unhumified substances and the humified remains of plant and animal tissues. The former is represented by the well-known classes of organic compounds, including carbohydrates, fats, waxes, and proteins. The soil, being a graveyard for the bodies of microorganisms, can be expected to contain essentially all the biochemical compounds synthesized by bacteria, actinomycetes, and fungi. Admittedly, many of these biochemicals will occur only in trace quantities.

The humified material, which represents the most active fraction of humus, consists of a series of highly acidic, yellow- to black-colored, high-molecular-weight polyelectrolytes referred to by such names as humic acid, fulvic acid, and so on. As noted later, these substances are formed by secondary synthesis reactions and have properties distinctly different from the biopolymers of living organisms, including the lignin of higher plants.[35–42]

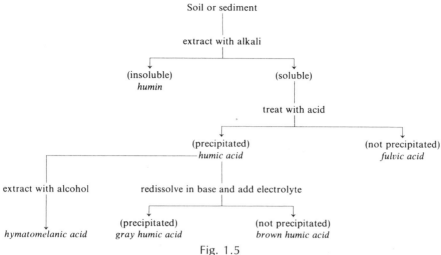

Fig. 1.5
Scheme for the fractionation of soil organic matter.

Extraction and Fractionation of Soil Humus

Humic substances are normally recovered from the soil by extraction with caustic alkali (usually 0.1–0.5 N NaOH), although in recent years use has been made of mild reagents, such as neutral sodium pyrophosphate.[36,42–46] The following fractions, based on solubility characteristics, are subsequently obtained: *humic acid,* soluble in alkali, insoluble in acid; *fulvic acid,* soluble in alkali, soluble in acid; *hymatomelanic acid,* alcohol-soluble part of humic acid; and *humin,* insoluble in alkali.

German scientists further divide humic acids into two groups by partial precipitation with electrolyte (salt solution) under alkaline conditions. The first group, the brown humic acids (*Braunhuminsäure*), are not coagulated by an electrolyte and are characteristic of humic acids in peat and Spodosols. The second group, the gray humic acids (*Grauhuminsäure*), are easily co-agulated and are characteristic of humic acids in Mollisols. In the older literature, considerable attention was given to the so-called *apocrenic* and *crenic* acids, which were light-yellow fulvic acid-type substances. A complete fractionation scheme is given in Fig. 1.5.

The *fulvic acid fraction* has a straw-yellow color at low pH values and turns to wine red at high pH's, passing through an orange color at a pH near 3.0. There is little doubt that compounds of a nonhumic nature are present. The term *fulvic acid* should be reserved as a class name for the pigmented components of the acid-soluble fraction.

Modern View of Humus Formation

It is now generally accepted that humic and fulvic acids are formed by a multiple-stage process that includes (1) decomposition of all plant components, including lignin, into simpler monomers; (2) metabolism of the monomers with an accompanying increase in the soil biomass; (3) repeated recycling of the biomass C (and N) with synthesis of new cells; and (4) concurrent polymerization of reactive monomers into high-molecular-weight polymers.[35,36]

Although many different types of reactions can lead to the production of dark-colored pigments (e.g., humic acids), the main pathway appears to be through condensation reactions involving polyphenols and quinones.[13,35-42] According to present-day concepts, polyphenols derived from plants (e.g., lignin), or synthesized by microorganisms, are enzymatically converted to quinones, which undergo self-condensation or combine with amino compounds to form N-containing polymers. An overall scheme for the formation of humic acid from polyphenols is shown in Fig. 1.6.

Flaig's[35] concept of humus formation is as follows:

1   Lignin, freed of its linkage with cellulose during decomposition of plant residues, is subjected to oxidative splitting with the formation of primary structural units (derivatives of phenylpropane).
2   The side-chains of the lignin-building units are oxidized, demethylation occurs, and the resulting polyphenols are converted to quinones by polyphenoloxidase enzymes.
3   Quinones arising from lignin (as well as from other sources) react with N-containing compounds to form dark-colored polymers.

The role of microorganisms as sources of polyphenols has been emphasized by Kononova,[36] who gives a detailed account of research in which histological microscopic techniques and chemical methods were used to study the decomposition of plant residues. Stages leading to the formation of humic substances were postulated to be as follows:

Stage 1   Fungi attack simple carbohydrates and parts of the protein and cellulose in the medullary rays, cambrium, and cortex of plant residues.
Stage 2   Cellulose of the xylem is decomposed by aerobic myxobacteria. Polyphenols synthesized by the myxobacteria are oxidized to quinones by polyphenoloxidase enzymes, and the quinones subsequently react with N compounds to form brown humic substances.
Stage 3   Lignin is decomposed. Phenols released during decay also serve as source materials for humus synthesis.

The relative importance of lignins and microorganisms as sources of po-

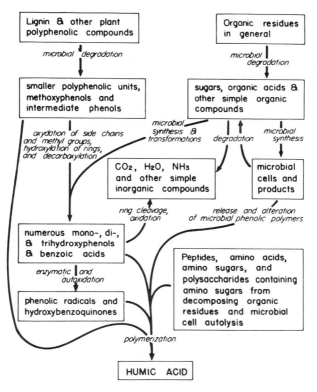

Fig. 1.6
Overall scheme for the formation of humic acid from polyphenols derived from lignin and synthesized by microorganisms.[10] (Reproduced from *Soils for the Management of Organic Wastes and Waste Waters* (1977) by permission of the American Society of Agronomy.)

lyphenols for humus synthesis is unknown, this may depend upon environmental conditions in the soil. Because lignins are relatively resistant to microbial decomposition and a major plant constituent, they are sometimes considered to be the major, if not the primary, source of phenolic units. However, some of the microscopic fungi that decompose lignin in soil produce humic acidlike substances in which the phenolic units originate from both lignin and through biosynthesis by fungi.

According to Martin and Haider,[40] microscopic fungi of the Imperfecti group play a significant role in the synthesis of humic substances in soil. Their studies have shown that such fungi as *Aspergillus sydowi, Epicoccum nigrum, Hendersonula toruloidea, Stachybotrys atra,* and *S. chartarum* degrade lignin as well as cellulose or other organic constituents and in the process synthesize appreciable amounts of humic acidlike polymers. Phe-

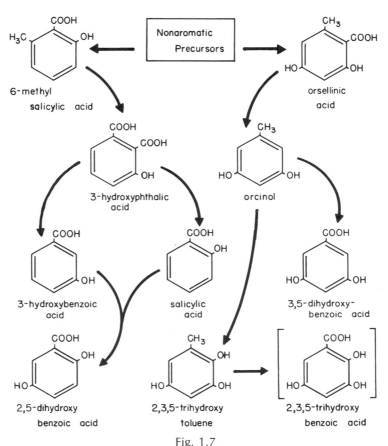

Fig. 1.7

Synthesis and possible transformations of polyphenols by *H. toruloidea*.[41] (Reproduced from *Soil Sci. Soc. Amer. Proc.* **36,** 311–315 (1972) by permission of the Soil Science Society of America.)

nolic units making up the polymer originated from lignin, as well as through synthesis by the fungi.

In recent studies, Haider and Martin[37,38] and Martin and Haider[39,40] investigated the synthesis of humic acid-type substances from several fungi belonging to the Imperfecti group. The synthesis and possible transformations of polyphenols by *H. toruloidea* are illustrated in Fig. 1.7. The first products were orsellinic acid, 6-methylsalicylic acid, and an unidentified phenol. In a short time, substantial amount of orcinol and some 3,5-dihydroxybenzoic acid and 2,3,5-trihydroxytoluene were identified in the culture solution. Additonal phenols were formed with time, including *m*-hydroxybenzoic acid, 2,5-dihydroxybenzoic acid, and small quantities of other compounds.

The quantities of humic acid synthesized by fungi can be appreciable.

For example, Martin et al.[41] found that as much as one-third of the substances synthesized by *H. toruloidea,* including the biomass, consisted of humic acid. Furthermore, a humic acid-type polymer could be recovered from the mycelium tissue by extraction with 0.5 N NaOH.

The production of humic substances by microorganisms may be partly an extracellular process. Following synthesis, the polyphenols are secreted into the external solution, where they are enzymatically oxidized to quinones, which subsequently combine with other metabolites (e.g., amino acids and peptides) to form humic polymers.[42] Reactions postulated to occur between quinones and amino acids are outlined in Chapter 5.

Humic Substances as a System of Polymers

A particularly useful concept that has evolved over the years and that has been popularized by Kononova[36] is that the various humic fractions represent a *system of polymers* that vary in a systematic way in elemental content, acidity, degree of polymerization, and molecular weight.[42-44] The proposed interrelationships are shown in Fig. 1.8. No sharp difference exists between the two main fractions (humic and fulvic acids) or their subgroups. The humin fraction (material not extracted with alkali) is not represented, but this component may consist of a mixture of (1) humic acids so intimately bound to mineral matter that the two cannot be separated and (2) highly condensed humic matter with a high C content ($>60\%$) and thereby insoluble in alkali.

All soils would be expected to contain a broad spectrum of humic substances, as depicted in Fig. 1.8. However, distribution patterns will vary from soil to soil and with depth in the soil profile. The humus of forest soils (Alfisols, Spodosols, and Ultisols) is characterized by a high content of fulvic acids; that of peat and grassland soils (Mollisols) contains high amounts of humic acid. As noted earlier, the humic acids of forest soils are mostly of the brown humic acid type; those of grassland soils are of the gray humic acid type.

Chemical Properties and Structures of Humic and Fulvic Acids

From Fig. 1.8 it can be seen that humic substances consist of a heterogenous mixture of compounds, with each fraction (humic acid, fulvic acid, etc.) being made up of molecules of different sizes. In contrast to humic acids, the low-molecular-weight fulvic acids contain higher oxygen but lower C contents, and they contain considerably more acidic functional groups, particularly COOH.[42-48] Another important difference is that practically all the oxygen in fulvic acids can be accounted for in known functional groups (COOH, OH, C=O); a high portion of the oxygen in humic acids occurs as a structural component of the nucleus (e.g., in ether or ester linkages).

The distribution of oxygen-containing functional groups in humic and ful-

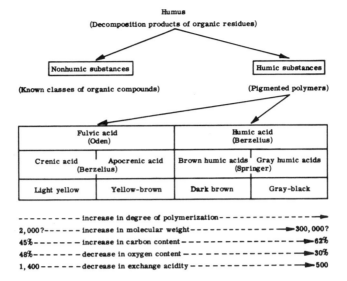

Fig. 1.8

Classification and chemical properties of humic substances. (Reprinted with permission from F. J. Stevenson and J. H. A. Butler[44] as modified from Scheffer and Ulrich,[47] in *Organic Geochemistry,* © 1969, Springer-Verlag.)

vic acids, as recorded in the recent literature, is summarized in Table 1.5. For any specific group, a considerable range of values is apparent, even with preparations obtained from the same soil type. The total acidities of the fulvic acids (640–1,420 meq/100 g) are unmistakably higher than those of the humic acids (560–890 meq/100 g). Both COOH and acidic OH groups (presumed to be phenolic OH) contribute to the acidic nature of these substances, with COOH being the most important. The concentration of acidic functional groups in fulvic acids would appear to be substantially higher than that in any other naturally occurring organic polymer.

According to current concepts, a "type" molecule for humic acid consists of micelles of a polymeric nature, the basic structure of which is an aromatic ring of the di- or trihydroxy-phenol type bridged by —O—, —CH_2—, —NH—, —N=, —S—, and other groups and containing both free OH groups and the double linkages of quinones. Some of the common chromophoric groups that may be responsible for the dark color of humic substances are as follows:

Table 1.5

Distribution of Oxygen-Containing Functional Groups in Humic and Fulvic Acids Isolated from Soils of Widely Different Climatic Zones (meq/100 g)[a]

Functional Group	Climatic Zone					
	Cool, Temperate					
	Arctic	Acid Soils	Neutral Soils	Subtropical	Tropical	Range
			Humic Acids			
Total acidity	560	570–890	620–660	630–770	620–750	560–890
COOH	320	150–570	390–450	420–520	380–450	150–570
Acidic OH	240	320–570	210–250	210–250	220–300	210–570
Weakly acidic + alcoholic OH	490	270–350	240–320	290	20–160	20–490
Quinone C=O	230	{10–180	{450–560	{80–150	{30–140	{10–560
Ketonic C=O	170					
OCH₃	40	40	30	30–50	60–80	30–80
			Fulvic Acids			
Total acidity	1,100	890–1,420	–	640–1,230	820–1,030	640–1,420
COOH	880	610–850	–	520–690	720–1,120	520–1,120
Acidic OH	220	280–570	–	120–270	30–250	30–570
Weakly acidic + alcoholic OH	380	340–460	–	690–950	260–520	260–950
Quinone C=O	200	{170–310		{120–260	30–150	{120–420
Ketonic C=O	200				160–270	
OCH₃	60	30–40		80–90	90–120	30–120

[a] From Stevenson[42] as adapted from the summary of Schnitzer.[45]

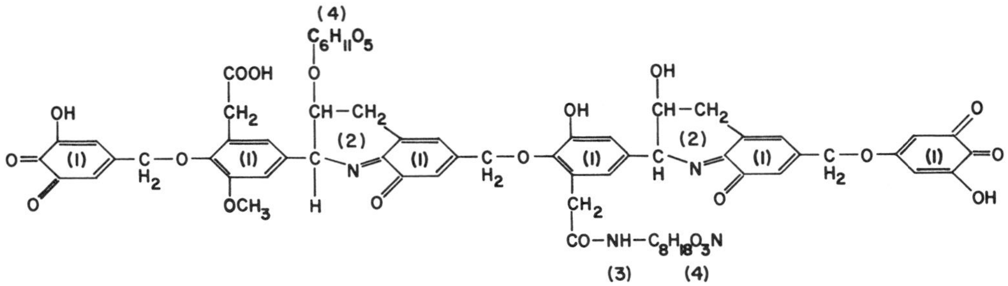

Fig. 1.9

Dragunov's structure of humic acid as recorded by Kononova.[36] (1) Aromatic ring of the di- and trihydroxybenzene type, part of which has the double linkage of a quinone group. (2) Nitrogen in cyclic forms. (3) Nitrogen in peripheral chains. (4) Carbohydrate residue.

"Type" structures, based on the concepts mentioned previously, have been proposed for humic acids, none of which can be considered entirely satisfactory. The structures shown in Figs. 1.9 and 1.10 show the presence of aromatic rings of the di- and trihydroxybenzene type, as well as the presence of the quinone group. The data of Table 1.5 suggest that Dragunov's structure (Fig. 1.9) does not contain a sufficient number of COOH groups (relative to phenolic OH). The two structures show N as a structural component, and they indicate the occurrence of carbohydrate and protein residues.

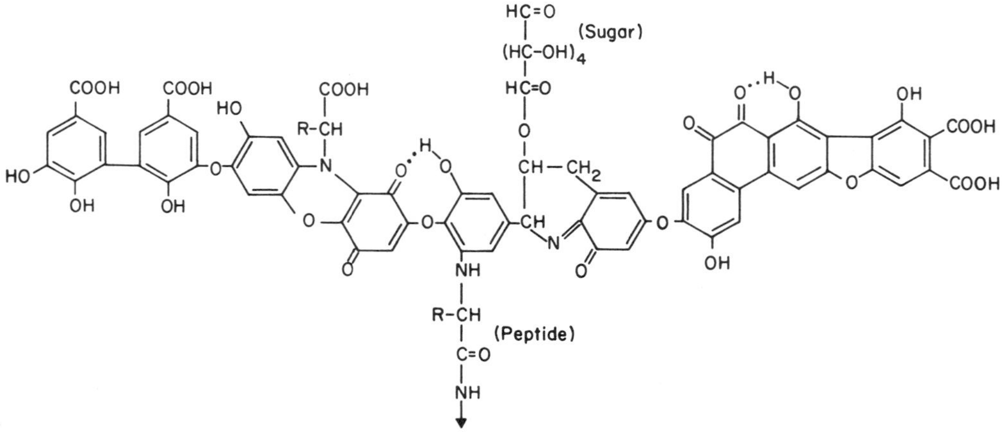

Fig. 1.10

Hypothetical structure of humic acid showing free and bound phenolic OH groups, quinone structures, oxygen as bridge units, and COOH groups variously placed on the aromatic ring. (From Stevenson.[42])

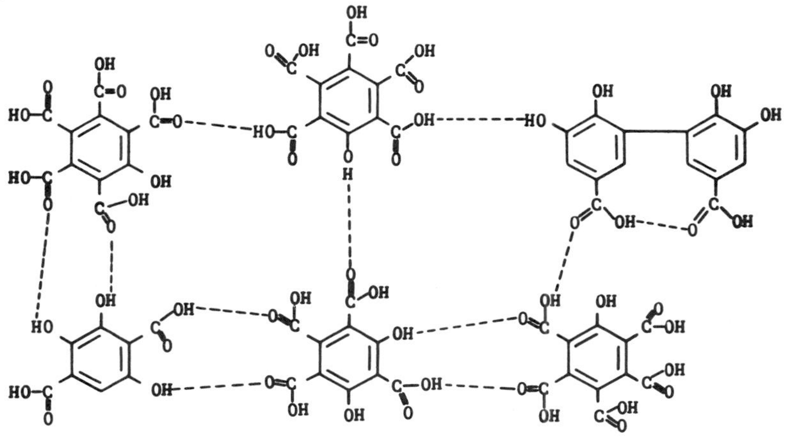

Fig. 1.11

Type structure of fulvic acid as proposed by Schnitzer and Khan.[46]

Schnitzer and Khan[46] concluded that fulvic acids consist in part of phenolic and benzenecarboxylic acids held together through H-bonds to form a polymeric structure of considerable stability (Fig. 1.11). Buffle's[49] model structure of fulvic acid (Fig. 1.12) contains aromatic and aliphatic components extensively substituted with oxygen-containing functional groups. Both structures show an abundance of COOH groups.

Nonhumic Substances in Soil

No more than two-thirds of the organic matter in most soils can be accounted for in the various humic fractions (humic acid, fulvic acid, or humin). Much of the remainder occurs as carbohydrates, proteinaceous constituents (e.g., peptides), and lipids.[42] The two main classes of organic compounds, consisting of humic and nonhumic substances, are not easily separated from one another, for the reason that a portion of the nonhumic material, such as carbohydrates, may be covalently bound to the humic matter, as shown by Fig. 1.10 and the following diagram.[50]

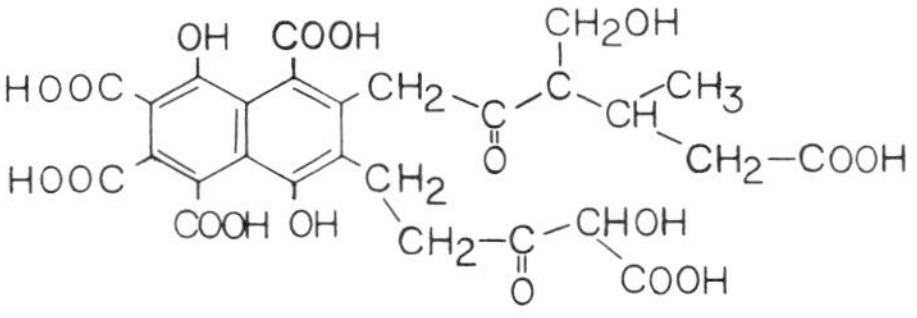

Fig. 1.12

Type structure of fulvic acid as proposed by Buffle.[49]

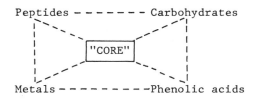

Carbohydrates

From 5 to 25% of the soil organic matter occurs in the form of carbohydrates.[42] Plant remains contribute carbohydrates in the form of simple sugars, hemicellulose, and cellulose, but these are more or less completely decomposed by bacteria, actinomycetes, and fungi, which in turn synthesize polysaccharides and other carbohydrates of their own. These latter materials make up the major part of the polysaccharides found in most soils.

The significance of carbohydrates in soil arises largely from the ability of complex polysaccharides to bind inorganic soil particles into stable aggregates. Carbohydrates are also capable of forming complexes with metal ions. Other soil properties affected by polysaccharides include cation exchange capacity (attributed to COOH groups of uronic acids) and biological activity (e.g., energy source for microorganisms). In certain submerged soils, the production of microbial gums and mucilages may lead to an undesirable reduction in permeability because of blocking of soil pores.

The carbohydrate material in soil occurs as (1) free sugars in the soil solution, (2) complex polysaccharides that can be extracted and separated from other organic constituents, and (3) polymeric molecules of various sizes and shapes that are so strongly attached to clay and/or humic colloids that they cannot be easily isolated and purified.[42] Stability of polysaccharides is due to a combination of several factors, including structural complexity, which makes them resistant to enzymatic attack, adsorption on clay minerals or oxide surfaces, formation of insoluble salts or chelate complexes with polyvalent cations, and binding by humic substances. In the latter case, the polysaccharides may occur as an integral component of humic and fulvic acids, being bound through an ester linkage (R_1COOR_2).

Lipids

The class of organic compounds designated as *lipids* represents a convenient analytical group rather than a specific type of compound, the common property being their solubility in various organic solvents (benzene, methanol, ethanol, acetone, chloroform, ether, etc.) or organic solvent mixtures. They represent a diverse group of materials ranging from relatively simple compounds such as fatty acids to more complex substances such as fats, waxes, and resins. The latter group of substances accounts for the bulk of the lipid material in most soils.[42]

From 1.2 to 6.3% of the organic matter in most soils occurs as fats, waxes, and resins, with somewhat higher values being reported for acid forest soils and peats.[42]

Lipids are of considerable interest because many compounds (aldehydes, ketones, phenolic acids, coumarins, glycosides, and certain aliphatic acids) are phytotoxic; other compounds (e.g., B vitamins) stimulate plant growth. Waxes and similar materials may be responsible for the water-repellent condition of certain sands.

USE OF ^{14}C IN SOIL ORGANIC MATTER STUDIES

The traditional manner of following the decomposition of plant residues in soil has been to measure the loss of C (as CO_2) from soil and plant residues incubated together and to subtract from this the loss of C from soil incubated in the absence of residues. This approach has serious limitations, including analytical errors in measuring long-time release of CO_2, particularly when the amount of plant residue added is kept small relative to the amount of native soil organic matter. Also, the assumption is made that addition of plant residues to the soil does not alter the decomposition rate of the native organic matter.

The use of substrates labeled with ^{14}C has made it possible to follow the decomposition of added residues with considerable accuracy, even in the presence of relatively large amounts of native organic matter. It has been possible also to identify the plant C as it becomes incorporated into fractions of the soil humus. A few studies have been carried out using the stable isotope ^{13}C, but most work has been done with the radioactive isotope ^{14}C.

The intitial studies using ^{14}C appear to be those of Bingeman et al.[51] and Hallam and Bartholomew,[52] who examined the effect of organic matter additions on the decomposition of native humus. Since then nearly 100 papers have been published in which ^{14}C has been used and many facets of organic matter decomposition have been covered. Only those aspects related to soil C transformations are discussed herein. For additional details, the reader is referred to the reviews of Jenkinson,[2] Paul and van Veen,[1] and others.[4,53–55]

Techniques and Approaches in ^{14}C Studies

Carbon-14 is a weak β-emitter having a half-life of about 5,730 years. Like other radioactive tracers, quantitative determination is based on the ionization or excitation of matter by radiations emitted by radioactive bodies. In early work, assay was done using equipment designed for ionization of gases, for which the Geiger–Müller tube was developed. A scintillation counter is now widely used, which measures the emission of light resulting from interaction of ionizing radiation with a fluorescent substance. For additional information on ^{14}C assay, a technical publication of the International Atomic Energy Agency[56] is recommended.

For C transformations in soil, a given quantity of ^{14}C-labeled plant tissue, or biochemical compound, is applied to a known amount of soil and the sample is incubated at constant temperature and humidity. The evolved CO_2 is trapped at intervals, usually in alkali, and assayed for ^{14}C content. Since the specific activity of ^{14}C in the substrate is known, the evolved CO_2 can be partitioned between soil organic matter and the added C source. For experiments carried out under field conditions, where quantitative recovery of CO_2 cannot easily be accomplished, the ^{14}C activity of C remaining in the soil is obtained by combustion of the soil residue and analysis of the liberated CO_2 for ^{14}C.

At the termination of an experiment, or after any given time interval, organic matter can be extracted and fractionated, each component being analyzed for ^{14}C.

Preparation of Plant Material Tagged with ^{14}C

In using ^{14}C to follow the decomposition of plant remains, it is imperative to use plant tissue that has been uniformly labeled. To insure that this is achieved, the growing plant must be maintained in an atmosphere containing CO_2 of constant ^{14}C composition during its entire growth period. This precludes use of soil as a growth media, which would contribute considerable quantities of $^{12}CO_2$ through decomposition of native organic matter by microorganisms. Another major problem is proper control of humidity.

Several types of growth chambers have been used with apparent success to obtain uniformly labeled ^{14}C plant material.[54,57]

Laboratory Incubations

Laboratory incubations provide a convenient and inexpensive way of following the pathway of organic matter decomposition in soil. In a typical incubation experiment, the treated soil is placed in an aeration train with a trap for collecting the evolved CO_2. Several types of incubation units have been described in the literature, some of which are rather sophisticated in that temperature and humidity are automatically controlled.

A simple incubation unit that permits collection of CO_2 during the decomposition of organic residues in soil is shown in Fig. 1.13. For optimum results, incubation must be carried out at constant temperature. The total quantity of CO_2 collected in the dilute NaOH solution is determined by titration with standardized HCl following addition of $BaCl_2$, according to the reactions:

$$2NaOH + CO_2 \rightarrow Na_2CO_3 + H_2O$$

$$Na_2CO_3 + BaCl_2 \rightarrow BaCO_3 + 2\,NaCl$$

$$NaOH\ (excess) + HCl \rightarrow NaCl + H_2O$$

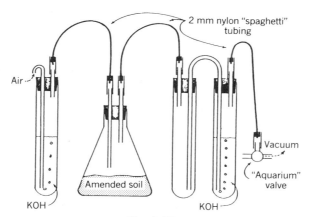

Fig. 1.13

Apparatus for studying the decomposition of organic substances in soil. The base is often a dilute NaOH solution (see text), (Provided through the courtesy of J. P. Martin.)

The ^{14}C content of the evolved CO_2 is determined on an aliquot of the NaOH solution by liquid scintillation counting or on the insoluble $BaCO_3$ using a Geiger–Müller tube.

Field Experiments

For these studies, measurements are made for the ^{14}C left behind in the soil from the labeled C source. The experiments are normally carried out in small cylinders, 10 to 30 cm in diameter, driven into the soil to a depth of 15 cm or more to prevent lateral movement of soil and added plant material and to prevent contamination of surrounding areas. The application rate for labeled plant material should approximate the natural rate. To obtain uniform mixing, the soil is removed from the 0- to 15-cm depth and the labeled material is incorporated while continuously stirring in a large container. Before returning the amended soil to the cylinder, the exposed subsoil is covered with a 6-mm mesh screen, which serves to delineate the amended soil while allowing for movement of roots and soil organisms.

At periodic intervals, soil is removed to the depth of the screen and subsamples are analyzed for ^{14}C after oxidation to CO_2. The ^{14}C content remaining in specific soil fractions or components can be determined through any of the fractionation schemes reported elsewhere in the text, such as humic acid, fulvic acid, carbohydrates, amino acids, and so on.

Decomposition of Simple Substrates and Establishment of Tagged Microbial Tissue

As noted earlier, the decomposition of plant remains in soil leads to the synthesis of microbial tissue. A major objective of studies using ^{14}C-labeled

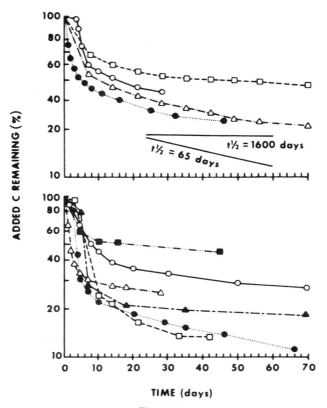

Fig. 1.14
Stabilization of ^{14}C-labeled glucose and acetate when added to soil in the laboratory. Each line represents the result of a separate experiment as recorded in the literature. (From Paul and McLaren,[58] p. 25, courtesy of Marcel Dekker, Inc.)

metabolites as a C source is to build up fresh microbial tissue (the biomass) tagged with ^{14}C. This is done by adding an energy source that will be completely and rapidly transformed into microbial tissue and CO_2. Glucose and acetate are two substrates commonly used for this purpose. With these compounds decomposition is normally complete within a matter of hours or days, with incorporation of up to two-thirds or more of the C into microbial tissue. A summary of C retention values as recorded in the literature is summarized in Fig. 1.14.

A model for decomposition of a simple substrate in soil is shown in Fig. 1.15. The diagram illustrates how recycling affects the apparent decomposition rate as measured by CO_2 production. As can be seen from Fig. 1.16, accumulation of biomass and microbial products commences soon after decomposition of the substrate; eventually, all of the substrate C remaining

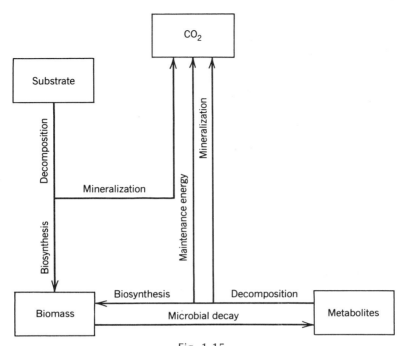

Fig. 1.15

Model for the decomposition of a simple substrate in soil. (Adapted from Paul and van Veen.[1])

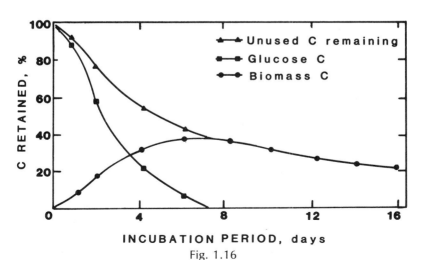

Fig. 1.16

Microbial utilization of glucose and incorporation of substrate C into the biomass. Incubation in the presence of $^{15}NH_4^+$ leads to labeling of the biomass with ^{15}N (see Chapter 5).

27

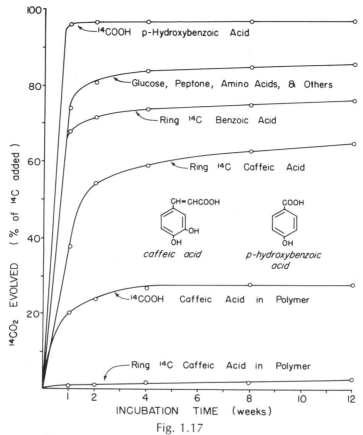

Fig. 1.17

Decomposition of several simple organic compounds, of specifically labeled ^{14}C-benzoic and caffeic acids, and of caffeic acid linked into phenolic polymers.[37] The soil was Greenfield sandy loam. (Reproduced from *Soil Sci. Soc. Amer. Proc.* **39,** 657–662 (1975) by permission of the Soil Science Society of America.)

behind in the soil occurs in synthesis products. A continued slow release of $^{14}CO_2$ occurs soon after disappearance of the substrate and can be attributed to decay of secondary microbial substances.

Phenolic acids of various types are produced in soil by numerous microorganisms, and they are present in combined form in lignins, tannins, microbial melanins, and other metabolites. Haider and Martin[37] conducted incubation experiments on specifically ^{14}C-labeled phenolic carboxylic acids. Parts of the molecule that were labeled included the aromatic COOH group, the three-C side chain, and the aromatic ring. A summary of the findings is shown in Fig. 1.17. With the exception of COOH attached to the aromatic ring, C of the phenolic derivatives (e.g., caffeic acid and *p*-hydroxybenzoic acid) was not lost as CO_2 to the same extent as from some simple organic compounds such as sugars, organic acids, and peptides. When linked into polymers, the ^{14}COOH and ring ^{14}C of caffeic acid were highly resistant to decomposition, particularly the ring ^{14}C.

Various attempts have been made to fractionate the C remaining in the soil after decomposition of ^{14}C-labeled substrates. In an experiment using radioactive acetate, Ivarson and Stevenson[59] found that residual activity was widely distributed in the different humus fractions, from which it was concluded that structural units of humus can be synthesized in a rather short time from simple organic compounds. A somewhat similar result was obtained by Wagner (see IAEA[55]) using ^{14}C-labeled glucose as substrate. The high stability of the C when transformed into microbial tissues and metabolites also has been demonstrated in studies by Hurst and Wagner[14] and Shields et al.[60]

A major difficulty in interpreting results of organic matter fractionation studies is that chemical changes may occur during extraction, with conversion of metabolic products into the so-called stable humus forms. This aspect of ^{14}C research requires further study.

Decomposition of ^{14}C-Labeled Plant Residues

The decay of natural products follows a much more complex pattern than was noted earlier for simple substrates. The model shown by Fig. 1.18 depicts decomposition for plant residue material (e.g., straw) made up of four components that decompose at different rates. As was the case noted earlier, actual decomposition rate will be higher than the one measured by CO_2 evolution. However, the substrate C will persist in the soil in an unmodified form for a longer time period (compare Fig. 1.19 with earlier Fig. 1.16).

In general, studies on the decomposition of ^{14}C-labeled plant residues have shown that the residues are attacked rapidly at first but after a few months the rate slows down to a very low value despite the fact that considerable amounts of plant C remain behind in the soil. This same pattern has been observed in laboratory incubations where the environment has been kept constant through careful control of temperature and humidity. The slowdown in decomposition with time is not believed to be due entirely to differences in the rate at which the various plant components decompose (lignin is attacked very slowly); it may be because part of the C of the more easily decomposable constituents (sugars, cellulose, etc.) is resynthesized into microbial components more resistant to decomposition than the initial plant material.

Field Trials

Several C cycling investigations have been performed now under field conditions. Results obtained for C retention after the first year (or first growing

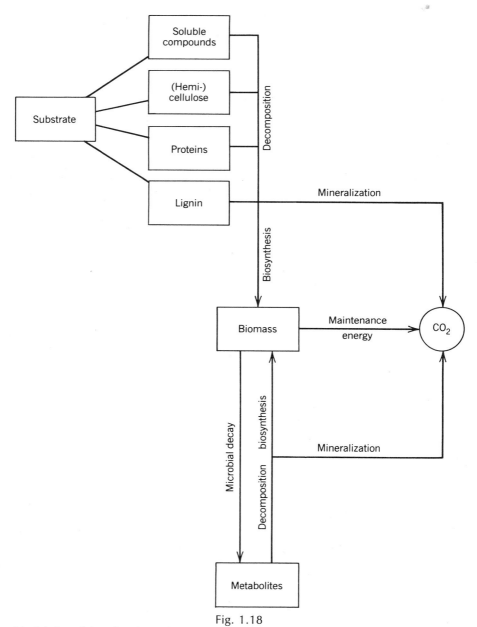

Fig. 1.18

Model describing the decomposition of a complex substrate. For comparison with a simple substrate see Fig. 1.15. (Adapted from Paul and van Veen.[1])

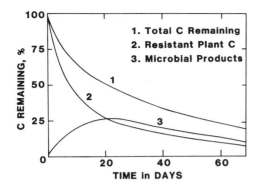

Fig. 1.19
Calculated decomposition of a complicated substrate. For comparison with a simple substrate see Fig. 1.16. (Adapted from Paul and van Veen.[1])

season) are summarized in Table 1.6. From 20 to 74% retention of applied C has been observed, depending on soil conditions and plant part involved. The highest values (63 to 74%) were obtained for ^{14}C-labeled roots of blue grama grass. It should be noted that most of the experimental data recorded in Table 1.6 are for temperate zone soils and that considerably different values for C retention would be expected for soils of colder and warmer regions. Major factors influencing decomposition in the natural soil include positioning of the plant material beneath or on the soil surface, soil water regime, temperature, and length of growing season as affected by climate.

Results of field studies with ^{14}C-labeled plant residues show that the C becomes increasingly resistant to decomposition with time. In the experi-

Table 1.6
Carbon Retained from ^{14}C-Labeled Plant Material Applied to Field Soils

Location	Type	Carbon Retained, %	Reference
Rothamsted, England	Ryegrass tops and roots	Approximately 33% first year irrespective of soil type or plant material	Jenkinson[21,61]
West Germany	Wheat straw and chaff	31% after first year for fallow and cropped soil	IAEA[55]
Austria	Maize (corn)	47% after first year when applied in August and 33% when applied in October	IAEA[55]
Saskatchewan, Canada	Wheat straw	35–45% after first growing season	Shields and Paul[62]
Colorado, US	Blue grama a. Herbage b. Roots	43–46% after 412 days 63–74% after 412 days	Nyhan[63]
Nigeria	Ryegrass	20% after first year and 14% after two years	Jenkinson and Ayanaba[64]

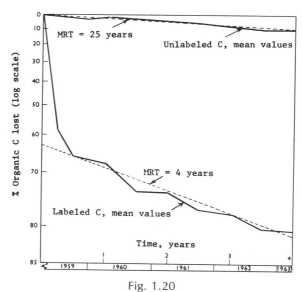

Fig. 1.20

Long-time C loss from organic residues as estimated from studies with ^{14}C-labeled ryegrass. (From Jenkinson.[65])

ments conducted by Jenkinson,[65] approximately 33% of the applied C re-mained behind in the soil after the first year, of which about one-third was believed to occur in microbial cells. After 4 years, about 20% of the labeled C still remained in the soil and only about 20% of this C was in microbial cells. The original residues decomposed rapidly with an estimated half-life of 14–30 days; after the first year, the residual C had a half-life of about 4 years. Under the conditions specified, the native humus had a half-life of about 25 years. These relationships are shown in Fig. 1.20. It should be noted that under natural soil conditions the mean residence time of organic matter is variable but usually somewhat longer then 25 years (discussed later).

Jenkinson and Rayner[66] developed a model for C transformations in soils based on the classical long-term rotation plots at the Rothamsted Experiment Station. Using half-lives for the various fractions of soil organic matter, they concluded that an annual input of 1 ton (907 kg) of plant C per ha for 10,000 years would lead to 11.3 tons of C (10,250 kg) in physically stabilized forms and 12.2 tons (11,065 kg) in chemically stabilized forms. The equivalent ^{14}C age of the organic matter would be 1,240 years. Clark and Paul[67] concluded that organic components of soil existed in the following major fractions when considered on a dynamic basis:

1   Decomposing plant residues and associated biomass, which turn over at least once every few years.

2 Microbial metabolites and cell wall constituents that become stabilized in soil and possess a half-live of 5–25 years.

3 The resistant fractions, which range in age from 250 to 2,500 years.

Incorporation of ^{14}C into Soil Organic Matter Fractions

Sørensen[68] fractionated the soil organic matter after allowing ^{14}C-labeled barley straw to decompose in soil. In agreement with the observations noted earlier, approximately one-third of the C remained behind in the soil after decomposition (100 days incubation at 20°C). A study was made of the distribution of ^{14}C in fractions of the soil organic matter, with the following result:

	Recovery of Residual ^{14}C
Fulvic acid	11.8–18.9%
Humic acid	19.5–26.4%
Humin	49.4–54.7%
Unaccounted for	5.0–12.9%

These results, together with studies using the hemicellulose and cellulose components of the straw, suggested that substances other than lignin contributed to the formation of humic substances. A word of caution was sounded by Sauerbeck and Führ (IAEA[55]), who obtained results that indicated that chemical reactions can occur between soil humus and soluble plant constituents during alkali extraction.

Sauerbeck (IAEA[55]) found that newly formed humic acids were considerably more stable than fulvic acids and that there was considerable interconversion among the different humus fractions with time.

THE PRIMING ACTION

Newly applied plant residues can either stimulate or retard decomposition of native humus in the soil. This change in decomposition rate is described as "priming" and is usually positive. The effect is shown in Fig. 1.21, where the amount of native C lost through "priming" is taken as the difference between the amount of soil CO_2 evolved in the presence and absence of the substrate.

In considering the so-called priming effect, it should be noted that certain errors can arise when using ^{14}C-labeled plant materials to follow decomposition processes in soil. They include:

1 Some of the labeled CO_2 may undergo exchange with carbonate C of the soil.

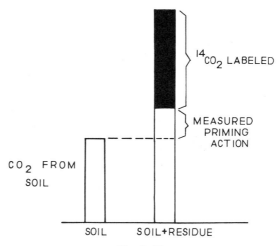

Fig. 1.21
Influence of applied plant residues labeled with ^{14}C on loss of native soil C as CO_2.

2   Plant C may not be uniformly labeled with ^{14}C.

3   Differences in the microbiological environment of the soil may result from the organic matter addition, such as pH, O_2 content, moisture, and others.

4   Calculation errors can be made in analyzing the data.

5   Accelerated decomposition of organic materials may arise (e.g., because of difference in wetting and drying the soil).

For a discussion of priming action the reader is referred to articles by Jenkinson, Sauerbeck, and Smith (see IAEA[55]).

One of the first investigators to suggest a priming action by addition of organic materials to soil was Löhnis.[69] Based on results of green manuring experiments in the field and greenhouse, he concluded that intensified bacterial activity accompanying incorporation of green manures in soil increased mineralization of the native humus N. This view was criticized by Pinck and Allison[70] and has yet to be resolved (see Chapter 5).

The largest priming actions ever recorded appear to be those of Broadbent and Bartholomew[71] and Broadbent and Norman,[72] who concluded that decomposition of native humus was increased as much as 10-fold by additions of tagged sudan grass tissues to soil. Broadbent and Norman[72] likened the process to a "forced draught on the smoldering bacterial fires of the soil." In studies carried out by Hallam and Bartholomew,[73] less total C remained in the soil after incubation with plant material than in soil incubated alone, from which they concluded that large, infrequent additions of plant residues were best from the standpont of organic matter maintenance.

Jenkinson (see IAEA[55]) concluded that explanations of the priming action could be placed in the following groups: (1) those setting out models that give rise to an apparent priming action even though decomposition rate of the native humus is unaltered, (2) those based on errors in relating ^{14}C evolved to the amount of labeled plant C decomposed, (3) those based on differences in the environment between soil incubated with and without organic additions, and (4) postulating mechanisms by which added organic matter can increase decomposition rate of the native soil organic matter.

Assuming that extra loss of soil C is caused by a priming action, the probable explanation is that there is a buildup of a large and vigorous population of microorganisms when energy material is added to the soil and that these microorganisms subsequently produce enzymes that attack the native soil organic matter. The extent of organic matter loss will depend upon a variety of factors, including the size and activity of the microflora. Presumably, the stimulating effect would last so long as plant residues constitute an important part of the decomposing mass. Easily decomposable plant residues, such as young, succulent plant tissues, would be particularly effective in accelerating C loss.

TRANSFORMATIONS IN WET SEDIMENTS

The special conditions that exist in wet sediments alter considerably the activities of micro- and macrofaunal organisms.[74-78] Accordingly, many of the concepts developed earlier for well-drained soils do not apply to waterlogged environments, such as lakes, streams, flooded soils, peats, marshes, and so on. Not only do wet sediments support a different flora of microorganisms, but decomposition of organic matter proceeds at a greatly reduced rate; furthermore, the end products of metabolism are different.

The dominant microorganisms of aquatic environments are the anaerobic bacteria, although facultative forms are present when NO_3^- is available to act as an electron acceptor during respiration. Fungi and actinomycetes are of little or no consequence; these organisms require O_2 for carrying out their life processes.

Plant remains are not completely metabolized in waterlogged environments. Intermediate decomposition products accumulate, and abundant amounts of such gases as methane (CH_4), hydrogen sulfide (H_2S), and H_2 are evolved. Since the energy of anaerobic fermentation is lower than that for aerobic digestion, less biomass per se is generated per unit of organic matter metabolized, even though more of the C remains behind in a reduced state. Several studies have attested to the slow rate of plant residue decay under anaerobic conditions, and to the greater preservation of C when the

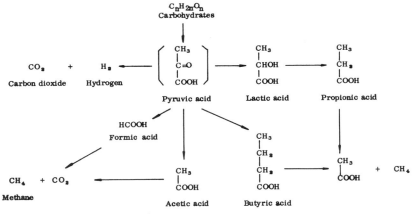

Fig. 1.22
Reactions of pyruvate in anaerobic systems, showing major organic acid and CH_4 formation. Carbon dioxide produced during fermentation serves as an electron (hydrogen) acceptor, with formation of CH_4 (see text). (Reprinted from Yoshida,[78] p. 101, by courtesy of Marcel Dekker, Inc.)

decay process has run its course. The great accumulations of organic matter in bog and marsh soils attest to the slow and incomplete mineralization of plant residues.

The basic metabolism of anaerobic bacteria differs from that of aerobic forms in two main ways. First, the end product of glycolysis (pyruvate) is disposed of by means other than by oxidation to CO_2 through the well-known tricarboxylic acid (Krebs) cycle. Specifically, pyruvate is transformed to other products, including a variety of organic acids and alcohols (e.g., ethanol). Fermentation of organic matter by the microflora is often characterized by the temporary accumulation of organic acids, such as acetic, formic, lactic, and butyric. Typical pyruvate reactions in anaerobic systems are shown in Fig. 1.22.

A second difference between anaerobic and aerobic metabolism is the way in which electrons or hydrogen are eliminated from the system. Biological oxidations result in the production of hydrogen (through dehydrogenation) or electrons that must be disposed of. In aerobic systems the terminal acceptor is O_2. In anaerobic metabolism other substances function as electron acceptors, such as NO_3^- for certain denitrifying species, SO_4^{2-} for *Desulfovibrio*, and CO_2 for methane-producing bacteria. The end products are N_2, H_2S, and CH_4, respectively.

The substrate for the methane-producing bacteria are organic acids of the type mentioned earlier, with CO_2 acting as the electron (or hydrogen) acceptor. Reactions leading to the formation of CH_4 with acetic acid as substrate are as follows:

$$CH_3COOH \ + \ 2H_2O \longrightarrow 2CO_2 \ + \ 8(H)$$

$$8(H) \ + \ CO_2 \longrightarrow CH_4 \ + \ 2H_2O$$

$$CH_3COOH \longrightarrow CH_4 \ + \ CO_2$$

Methane

Other organic acids can also be used for CH_4 production, as well as various alcohols.

Thus, in contrast to aerobic systems, where CO_2 and H_2O are the main decomposition products, anaerobic metabolism is characterized by the formation of organic acids, CH_4, H_2S, and a variety of secondary decomposition products, including mercaptans (e.g., methyl mercaptan CH_3SH), primary and secondary amines, aldehydes, and ketones. Many of these products are harmful to plant growth. Butyric acid, for example, is believed to contribute to the poor growth often observed in rice paddy fields.[79,80] Ethylene in toxic amounts has been reported to be produced by decay of organic residues under anaerobic soil conditions.[81-83] According to Kilham and Alexander,[84] accumulations of organic matter in some flooded soils is due to the inhibitory effect of organic acids and hydrogen sulfide on microbial activity.

A list of compounds peculiar to wet sediments is given in Table 1.7. Brief discussions of individual items follow.

Protein Decomposition

In contrast to decomposition in aerobic soils, where the major end products are CO_2, NO_3^-, SO_4^{2-}, and H_2O, foul-smelling compounds are released by

Table 1.7
Organic Compounds Peculiar to Wet Sediments[a]

Class	Comments
Fermentation products	Incomplete oxidation leads to production of CH_4, organic acids, amines, mercaptans, aldehydes, and ketones.
Modified or partially modified remains of plants	In addition to slightly altered lignins, carotenoids, sterols, and porphyrins of chlorophyll origin are preserved.
Synthetic organic chemicals[a]	Many man-made chemicals (e.g., DDT) decompose slowly if at all under anaerobic conditions.
Carcinogenic compounds[a]	Synthesis of methylmercury, dimethylarsine, dimethylselenide, and nitrosamines of various types.

[a] Environmental aspects of the C cycle are discussed in Chapter 3.

the anaerobic degradation (putrification) of protein-rich substances. They include amines, organic acids, indole, skatole, mercaptans, and H_2S.

Cellulose Decomposition

Several microorganisms are capable of decomposing cellulose in the absence of O_2, the most common being species of *Clostridium*. Some species are specific in their requirement for cellulose. The main end products are CO_2, H_2, ethanol, and various organic acids. Methane is also a byproduct, but this gas is apparently produced not by the *Clostridium* but by bacteria that metabolize the organic acids liberated during the primary stage.

Modification in Lignin

Mention was made earlier of the significance of fungi in lignin decomposition. Since these organisms are not found in aquatic sediments, the lignin (in a modified form) tends to accumulate. The preservation of organic matter in peats and poorly drained soils is well known.

The main modification of lignin in aqueous systems appears to be enzymatic cleavage of methoxyl (OCH_3) groups, presumably by the anaerobic bacteria. The fate of the remainder of the lignin molecule is uncertain.

Preservation of Other Plant Components

Various lipid components appear to be less subject to decomposition in anaerobic systems than in aerobic soils. Carotenoids, certain sterols, and porphyrins derived from chlorophyll are generally prevalent in wet sediments, whereas they occur in well-drained soils in trace amounts only.

Changes Brought About in Soils and Sediments During Submergence

The consideration previously outlined suggests that considerable changes occur when soils or sediments are submerged, such as through flooding by natural or artifical means. The major changes can be summarized as follows:

1 The utilization of organic matter by microorganisms results in a marked drop in dissolved O_2 and an increase in CO_2 content. The net result is a concurrent drop in the oxidation-reduction potential (E_h). Microorganisms cause the change in E_h through the consumption of O_2 and the liberation of reduced products.
2 There may be changes in pH as well as in electrolyte concentration in general.
3 Denitrification may occur with escape of NO_3^- as N_2 and N_2O. On the other hand, NH_3 (as NH_4^+) may accumulate.
4 Intermediate organic matter decomposition products may accumulate, such as organic acids.
5 Sulfate is reduced to H_2S.

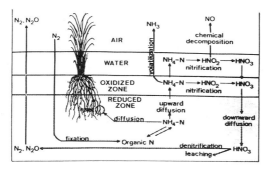

Fig. 1.23

Schematic representation of N transformations in a flooded soil ecosystem.[77] [Reproduced from *Plant Soil* **67**, 209–220 (1982) by permission of Martinus Nijhoff/Dr. W. Junk Publishers. Printed in the Netherlands.]

6 Manganese and iron may be reduced and their solubilities enhanced. These transformations are controlled to a considerable extent by microorganisms (change in E_h accompanying depletion of O_2 leads to changes in oxidation state). Various other inorganic constituents, such as phosphate, may become more soluble.

Nitrogen Transformations

All aspects of the N cycle must be viewed in a new context when considered in terms of aquatic systems (see Chapters 4 and 5). For example, many of the organisms that typically fix N_2 in aerobic soils fail to function at low O_2 tensions. Thus, in contrast to agricultural soils, where aerobic organisms (*Rhizobium, Azotobacter, Beijerinckia,* and others) play a key role in N_2 fixation, most of the N fixed in water-saturated environments results from the activities of anaerobic organisms, such as *Clostridium,* the blue-green algae, and certain green and purple photosynthetic bacteria. Many investigators believe that N_2 fixation in submerged soils (e.g., rice paddy fields) is brought about mostly by blue-green algae. A schematic representation of N transformations in a flooded soil ecosystem is shown in Fig. 1.23.

The mineralization–immobilization reaction is also modified in zones where O_2 is absent. Mineralization does not proceed beyond the NH_3 stage, although further oxidation may occur at the surface-oxidized layer, where sufficient O_2 is present to meet the needs of nitrifying bacteria. The ammonification process (conversion of organic N to NH_3) may also be curtailed or otherwise modified. Since anaerobic decomposition provides little energy, larger amounts of substrate must be oxidized per unit of C assimilated, which suggests less assimilation of mineral N into microbial tissue than for aerobic decomposition. Accordingly, competition of microorganisms with plants for available N may be less severe in aquatic environments than in aerobic soils.

Nitrate, introduced into aqueous sediments in surface or ground waters, applied as fertilizer, or formed at the oxidized layer, is particularly subject to gaseous loss through denitrification. A quantitative measure of denitrification losses is considered basic to an understanding of the N balance in lakes and streams and for the proper utilization of fertilizer N in rice fields. The subject of denitrification is considered in greater detail in Chapter 4.

Oxidation-Reduction Reactions

As indicated earlier, the shift from aerobic to anaerobic metabolism in wet sediments results in depletion of dissolved O_2 with an accompanying change in the oxidation-reduction potential. The absolute value for E_h serves as an indication of the tendency of organic and inorganic substances to gain electrons (become reduced) or lose electrons (become oxidized). Thus a high E_h indicates an oxidizing environment, whereas a low or negative value suggests a reducing environment. Negative E_h values are common in paddy fields and aqueous sediments containing decomposable organic matter.

From an energy standpoint, microbes reduce NO_3^- before SO_4^{2-}; both NO_3^- and SO_4^{2-} are reduced before CH_4 formation. However, the kinetics and kind of reduction products are determined by a variety of factors, including amount of organic mater, pH, temperature, duration of flooding, and kind and amount of inorganic electron acceptors.

The ability of inorganic substances to act as electron acceptors—to be reduced—would be expected to follow the thermodynamic sequence shown in Table 1.8.

SUMMARY

The decomposition of plant and animal remains in soil constitutes a biological process that is carried out by bacteria, actinomycetes, and fungi. Part of the C is utilized for the synthesis of body tissue (the soil biomass); part is incorporated into stable humus. Biochemical transformations of organic matter result largely from the enzymatic activities of proliferating microorganisms active in the decay process.

Humic substances represent a complex mixture of molecules having various sizes and shapes, but no completely satisfactory scheme has been forthcoming for their isolation, purification, and fractionation. There is some evidence to indicate that, in any given soil, the various fractions obtained on the basis of solubility characteristics (e.g., humic acid, fulvic acid, and others) represent a system of polymers whose chemical properties (elemental composition, functional group content, etc.) change systematically with increasing molecular weight.

Considerable information has been obtained on decomposition processes

Table 1.8
Thermodynamic Sequence for Reduction of Inorganic Substances

Reaction	$E_h{}^a$
Disappearance of O_2	
$O_2 + 4H^+ + 4e \rightleftharpoons 2H_2O$	0.816 V
Disappearance of NO_3^-	
$NO_3^- + 2H^+ + 2e \rightleftharpoons NO_2^- + H_2O$	0.421 V
Formation of Mn^{2+}	
$MnO_2 + 4H^+ + 2e^- \rightleftharpoons Mn^{2+} + 2H_2O$	0.396 V
Reduction of Fe^{3+} to Fe^{2+}	
$Fe(OH)_3 + 3H^+ + e^- \rightleftharpoons Fe^{2+} + 3H_2O$	−0.182 V
Formation of H_2S	
$SO_4^{2-} + 10H^+ + 8e^- \rightleftharpoons H_2S + 4H_2O$	−0.215 V
Formation of CH_4	
$CO_2 + 8H^+ + 8e^- \rightleftharpoons CH_4 + 2H_2O$	−0.244 V

[a] At pH 7.0.

in recent years. As far as research with ^{14}C-labeled substrates is concerned, the following tentative conclusions can be drawn.

1 The concept of soil organic matter as an "inert," biologically resistant material has not been substantiated.

2 Addition of fresh organic residues to the soil may result in a small priming action on the native soil organic matter.

3 Humus can by synthesized in a rather short time by microorganisms living on rather simple substrates.

4 Plant residues decay rather rapidly in soil and are more or less completely transformed, even the lignin fraction.

5 Except possibly for lignin, freshly incorporated C first enters into microbial tissue (soil biomass), then into "labile" fractions of the soil organic matter, and finally into complex humic polymers during advanced stages of humification.

The conditions in wet sediments considerably alter the process whereby organic residues undergo decay in soils. Plant remains are not completely metabolized and intermediate decomposition products accumulate, some of which are toxic to plants. In addition, specific types of compounds (e.g., methylmercury, dimethylarsine, nitrosamines) can be produced that are carcinogenic.

REFERENCES

1   E. A. Paul and J. A van Veen, *Trans. 11th Intern. Congr. Soil Sci.* **3**, 61 (1978).

2   D. S. Jenkinson, *Soil Sci.* **111**, 64 (1971).

3   C. C. Delwiche, "Carbon Cycle," in R. W. Fairbridge and C. W. Finkle, Jr., Eds., *The Encyclopedia of Soil Science,* Part 1, Dowden, Hutchinson, and Ross, Stroudsburg, 1969, pp. 61–64.

4   W. B.McGill and C. V. Cole, *Geoderma* **26**, 267 (1981).

5   H. L. Bohn, *Soil Sci. Soc. Amer. J.* **40**, 468 (1976).

6   B. Bolin, "The Carbon Cycle," in B. Bolin and R. B. Cook, Eds., *The Major Biogeochemical Cycles and their Interactions,* Wiley, New York, 1983, pp. 41–45.

7   R. M. Garrels, F. T. MacKenzie, and C. Hunt, *Chemical Cycles and the Global Environment,* Kauffman, Los Altos, Calif., 1975.

8   B. Bolin, E. T. Degens, S. Kempe, and P. Ketner, *The Global Carbon Cycle,* SCOPE Report 13, Wiley, New York, 1979.

9   A. Burges, *Micro-Organisms in the Soil,* Hutchinson, London, 1958.

10 J. P. Martin and D. D. Focht, "Biological Properties of Soils," in L. F. Elliott and F. J. Stevenson, Eds., *Soils for Management of Organic Wastes and Waste Waters,* American Society of Agronomy, Madison, Wis., pp. 114–169.

11 W. Flaig, H. Beutelspacher, and E. Rietz, "Chemical Composition and Physical Properties of Humic Substances," in J. E. Gieseking, Ed., *Soil Components, Vol. 1. Organic Components,* Springer-Verlag, New York, 1975,pp. 1–211.

12 W. J. Schubert, *Lignin Biochemistry,* Academic Press, New York, 1965.

13 K. Haider, J. P. Martin, and Z. Filip, "Humus Biochemistry," in E. A. Paul and A. D. McLaren, Eds., *Soil Biochemistry,* Vol. 4, Dekker, New York, 1974, pp. 195–244.

14 H. M. Hurst and G. H. Wagner, *Soil Sci. Soc. Amer. Proc.* **33**, 707 (1969).

15 F. E. Clark, "Bacteria in Soil," in A. Burges and F. Raw, Eds., *Soil Biology,* Academic Press, New York, 1967, pp. 15–49.

16 D. S. Jenkinson and J. N. Ladd, "Microbial Biomass in Soil, Measurement and Turnover," in E. A. Paul and J. N. Ladd, Eds., *Soil Biochemistry,* Vol. 5, Dekker, New York, 1981, pp. 415–472.

17 E. A. Paul and R. P. Voroney, "Nutrient and Energy Flows Through Soil Microbial Biomass," in R. C. Ellwood et al., Eds., *Contemporary Microbial Ecology,* Academic Press, New York, 1980, pp. 215–237.

18 L. A. Babiuk and E. A. Paul, *Can. J. Microbiol.* **16**, 57 (1970).

19 J. P. E. Anderson and K. H. Domsch, *Soil Biol. Biochem.* **10**, 207, 215 (1978).

20 D. S. Jenkinson and J. M. Oades, *Soil Biol. Biochem.* **11**, 193, 201 (1979).

21 D. S. Jenkinson and D. S. Powlson, *Soil Biol. Biochem.* **8**, 167, 179, 209 (1976).

22 P. Nannipier, R. L. Johnson, and E. A. Paul, *Soil Biol. Biochem.* **10**, 223 (1978).

23 D. Parkinson and E. A. Paul, "Microbial Biomass," in A. L. Page, R. H. Miller, and D. R. Keeney, Eds., *Methods of Soil Analysis, Part 2. Chemical and Biological Properties,* 2nd ed., American Society of Agronomy, Madison, Wis., 1982, pp. 821–830.

24 W. N. Miller and L. E. Casida, *Can. J. Microbiol.* **16**, 299 (1970).

25 J. Skujins and L. E. Casida, *Can. J. Microbiol.* **16**, 299 (1970).

26 C. C. Lee, R. F. Harris, J. D. H. Williams, D. E. Armstrong, and J. K. Syers, *Soil Sci. Soc. Amer. Proc.* **35**, 82 (1971).

27 E. A. Paul and R. L. Johnson, *Appl. Environ. Microbiol.* **34**, 263 (1977).

28 S. Kiss, M. Drăgan-Bularda, and D. Rădulescu, *Adv. Agron.* **27**, 25 (1975).

29 V. F. Kuprevich and T. A. Schcherbakova, "Comparative Enzymatic Activity in Diverse Types of Soil," in A. D. McLaren and J. Skujins, Eds., *Soil Biochemistry,* Vol. 2, Dekker, New York, 1971, pp. 167–201.

30 M. A. Tabatabai, "Soil Enzymes," in A. L. Page, R. H. Miller, and D. R. Keeney, Eds., *Methods of Soil Analysis,* Part 2, American Society of Agronomy, Madison, Wis., 1982, pp. 903–947.

31 J. A. Thornton, D. Crim, and A. D. McLaren, *J. Forensic Sci.* **20,** 674 (1975).

32 A. D. McLaren, *Chem. Sci.* **8,** 97 (1975).

33 D. J. Ross, *Soil Biol. Biochem.* **3,** 97 (1971).

34 J. Skujins, *Bull. Ecol. Res. Comm. NFR* **17,** 235 (1973).

35 W. Flaig, "The Chemistry of Humic Substances," in *The Use of Isotopes in Soil Organic Matter Studies,* Report of FAO/IAEA Technical Meeting, Pergamon, New York, 1966, pp. 103–127.

36 M. M. Kononova, *Soil Organic Matter,* Pergamon, Oxford, 1966.

37 K. Haider and J. P. Martin, *Soil Sci. Soc. Amer. Proc.* **39,** 657 (1975).

38 K. Haider and J. P. Martin, *Soil Biol. Biochem.* **2,** 145 (1970).

39 J. P. Martin and K. Haider, *Soil Sci.* **107,** 260 (1969).

40 J. P. Martin and K. Haider, *Soil Sci.* **111,** 54 (1971).

41 J. P. Martin, K. Haider, and D. Wolf, *Soil Sci. Soc. Amer. Proc.* **36,** 311 (1972).

42 F. J. Stevenson, *Humus Chemistry: Genesis, Composition, Reactions,* Wiley-Interscience, New York, 1982.

43 P. Dubach and N. C. Mehta, *Soil Fertilizers* **26,** 293 (1963).

44 F. J. Stevenson and J. H. A. Butler, "Chemistry of Humic Acids and Related Pigments," in G. Englinton and M. T. J. Murphy, Eds., *Organic Geochemistry,* Springer-Verlag, New York, 1969, pp. 534–557.

45 M. Schnitzer, "Recent Findings on the Characterization of Humic Substances Extracted from Soils," in *Proceedings Symposium on Soil Organic Matter,* FAO/IAEA, Vienna, 1977, pp. 117–131.

46 M. Schnitzer and S. U. Khan, *Humic Substances in the Environment,* Dekker, New York, 1972.

47 F. Scheffer and B. Ulrich, *Humus und Humusdüngung,* Ferdinand Enke, Stuttgart, 1960.

48 M. H. B. Hayes and R. S. Swift, "The Chemistry of Soil Organic Colloids," in D. J. Greenland and M. H. B. Hayes, Eds., *The Chemistry of Soil Constituents,* Wiley, New York, 1978, pp. 179–320.

49 J. A. E. Buffle, "Les Substances Humiques et leurs Interactions avec les Ions Mineraux," in *Conference Proceedings de la Commission d'Hydrologie Applique de l'AGHTM,* l'Universite d'Orsay, 1977, pp. 3–10.

50 R. D. Haworth, *Soil Sci.* **111,** 71 (1971).

51 C. W. Bingeman, J. E. Varner, and W. P. Martin, *Soil Sci. Soc. Amer. Proc.* **17,** 34 (1953).

52 M. J. Hallam and W. V. Bartholomew, *Soil Sci. Soc. Amer. Proc.* **17,** 365 (1953).

53 B. J. O'Brien and J. D. Stout, *Soil Biol. Biochem.* **10,** 309 (1978).

54 International Atomic Energy Agency (IAEA), *The Use of Isotopes in Soil Organic Matter Studies,* FAO/IAEA, 1966, pp. 199–203.

55 International Atomic Energy Agency (IAEA), *Isotopes and Radiation in soil Organic Matter Studies,* FAO/IAEA, Vienna, 1968, pp. 3–11, 57–66, 197–205, 241–250, 351–361.

56 International Atomic Energy Agency (IAEA), *Tracer Manual on Crops and Soils, Tech. Report Series 171,* FAO/IAEA, Vienna, 1976.

57 A. Anderson, G. Nielsen, and H. Sørensen, *Physiol. Plantarum* **14,** 378 (1961).

58 E. A. Paul and A. D. McLaren, "Biochemistry of the Soil Subsystem," in E. A. Paul and A. D. McLaren, Eds., *Soil Biochemistry,* Vol. 3, Dekker, New York, 1975, pp. 1–36.

59 K. D. Ivarson and I. L. Stevenson, *Can J. Microbiol.* **10,** 677 (1964).

60 J. A. Shields, E. A. Paul, and W. E. Lowe, *Soil Biol. Biochem.* **6,** 31 (1974).

61 D. S. Jenkinson, *J. Soil Sci.* **17,** 280 (1966).

62 J. A. Shields and E. A. Paul, *Can. J. Soil Sci.* **53,** 297 (1973).

63 J. W. Nyhan, *Soil Sci. Soc. Amer. Proc.* **39,** 643 (1975).

64 D. S. Jenkinson and A. Ayanaba, *Soil Sci. Soc. Amer. J.* **41,** 912 (1977).

65 D. S. Jenkinson, *J. Soil Sci.* **16,** 104 (1965).

66 D. S. Jenkinson and J. H. Rayner, *Soil Sci.* **123,** 298 (1977).

67 F. E. Clark and E. A. Paul, *Adv. Agron.* **22,** 375 (1970).

68 H. Sørensen, *Soil Sci.* **95,** 45 (1963).

69 F. Löhnis, *Soil Sci.* **22,** 355 (1926).

70 L. A. Pinck and F. E. Allison, *Soil Sci.* **71,** 67–75.

71 F. E. Broadbent and W. V. Bartholomew, *Soil Sci. Soc. Amer. Proc.* **13,** 271 (1948).

72 F. E. Broadbent and A. G. Norman, *Soil Sci. Soc. Amer. Proc.* **11,** 264 (1946).

73 M. J. Hallam and W. V. Bartholomew, *Soil Sci. Soc. Amer. Proc.* **17,** 365 (1953).

74 W. H. Patrick, Jr., "Nitrogen Transformations in Submerged Soils," in F. J. Stevenson, Ed., *Nitrogen in Agricultural Soils,* American Society of Agronomy, Madison, Wis., 1982, pp. 449–465.

75 F. N. Ponnamperuma, *Adv. Agron.* **24,** 29 (1972).

76 N. K. Savant and S. K. De Datta, *Adv. Agron.* **35,** 241 (1982).

77 K. R. Reddy, *Plant Soil* **67,** 209 (1982).

78 T. Yoshida, "Microbial Metabolism of Flooded Soils," in E. A. Paul and A. D. McLaren, Eds., *Soil Biochemistry,* Vol. 3, Dekker, New York, 1975, pp. 83–122.

79 Y. Takai, *J. Sci. Soil Manure* **28,** 40, 138 (1967).

80 Y. Takijima, *Soil Sci. Plant Nutr.* **10,** 7, 14 (1964).

81 G. Goodlass and K. A. Smith, *Soil Biol. Biochem.* **10,** 201 (1964).

82 K. A. Smith and S. W. F. Restall, *J. Soil Sci.* **22,** 430 (1971).

83 K. A. Smith and R. S. Russell, *Nature* **222,** 769 (1969).

84 O. W. Kilham and M. Alexander, *Soil Sci.* **137,** 419 (1984).

TWO

CARBON BALANCE OF THE SOIL AND ROLE OF ORGANIC MATTER IN SOIL FERTILITY

All soils contain carbon (C) in the form of organic matter, or humus, the two terms being used synonymously. Absolute amounts of C vary considerably from one soil to another, being as low as 1% or less in coarse-textured soils (sands) to as much as 3.5% in prairie grassland soils (e.g., Mollisols). Poorly drained soils (Aquepts) often have C contents approaching 10%. Many tropical soils, such as the Oxisols, are notoriously low in organic C.

For all practical purposes, the organic C content of the soil parallels that for N. The latter, being the more easily measured parameter, is frequently used as an index of organic matter or C content. The C/N ratio of the soil varies somewhat, but the usual range is from 10 to 12 (Chapter 5).

The factor 1.724 is normally used to convert organic C to organic matter and is based on an assumed C content of 58% for the organic matter. It should be noted, however, that the C content of soil organic matter can vary considerably from this value and that the percentage ususally decreases with depth in the profile.

Topics discussed in this chapter include (1) factors affecting organic C content, including effect of cropping; (2) ^{14}C-dating; (3) role and function of humus; and (4) value of commercial humates as soil amendments. For additional details on factors affecting organic matter levels, the work of Jenny[1-5] is recommended. Several reviews are also available on this subject,[6-9] as well as on ^{14}C dating[10] and the role and function of organic matter.[11-13]

FACTORS AFFECTING LEVELS OF ORGANIC C IN SOILS

Organic C does not accumulate indefinitely in well-drained soils, and with time an equilibrium level is attained that is governed by the soil-forming factors of climate (cl), topography or relief (r), vegetation and organisms (o), parent material (p), and age or time (t).[1-9]

$$C \text{ or } N = f(cl, r, o, p, t, \ldots) \qquad (1)$$

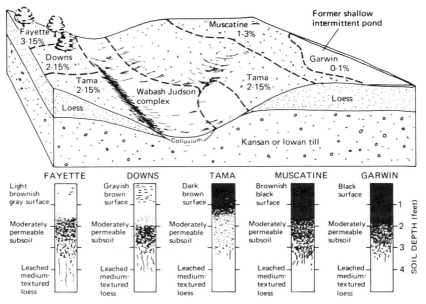

Fig. 2.1

Variability in soil organic matter content within a soil association as influenced by such factors as topography, drainage, parent material, and vegetation. Two of the soils are Alfisols (Fayette and Downs) while three are Mollisols (Tama, Muscatine, and Garwin). (Reprinted with permission of Macmillan Publishing Company from *The Nature and Properties of Soils* by N. C. Brady. Copyright © 1974 by Macmillan Publishing Company.)

From the preceding, a set of individual equations can be developed, for example,

$$C \text{ or } N = f(\text{climate})_{r,o,p,t} \qquad (2)$$

where subscripts indicate that the remaining factors do not vary.

As suggested by the preceding equation, evaluation of any given factor requires that all other factors remain constant, which seldom, if ever, occurs under natural conditions. The numerous combinations under which the various factors operate account for the great variability in the organic C content of soils, even in a very localized area. Figure 2.1 depicts differences in organic matter content within a soil association as influenced by such factors as topography, drainage, parent material, and vegetation.[14]

Jenny and his coworkers[1-5] attempted to evaluate the importance of the soil-forming factors on the N content of the soil by treating each one as an independent variable. This approach can be criticized on the grounds that an alteration of any one factor produces changes in one or more of the remaining factors. Nevertheless, his studies have contributed substantially

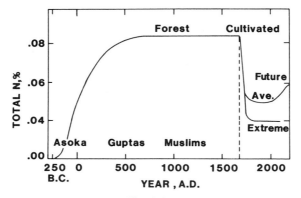

Fig. 2.2

Idealized time function for N accumulations in loam soils of the Indo-Gangetic Divide in India as envisioned by Jenny and Raychaudhuri.[5]

to our understanding of the factors influencing the N and organic C contents of soil. According to Jenny,[1] the order of importance of the soil-forming factors in determining the N content of loamy soils within the United States as a whole is as follows: climate > vegetation > topography = parent material > age.

The Time Factor of Soil Formation

Information on the influence of time on soil organic matter (C and N) levels has come from studies of time sequences (or chronosequences) on mud flows, spoil banks, sand dunes, road cuts, and the moraines of receding glaciers. This work, reviewed elsewhere,[8,9] has shown that organic matter levels increase rather rapidly during the first few years of soil formation, subsequently slows down, and attains an equilibrium level characteristic of the environment under which the soil was formed. In the case of moraines of the Alaskan glaciers, equilibrium levels of C and N were attained within about 110 years.[15] Similar results have been attained during soil development on other landscapes, including strip mine spoils.[16,17]

Under drought conditions, rather long periods of time may be required to attain equilibrium levels of C in soil. Thus Syers et al.[18] found that organic C was still increasing after 10,000 years of soil formation on wind-blown sand in New Zealand. Over extended periods, of the order of geological time, the C level may undergo further change because of variations in climate or of alteration in soil composition through pedogenic processes.

An idealized diagram showing the effect of time on the N (C) content of loam soils of the Indo–Gangetic Divide in India, as envisioned by Jenny and Raychaudhuri,[5] is shown in Fig. 2.2. In this case, steady-state levels of N during soil development on sediments deposited by flood waters during the

Asoka period (250 B.C.) were believed to have been attained sometime near the end of the Gupta Dynasty (500 A.D.) and to have remained in this primeval condition for a millennium until the reign of Shah Jahn of Taj Mahal fame (about 1650 A.D.), at which time the area was converted to agricultural use. Decreases in soil N levels were subsequently brought about through the activities of man, with establishment of new steady-state levels of organic C.

Although several reasons have been given for the establishment of equilibrium levels of organic matter in soil, none has proven entirely satisfactory. Included with the explanations are:

1 Organic colloids (e.g., humic acids) are produced that resist attack by microorganisms.
2 Humus is protected from decay through its interaction with mineral matter (e.g., polyvalent cations and clay).
3 A limitation of one or more essential nutrients (N, P, S) places a ceiling on the quantity of stable humus that can be synthesized.

The process of soil formation also leads to considerable migration of organic C into the lower soil horizons, often in association with clay or metal ions. Worm and root channels and ped surfaces become coated with dark-colored mixtures of humus and clay. Buol and Hole[19] observed that clay coatings on ped surfaces have considerably higher organic C contents than material within the peds. In some soils, streaks or tongues result from the downward seepage of humus. Illuvial humus also appears as coatings on sand and silt particles. Localized accumulations of sesquioxides (Fe and Al) and humus are common. In many soils, a secondary maximum in humus content coincides with an accumulation of clay in the B horizon. Transfer of organic matter into the lower soil horizons can continue for some time after equilibrium levels are reached in the surface layer; eventually, the total quantity in the profile stabilizes and remains essentially constant over time.

In addition to leaching, organic matter can be transported downward in soil through the action of soil animals. Earthworms, for example, can completely mix soil to depths of a meter or so, transferring organic matter downward in the process. Burrowing animals move soil material low in organic matter from the deeper horizons to the surface and vice versa.

Effect of Climate

Climate is the most important single factor that determines the array of plant species at any given location, the quantity of plant material produced, and the intensity of microbial activity in the soil; consequently, this factor plays a prominent role in determining organic matter levels. Considering climate in its entirety, a humid climate leads to forest associations and the development of Spodosols and Alfisols; a semiarid climate leads to grassland

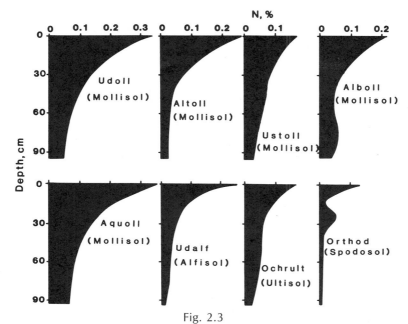

Fig. 2.3

Profile distribution of N in representative soils of the north central region of the United States (Adapted from Stevenson.[9])

associations and the development of Mollisols. Grassland soils exceed all other well-aerated soils in organic matter content; desert, semidesert, and certain tropical soils have the lowest. The profile distribution of N in soils representative of several great soil groups in the North Central region of the USA is shown in Fig. 2.3.

Soils formed under restricted drainage (Histosols and Inceptisols) do not follow a climatic pattern. In these soils, O_2 deficiency prevents the complete destruction of plant remains by microorganisms over a wide temperature range.

Extensive studies were made by Jenny and his coworkers[1-5] on the effect of climate on N levels in soil. Jenny's[1] results for north to south transects of the semiarid, semihumid, and humid regions of the USA are shown in Fig. 2.4. In each case the N content of the soil was two to three times lower for each rise of 10°C in mean annual temperature. Whereas relationships derived for USA soils cannot be extrapolated directly to other areas, it is well known that many soils of the warmer climatic zones have very low N (and C) contents. However, some tropical soils have C contents that compare favorably with those of temperate regions.[7]

Several explanations have been given for the decrease in soil N (and C) levels with an increase in mean annual temperature. Jenny[1-5] explained this

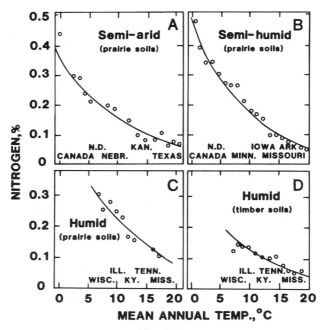

Fig. 2.4

Average total N content as related to mean annual temperature for soils along north to south transects of the semiarid, semihumid, and humid regions of the central United States.[1] (Reproduced from *Nitrogen in Agricultural Soils* (1982) by permission of the American Society of Agronomy.)

relationship to temperature effects on biological activity, as expressed by the van't Hoff equation. Thus soil N content was described by the formula

$$N = \frac{a}{1 + Ce^{kt}}$$

where t is the mean annual temperature, e is base of the natural logarithm, and a, C, and k are constants.

A major criticism of Jenny's use of the van't Hoff equation is that the theory only takes into account the influence of temperature on the activities of soil microorganisms; temperature effects on photosynthesis (production of raw material for humus synthesis) are ignored. Also no allowance is made of time–temperature relationships. In cold climates, for example, the soil is frozen for a good portion of the year and biological activity is nil. A differential effect of temperature on the activities of microorganisms and higher plants has been pointed out by Senstius.[20] Specifically, as the temperature is raised, the activities of microorganisms increase more so than does the

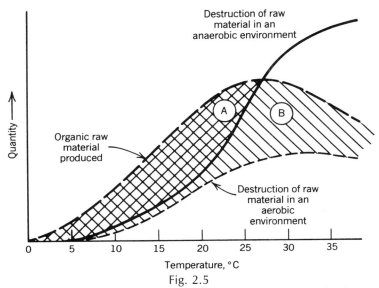

Fig. 2.5

Influence of temperature on organic raw material production through photosynthesis and organic matter destruction by microorganisms. Accumulations of humus under aerobic soil conditions are confined to zone A; accumulations under anaerobic conditions occur over the entire temperature range (A + B). (Adapted from Senstius.[20])

photosynthetic process of higher plants. Furthermore, the life activities of higher plants start out at lower temperatures than do those of microorganisms (0°C versus 5°C), and the optimum is lower (25°C versus 30°C). On this basis, temperatures below about 25°C should favor the production and preservation of humic substances; higher temperatures should favor destructive processes. Inability to maintain organic matter at high levels in tropical soils was attributed by Senstius[20] to the higher activities of microorganisms at the warmer temperatures, as depicted in Fig. 2.5.

Enders[21] presented a unique concept concerning humus accumulations in soil. He concluded that the best soil conditions for the synthesis and preservation of humic substances were frequent and abrupt changes in the environment (e.g., humidity and temperature); consequently, soils formed in harsh continental climates should have high C and N contents. Harmsen[22] used this same theory to explain the greater synthesis of humic substances in grassland soils, as compared to arable land, claiming that in the former the combination of organic substrates in the surface soils and frequent and sharp fluctuations in temperature, moisture, and irradiation leads to greater synnthesis of humic substances. According to Harmsen,[22] the extreme surface of the soil (upper few millimeters) is the site of synthesis of humic substances and fixation of N.

The effect of increasing rainfall (moisture component of climate) on soil

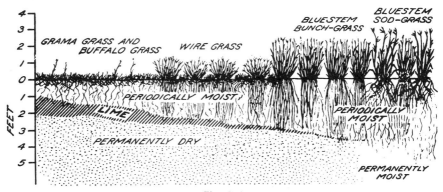

Fig. 2.6

Relationship between vegetative growth and moisture supply along a west to east transect to the Great Plains. (From Shantz.[23])

organic matter content is to promote greater plant growth and the production of larger quantities of raw material for humus synthesis. The quantity of plant material produced, and subsequently returned to soil, can vary from a trace in arid and arctic regions to several metric tons per hectare in warm climates where plant growth occurs throughout the year. Both roots and tops serve as energy sources for humus synthesis.

Total rainfall is not a satisfactory index of soil moisture, because of great variations in evaporation. Jenny[1,2] has used what is known as the NS quotient, which is the ratio of precipitation (in mm) to the absolute saturation deficit of the air (in mm of Hg).

For grassland soils along a west to east transect of central USA, a definite correlation exists between the depth of penetration of the root system (and thickness of the grass cover) and the amount of N contained in the surface layer (compare Fig. 2.6 with Fig. 2.7).

Vegetation

It is a well-known fact that, other factors being constant, the C content of grassland soils (e.g., Mollisols) is substantially higher than that for forest soils (e.g., Alfisols). Some of the reasons given for this are as follows:

1   Larger quantities of raw material for humus synthesis are produced under grass.

2   Nitrification is inhibited in grassland soils, thereby leading to the preservation of N (and C).

3   Humus synthesis occurs in the rhizosphere, which is more extensive under grass than under forest vegetation.

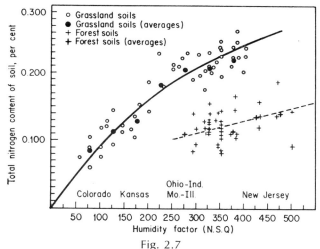

Fig. 2.7

Relationship between N content and humidity factor for soils along a west to east transect of the central United States. (From Jenny.[1])

4 Inadequate aeration occurs under grass, thereby contributing to organic matter preservation.

5 The high base status of grassland soils promotes the fixation of NH_3 by lignin and thereby conservation of N and C.

A combination of several factors is probably involved, with item (3) being of major importance. In a study involving transformation of ^{14}C-labeled organic substrates, Reid and Goss[24] found that decomposition rates for applied organic matter were reduced in the rhizosphere of corn and perennial rye.

In the case of forest soils, differences in the profile distribution of C and N occur by virtue of the manner in which the leaf litter becomes mixed with mineral matter. In soils formed under deciduous forests on sites that are well drained and well supplied with Ca (Alfisols), the litter becomes well mixed with the mineral layer through the activities of earthworms and other faunal organisms. In this case, the top 10–15 cm of soil becomes coated with humus. On the other hand, on sites low in available Ca (Spodosols), the leaf litter does not become mixed with the mineral layer but forms a mat on the soil surface. An organic-rich layer of acid (mor) humus accumulates at the soil surface and humus accumulates only in the top few cm of soil.

Parent Material

Parent material affects the C content of the soil through its influence on texture. It is a well-known fact that, for any given climatic zone, and provided vegetation and topography are constant, C and N contents depend on

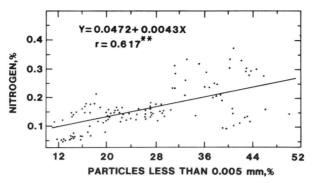

Fig. 2.8

Relationship between N content and percent of particles less than 0.005 mm for the surface soils of virgin sod from 13 locations in the Great Plains region of the United States. (From Haas et al.[25])

textural properties. The fixation of humic substances in the form of organic–mineral complexes serves to preserve organic matter. Thus heavy-textured soils have higher C contents than loamy soils, which in turn have higher C contents than sandy soils. A typical result showing a positive correlation between organic matter (N) content and particles less than 0.005 mm is given in Fig. 2.8.

Organic matter has several characteristics, such as resistance to attack by microorganisms and to removal by chemical extractants, which suggest that it occurs in intimate association with mineral matter. Retention may also be affected by the type of clay mineral present. Montmorillonitic clays, which have high adsorption capacities for organic molecules, are particularly effective in protecting nitrogenous constituents against attack by microorganisms.

Topography

Topography, or relief, affects C content through its influence on climate, runoff, evaporation, and transpiration. Local variations in topography, such as knolls, slopes, and depressions, modify the plant microclimate, defined by Aandahl[26] as the climate in the immediate vicinity of the soil profile. Soils occurring in depressions, where the climate is "locally humid," have higher C contents than those occurring on the knolls, where the climate is "locally arid."

Plant remains are not completely metabolized in water-logged environments (see Chapter 1). Thus naturally moist and poorly drained soils are usually high in organic matter, because the anaerobic conditions that prevail during wet periods of the year prevent destruction of organic matter. Accumulations are possible over the entire temperature range (see Fig. 2.5).

Accumulations of organic matter are especially evident in Histosols (peats and mucks), but many swamps are also rich in organic matter. A serious economic problem is encountered when Histosols are drained for agricultural use, because of loss of soil through subsidence.[27] The major cause is enhanced microbial oxidation of the soil organic matter, with accompanying conversion of organic C to CO_2.

EFFECT OF CROPPING ON SOIL ORGANIC MATTER LEVELS

Marked changes are brought about in the C (and N) content of the soil through the activities of man. Usually, but not always, organic C levels decline when soils are first placed under cultivation. For extensive documentation of this work, the reviews of Stevenson[9] and Campbell[28] are recommended.

Early findings, such as those obtained by Salter and Green[29] for some rotation experiments at the Ohio Agricultural Experiment Station showed that loss of N was least with rotations containing a legume and greatest under continuous row cropping. An item of some significance was that losses were more or less linear over the 31-year period of the experiment. Therefore unless a change occurred, organic matter content would ultimately reach an absolute minimum (e.g., approach zero). The following equation was derived by Salter and Green[29] to describe N loss:

$$N = N_0 e^{-rt} \quad \text{or} \quad \frac{dN}{dt} = -rN \tag{3}$$

where N_0 is the initial N content and r is the fraction of the N remaining after a single year's cropping.

The conclusion that virtually all the organic matter (and N) might eventually be lost from soil by cropping caused considerable alarm in agronomic circles. It was feared that unless drastic measures were taken to maintain organic matter reserves, many soils would become unproductive.

Jenny's[30] early results for losses of soil N over a 60-year period under average farming conditions in the corn-belt section of the United States were in general agreement with the findings of Salter and Green in that cropping caused a decline in organic matter levels. However, his findings suggested that destruction of organic matter would not be complete but that new *equilibrium levels* would be attained. For soils of the corn belt, about 25% of the N was found to be lost the first 20 years, 10% the second 20 years, and 7% the third 20 years. The data suggested that new equilibrium levels would be attained within a 60- to 100-year time period. As shown earlier, a similar time period was required to attain equilibrium levels of organic matter during soil formation.

The validity of Jenny's equilibrium concept has been established in sev-

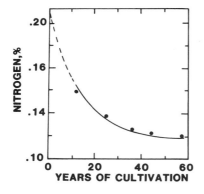

Fig. 2.9
Decline of soil N content as influenced by years of cropping at Hays, Kansas. (From Hobbs and Brown.[32])

eral long-time cropping systems (e.g., see Campbell[28]). Results obtained for the soils at Hays, Kansas, and for the Morrow Plots at the University of Illinois are given in Figs. 2.9 and 2.10, respectively. In each case, N losses during the early years were much more rapid than those in later years. In the case of the Morrow plots, losses were greatest with the continuous corn

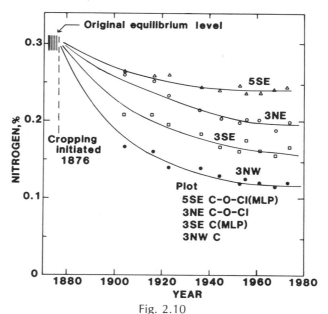

Fig. 2.10
Effect of long-time rotations on the N content of select soils from the Morrow plots at the University of Illinois.[9] C = corn; O = oats, Cl = clover; MLP = manure-lime-phosphate. (Reproduced from *Nitrogen in Agricultural Soils* (1982) by permission of the American Society of Agronomy.)

plot and least with the corn–oats–clover rotation plot amended with manure, lime, and phosphate.

Jenny attempted to correct the deficiency in Salter and Green's equation by including a factor for the annual return of N to the soil. Thus equation 3 becomes:

$$\frac{dN}{dt} = -rN + A \tag{4}$$

where A is the annual rate of N addition.

This equation was transformed to the following form by Bartholomew and Kirkham:[31]

$$N = \frac{A}{r} - \left(\frac{A}{r} - N_0\right)e^{-rt} \tag{5}$$

A plot of N versus e^{-rt}, therefore, should yield a straight line in which the y intercept (A/r) would describe the expected equilibrium value and the $(N_0 - A/r)e^{-rt}$, the change process. The change in the magnitude of the latter with time provides a measure of the rate of establishment of equilibrium.

Graphical methods were used by Bartholomew and Kirkham[31] to obtain the constants A and r for the experimental plots of several long-time rotation experiments. In brief, their findings were in agreement with Jenny's observation that equilibrium levels can be expected to be attained within a 60- to 100-year period of cultivation.

Crop yields have a potential effect on organic N levels through a feedback effect (greater return of plant residues as yields increase). Russell[33] used computer-based numerical methods to predict long-term effects of increased yields on N levels for the Morrow Plots at the University of Illinois and the Sanford Field at the University of Missouri. The basic equation was:

$$\frac{dN}{dt} = -K_1(t)*N + K_2 + K_3(t)*Y(t) \tag{6}$$

where $Y(t)$ is plant yield at time t, $K_1(t)$ is the decomposition coefficient, K_2 represents addition to soil organic matter from noncrop sources (including manures), and $K_3(t)$ is a coefficient related to the specific crop at time t. This equation permits estimates to be made of the effect of crop yield within a rotation on soil N levels.

For the Morrow Plots, increasing corn yields in a continuous corn system had negligible effects on soil N levels, but strong positive effects were noted for oats and clover. All crops in the Sanford Field had some feedback on soil N levels.

The decline in the organic matter content of the soil when land is cultivated cannot be attributed entirely to a reduction in the quantity of plant residues available for humus synthesis. A temporary increase in respiration rate occurs each time an air-dried soil is wetted,[34,35] and since considerable amounts of fresh soil are subjected to repeated wetting and drying through cultivation, losses of organic matter by this process could be appreciable. Still another effect of cultivation in stimulating microbial activity may be exposure of organic matter not previously accessible to microorganisms.

Agricultural practices that may accelerate decomposition and mineralization of soil humus are as follows:

Cultivation   Improves aeration and moisture, thereby increasing microbial activity and the release of organic compounds to soluble forms.

Irrigation   Improves the moisture status of the soil, with enhancement in the activities of microorganisms.

Liming   Increases the activities of earthworms and other faunal organisms; encourages actinomycetes, which may be more effective decomposers than bacteria or fungi; facilitates the precipitation of metallic cations that are effective in stabilizing humic substances.

Green manuring   Greatly increases the numbers of microorganisms and thereby the rate of oxidation of organic matter.

For many cultivated soils, particularly the Mollisols, organic matter and N can only be maintained at a level approaching that of the native uncropped soil by inclusion of a sod crop in the rotation or by frequent and heavy applications of manures and crop residues. Increases in soil N (and C) levels have been observed by returning previously tilled soil to a grass sod,[36-39] a typical result being shown in Fig. 2.11. Introduction of legumes on soils initially low in N has been reported to lead to increased levels of soil N.[40] Increases in soil N levels have been also observed through zero tillage.[40-42]

In contrast to the work alluded to earlier, indicating an initial rapid loss of soil N following a change in cultural practice, linear changes were observed by Rasmussen et al.[43] for a wheat–fallow cropping sequence on a Pacific Northwest semiarid soil (45-year cropping period). As shown in Fig. 2.12, linear declines in soil N occurred with or without return of wheat straw residues to the soil; a linear increase occurred when manure was applied at a high rate (22.4 metric tons/ha·yr). The greatest loss of N occurred when the wheat straw was burned rather than returned to the soil. For this soil, well over 100 years will be required for establishment of new equilibrium levels of organic matter and N.[43]

Jenkinson and Johnson[44] demonstrated a pronounced increase in soil N content through long-time applications of manure (35 tons/ha·yr) to soil under permanent barley at the Rothamsted Experimental Station in England

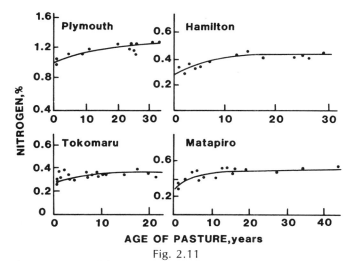

Fig. 2.11

Changes in the N content of the surface 7.6 cm of soil versus time after establishment of permanent pasture on four New Zealand soils. (From Jackman.[38])

(Fig. 2.13). The plots were established in 1852 and equilibrium levels had still not been attained in the manured plot by the time they were sampled 123 years later (in 1975). Only minor changes in soil N levels occurred in the unmanured plot, while a slow but continuous decline occurred in a plot receiving farmyard manure annually between 1852 and 1871 but not there-

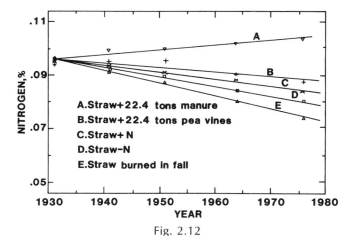

Fig. 2.12

Soil N levels (30 cm depth) in a silt loam soil near Pendleton, Oregon, as related to crop residue treatment in a wheat–fallow rotation system. (From Rasmussen et al.[43])

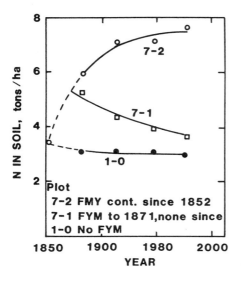

Fig. 2.13
Nitrogen in the top 23 cm of a soil under continuous barley at the Rothamsted Experimental Station. (From Jenkinson and Johnson.[44])

after. Anderson and Peterson[45] observed increases in soil N content by long-time applications of manure (27 metric tons/ha·yr) to a Nebraskan soil.

Results obtained by Larson et al.[46] indicated that application of plant residues at a rate of 6 tons/ha·year was needed for maintenance of organic matter in a Typic Haplaquoll soil in Iowa (Marshall silty clay loam); higher application rates led to increases in soil N. Rasmussen et al.[43] predicted that an annual application of 5 metric tons of mature crop residue per hectare would be required to maintain soil organic matter at present levels in a wheat–fallow rotation on Pacific Northwest semiarid soils. For a wheat-fallow system in a cooler climate, a smaller quantity of plant residues may be needed.[47] Hobbs and Brown[32] concluded that 56–67 metric tons of farmyard manure every three years were necessary to prevent further loss of organic matter from some prairie soils of Kansas. In other work, Sivapalan[48] observed that residues rich in polyphenols promote the synthesis of large amounts of "true humic matter" with a high N content.

The effectiveness of organic residues in maintaining soil organic matter reserves depends upon such factors as rate of application, kind of residue and its N content (C/N ratio), manner of incorporation into the soil, soil characteristics, and seasonal variations in temperature and moisture. Select references on soil management factors affecting organic matter levels in soils from widely different climatic regions of the earth can be found in several reviews.[8,9,28] Factors affecting the fate of organic wastes when applied at high disposal rates to soils of the various climatic regions of the United States (arid, cool subhumid and humid, hot humid) are discussed in the volume by Elliott and Stevenson.[49]

PALEOHUMUS

Remnants of plant and animal life, as well as products produced from them through humification, have been observed in buried soils (paleosols) of all ages.[50-54] This organic matter, or paleohumus, is of interest in geology and pedology because of its importance as a stratigraphic marker and as a key to the environment of the geologic past. Thus the occurrence of dark-colored humus zones, when used in conjunction with other pedological observations, has served as a basis for establishing the identification of buried soils, from which it has been possible to draw conclusions relative to climate, vegetative patterns, and the morphology of former land surfaces.[50]

Humus is the fraction of the soil most susceptible to change. Thus, in most instances, the humus found in buried soils represents only a small fraction of the amount initially present. The degree to which humus has been

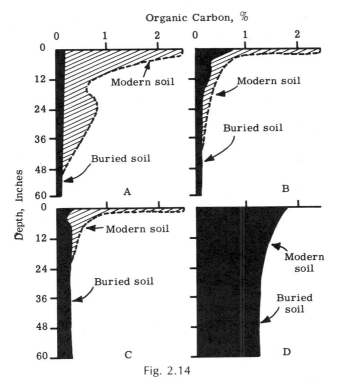

Fig. 2.14

Comparison of organic C contents of some paleosols with their supposed modern counterparts (from Stevenson[50] as prepared from literature data. © 1969 The Williams & Wilkins Co., Baltimore). (A) Yarmouth "Planosol" (Simonson[51,52]). (B) Yarmouth "Gray-brown Podzolic" (Simonson[51]). (C) Farmdale "Gray-brown Podzolic" (Hogan and Beatty[53]). (D) Early-Wisconsinan "Regosol" (Ruhe et al.[54]).

lost will depend on such factors as the change in environment that preceded the new cycle of sedimentation, the circumstances under which the soil was buried, and activities of living organisms in the buried sediment.

In Fig. 2.14 the organic carbon contents of four paleosols are compared with those of their modern counterparts. Except for the Wisconsinan "Regosol" (profile D) much of the humus had been lost, particularly from the surface layer.

^{14}C DATING OF SOILS

Absolute ages cannot be determined for the organic matter of soils or recent sediments because of continued decomposition of old humus and resynthesis of new humus by microorganisms. The term *mean residence time* has been used to express the results of ^{14}C measurements for the *average age* of modern humus. The resistance of humus to biological decomposition has long been known but not until the advent of ^{14}C dating was it possible to express average age on a quantitative basis.

Principles of ^{14}C Dating

Organic materials formed from atmospheric CO_2 acquire during their synthesis a small amount of cosmic ray–produced ^{14}C. When photosynthesis ceases, slow radioactive decay of the ^{14}C begins. This gradual loss of radioactivity provides a method for estimating the age of any residual C persisting in the environment. The ^{14}C-dating method, first discovered by Libby and his associates,[55] is suitable for accurate age estimates up to 50,000 years but extension to 70,000 years is possible.

Upon entry into the atmosphere, cosmic radiation produces neutrons, some of which are captured by N with the production of ^{14}C, according to the equation:

$$^{14}N + n \rightarrow {}^{14}C + H$$

The ^{14}C combines with atmospheric O_2 to form $^{14}CO_2$, which is carried to the earth's surface by convection and turbulent mixing in the troposphere. Through photosynthesis, plants acquire small amounts of this ^{14}C. When the plant (or the animal which consumes the plant) dies and CO_2 exchange with the atmosphere ceases, the ^{14}C content of the material (the $^{14}C/^{12}C$ ratio) diminishes with time because of radioactive decay. The basis for age estimate is detection of the charged beta particle. The number of decaying atoms per minute is proportional to the ^{14}C content of the material under investigation.

The ^{14}C radioactivity of a given sample decreases in a predictable manner with time, the half-life being 5,730 years. After this period, only one-half of the original ^{14}C activity will be left. After twice this period, or 11,460 years, only one-fourth will be left, and after another 5,730 years, or a total of 17,190 years, only one-eighth will be left, and so on. The formula relating age to ^{14}C activity is:

$$\log A = \log A_0 - \frac{\log 2}{t^{1/2}}t$$

where A is the number of radioactive nuclei remaining after time interval t, A_0 is the number of radioactive nuclei present at zero time, t is time or age since zero time, and $t^{1/2}$ is the half-life of the radioactive nuclide. The steady-state specific activity of ^{14}C (A_0) has been determined to be 15 disintegrations per minute per gram of C (d/m·g). A wood fragment found to have a specific activity of 5 d/m·g would have an estimated age of 9,100 years.

The fundamental assumption underlying the ^{14}C dating method is that production of ^{14}C by cosmic radiation has been constant for at least 70,000 years and that C exchange between the reservoirs (atmosphere, biosphere, and oceans) has followed the same rate. These conditions appear to have been fulfilled, although deviations are known to have occurred. Since about 1870, the distribution of ^{14}C has been complicated by consumption of fossil fuel, which has diluted the ^{14}C in the atmosphere by release of ^{12}C as CO_2. More recently, the explosion of thermonuclear bombs has added large amounts of ^{14}C to the atmosphere; in 1964, the ^{14}C level of the atmosphere was twice the natural level. At present, the ^{14}C content of the atmosphere is still much higher than the previous normal level. Enrichment of the atmosphere with "bomb" radiocarbon has been used as the basis for determining rates of movement and turnover of organic C in soils.[10,56]

Mean Residence Time (MRT) of Soil Organic Matter

Because of the very low levels of naturally occurring ^{14}C, special preparative techniques are required for age determinations and care must be taken to prevent contamination.

Typical MRTs for C in the surface layer of the soil are recorded in Table 2.1. While considerable variation in mean ages have been reported (250 to over 3,000 years), the findings attest to the high resistance of humus to microbial attack. In general, humus of the Chernozem (Mollisol) is more stable than that of the Gray-Brown Podzolic (Ultisol).

Results obtained for the MRTs of individual fractions of humus suggest that, within the organic pool of the soil, some components are more stable than others.

Campbell et al.[58] employed a rather complex fractionation scheme based on alkali extraction and acid hydrolysis of the isolated fractions to isolate components of the organic matter in a Chernozem (Mollisol) and a Gray-Wooded Podzolic (Ultisol) soil. The various fractions of the Chernozem soil were dated with MRTs that ranged from 25 to 1,400 years. A summary of

Table 2.1

Mean Residence Time (MRT) of Organic Matter in Some Representative Soils

Sample Description	MRT	Reference
Cheyenne grassland soil, North Dakota		
Virgin soil	1,175 ± 100	Paul et al.[57]
Clean cultivated area	1,900 ± 120	Ibid.
Manured orchard	880 ± 74	Ibid.
Bridgeport loam, Wyoming		
Surface sod layer and corn–fallow–corn plot	3,280	Ibid.
Clay fraction from virgin soil	6,690	Ibid.
Continuous wheat plot	1,815	Ibid.
Orthic Black Chernozem (Mollisol), Saskatchewan	1,000	Ibid.
Black Chernozemic (Mollisol), Saskatchewan	870 ± 50	Campbell et al.[58]
Grey-Wooded Podzolic (Alfisol), Saskatchewan	250 ± 60	
Black Chernozems (Mollisol), Saskatchewan, two catenas examined		
Crest of catena to depression	545 to present	Martel and Paul[59]
B horizons	700 to 4,000	
Buried horizons	5,900 and 8,410	

the main findings is shown in Table 2.2. The MRT of the unfractionated soil was 870 ± 50 years. Stability of the major humus fractions followed the order: humic acid > humin > fulvic acid. The material hydrolyzable in 6 N HCl, consisting of amino acids, carbohydrates, etc., had the lowest MRT. For the Ultisol (data not shown), fractions were obtained with MRTs of from 0 to 485 years; MRT for the unfractionated soil was 250 ± 60 years. Humic fractions from the Mollisol were more stable than their counterparts from the Ultisol.

Absolute Ages of Buried Soils

The humus component is seldom suitable for absolute age estimations of buried soils and this is usually attempted only as a last resort. More suitable materials for ^{14}C dating are charcoal, fossil wood, uncontaminated peaty layers, and well-preserved shells. Contamination of buried soils by more recent C is always a problem, such as by downward movement of soluble organics in percolating water and penetration of the buried soil by plant roots. Principles involved in dating pedogenic events in the Quaternary have been discussed by Ruhe.[60]

Table 2.2

Mean Residence Times (MRT) for Different Organic
Matter Fractions of a Chernozemic Black Soil
(Melfort silt loam)[a]

Component	MRT, years
Unfractionated soil	870 ± 50
Acid extract of soil	325 ± 60
Fulvic acid	495 ± 60
Humic acid	
Total sample	$1,235 \pm 60$
Acid hydrolysate	25 ± 50
Nonhydrolyzable	$1,400 \pm 60$
Humin	
Total sample	$1,140 \pm 50$
Acid hydrolysate	465 ± 50
Nonhydrolyzable	$1,230 \pm 60$

[a] Adapted from Campbell et al.[58]

Recognition of contamination may be difficult, particularly when there is no reason to suspect that the data obtained are in error. The likelihood of foreign C contamination is often indicated by widely divergent dates from the same stratigraphic unit or when an inversion occurs from known stratigraphic relationships (i.e., younger ^{14}C-dating age in material taken at a deeper depth).

Visible contaminants (e.g., root hairs) can be physically removed. The problem centers on elimination of younger humus material. A flow sheet showing a method of sample preparation and conversion of CO_2 to acetylene (C_2H_4) for counting of ^{14}C activity is given in Fig. 2.15.

Ruhe[60] cites examples where buried soils were dated and a comparison was made with dates of the NaOH-soluble (mobile) organic fraction. In one case the insoluble component had an age of $20,500 \pm 400$ years whereas the "mobile" fraction had an age of $13,400 \pm 600$ years, indicating that contamination had occurred. Agreement in the ages of the alkali-insoluble and soluble fractions would suggest little or no contamination and that the estimated date is valid.

In a recent study of radiocarbon dates obtained from soil-buried charcoals, Goh and Molloy[61] observed that the larger charcoal particles yielded a significantly older age than the finer particles. This was attributed to the presence of easily extractable fulvic acid in the smaller charcoal particles. The problem of removing organic contaminants from charcoal has been discussed by Goh.[63]

A new method has recently been proposed for dating sediments based on the racemization of amino acids. Living organisms synthesize amino acids

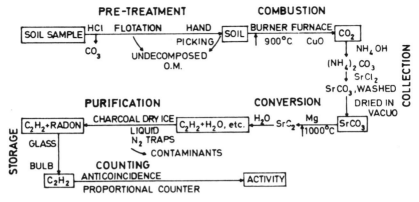

Fig. 2.15

Flow sheet showing sample preparation and conversion of CO_2 to acetylene (C_2H_4) for determination of ^{14}C activity. (From IAEA.[62])

that are largely in the L-stereoisomeric configuration. After burial, these compounds undergo slow conversion to the D-configuration. The degree of racemization provides an estimate of age and has been used for dating sediments and fossil calcareous shells. This work has been amply reviewed by Dungworth.[64]

ROLE AND FUNCTION OF SOIL HUMUS

Any discussion of the C balance of the soil would be incomplete without some consideration being given to the role of organic matter (humus) in crop production. Other factors being constant (climate, drainage, topography, parent material, etc.), soils rich in humus are generally more productive than those poor in humus. The high productivity of the dark-colored soils of the corn-belt section of the United States is usually attributed to their high content of stable humus.

Humus contributes to the fertility, or productivity, of the soil through its positive effects on the chemical, physical, and biological properties of the soil. It has a nutritional function in that it serves as a reservoir of N, P, and S for plant growth, a physical function in that it promotes good soil structure, and a biological function in that it serves as a source of energy for micro-organisms.

Nutrient Availability

The reviews of Stevenson[12,13] show that three interrelated organic matter fractions must be taken into account when considering nutrient–organic matter interactions in soil. They are (1) plant and animal residues, which upon decomposition provide N, P, and S for plant growth; (2) the microbial biomass, which serves as a temporary storage unit for nutrients (MRT of several months to a few years); and (3) the resistant humus fraction, which has a long MRT (250 to 1,000 + years). Under conditions where steady-state levels of organic matter have been attained, mineralization of native humus is compensated for by synthesis of new humus.

Conversion of organic N, P, and S to available mineral forms (NH_4^+, NO_3^-, PO_4^{3-}, SO_4^{2-}) occurs through the activity of microorganisms and is influenced by those factors affecting microbial activity (temperature, moisture, pH, etc.), as well as by the C/N, C/P, and C/S ratios of decomposing plant residues, as discussed in other chapters. The process is referred to as *mineralization* and is nearly always accompanied by conversion of mineral forms of the nutrients to organic forms, or *immobilization*.

$$\text{Organic N, P, S} \underset{\text{immobilization}}{\overset{\text{mineralization}}{\rightleftharpoons}} NH_4^+,\ NO_3^-,\ PO_4^{3-},\ SO_4^{2-}$$

The assumption is often made that 1 to 3% of the soil organic matter is mineralized during the course of the growing season, with net release of equal percentages of the N, P, and S. This statement must be accepted with reservation, for the reason that the humus content of most soils is in a state of quasi-equilibrium. A *net annual release* of nutrients occurs only when organic matter levels are declining, a condition that is to be avoided as exploitation generally leads to a reduction in the productive capacity of the soil. Any soil under constant management attains a balance between gains and losses of N, P, and S. Thereafter, nutrients continue to be liberated but the amounts released are compensated for by incorporation of equal amounts into newly formed humus. The possibility that plant nutrition can be improved on a sustaining basis by adoption of cultivation practices designed to promote the decomposition (and subsequent loss) of stable humus is limited.

In many countries, adequate quantities of plant food materials (N, P, and K fertilizers) are available to the farmer at reasonable cost. Undoubtedly, this has changed our concept of the role of humus in supplying nutrients for crop growth. Nevertheless, humus should be regarded as a positive asset to high crop production. Soil scientists generally agree that practices that deplete humus reserves lower the yield potential.

The C/N/P/S Ratio

While considerable variation is found in the C/N/organic P/total S ratio for individual soils, the mean for soils from different regions of the world is remarkably similar (see Chapter 7). As an average, the proportion of C/N/ P/S in soil humus is approximately 140:10:1.3:1.3. Ratios recorded for C and P are somewhat more variable than for C and N, or C and S. One explanation for the wider range of C/P ratios is that, unlike N and S, P is not a structural constituent of humic and fulvic acids.

The release of N, P, or S to available mineral forms can follow a number of pathways, the more common ones being (1) an initial net immobilization followed by net mineralization in later stages, (2) a steady rate of release with time, and (3) an initial net mineralization followed by net immobilization. The pattern of nutrient release has not been shown to be related to any given soil property, including the N, P, or S content of the soil.

A priori one might expect that the relative rates of mineralization of N, P, and S would be similar and that they would be released in the same ratios in which they occur in soil organic matter. However, this has not always been the case. Differences in the relative amounts of N, P, and S released from soil organic matter may be due to one or more of the following.

1 Nitrogen, P, and S occur in different organic compounds and organic matter fractions; consequently, they are not released in the same ratios.

2 The application of plant and animal residues results in differential mineralization–immobilization of nutrients. The N, P, and S contents of applied plant residues, as reflected through the C/N, C/P, and C/S ratios, play a major role in regulating the amounts of these nutrients that accumulate in available mineral forms at any one time (see Chapter 5). As a general rule, an initial *net mineralization* (gain in mineral forms of the nutrients) will occur at C/N ratios less than 20:1, C/P ratios less than 200:1, and C/S ratios less than 200:1. In contrast, an initial *net immobilization* of N will occur at C/N ratios of 30:1, of P at C/P ratios over 300:1, and of S at C/S ratios over 400:1. Intermediate ratios lead to neither a gain nor a loss of the nutrient.

3 The release of NH_4^+ and NO_3^- is obscured through fixation reactions involving NH_4^+ and NH_3 and by denitrification.

4 The release of PO_4^{3-} and SO_4^{2-} is obscured through formation of insoluble salts (e g., calcium phosphates and sulfates).

Indirect Effects

In addition to serving as a source of N, P, and S, organic matter influences nutrient supply in other ways. Organic matter is required as a source of energy for molecular N_2 fixation by microorganisms; accordingly, the amount of N_2 fixed by the free-living fixers will be influenced by the quantity of available energy in the form of carbohydrates. The process of denitrification is also affected by a supply of decomposable organic matter. Several investigators have demonstrated a direct correlation between denitrification rate and the content of soluble C in the soil solution (see Chapter 4).

The availability of phosphate in soil is often limited by fixation reactions, which convert the monophosphate ion ($H_2PO_4^-$) to various insoluble forms.

As will be pointed out in Chapter 7, addition of organic residues to the soil often enhances the availability of the native soil P to higher plants.

Aluminum toxicity is a major problem in many acid soils. However, acid soils rich in native organic matter, or amended with large quantities of organic residues, give low Al^{3+} concentrations in the soil solution and permit good growth of crops under conditions where toxicities would otherwise occur. The role of organic matter in alleviating toxicity effects of heavy metals is discussed in Chapter 9.

Source of Energy for Soil Organisms

Humus serves as a source of energy for both macro- and microorganisms. The numbers of bacteria, actinomycetes, and fungi in the soil are related in a general way to humus content; macrofaunal organisms are similarly affected.

The role played by the soil fauna has not been completely elaborated, but the functions they perform are multiple and varied. For instance, earthworms may be important agents in producing good soil structure. They construct extensive channels through the soil that serve not only to loosen the soil but also to improve drainage and aeration. Earthworms can flourish only in soils that are well provided with organic matter.

Growth of Higher Plants

Organic substances in soil have a direct physiological effect on plant growth. Some compounds, such as phenolic acids, have phytotoxic properties; others, such as the auxins, enhance the growth of higher plants. The condition of "soil sickness" has frequently been connected with the accumulation of organic toxins. Phytotoxic substances are believed to be responsible for the low yields of wheat under the stuble–mulch system of farming in semiarid regions of the world.

Soil Physical Condition

Humus has a profound effect on the structure of most soils. The deterioration of structure that accompanies intensive tillage is usually less severe in soils adequately supplied with humus. When humus is lost, soils tend to become hard, compact, and cloddy. Seedbed preparation and tillage operations are easier to carry out and are more effective when humus levels are high.

Aeration, water-holding capacity, and permeability are favorably affected by humus. The frequent addition of easily decomposable organic residues leads to the synthesis of complex organic compounds that bind soil particles into structural units called aggregates. These aggregates help to maintain a loose, open, granular condition. Water is then better able to enter and percolate downward through the soil.

Soil Erosion

Humus definitely increases the ability of the soil to resist erosion. It enables the soil to hold more water. Even more important is its effect in promoting soil granulation and maintaining large pores through which water can enter and percolate downward. In a granular soil, individual particles are not easily carried along by moving water. Loss of pore space in long-cultivated soils is mainly a loss of the larger pore spaces. As a result, the soil becomes dense and compact, water enters slowly, and surface runoff increases.

Buffering and Exchange Capacity

From 20 to 70% of the exchange capacity of the soil is due to colloidal humic substances. Exchange acidities of isolated fractions of humus range from 300 to 1,400 meq/100 g. Humus exhibits buffering over a wide pH range.

Miscellaneous Effects

In recent years, information relative to a number of additional effects of soil humus has been forthcoming. These newly recognized features have implications with respect to both the yields of crop plants and the quality of the environment. They include (1) interaction of humus with organic pesticides and (2) sorption of N and S oxides from the atmosphere.

COMMERCIAL HUMATES AS SOIL AMENDMENTS

From time to time, lignites or products derived from them have been marketed for use as organic soil amendments. In some cases, rather extravagant claims have been made for their beneficial effects on the physical, chemical, and biological properties of the soil. The main focus of this section will be on the potential value of such products for use on normally productive agricultural soils.[65] The conclusions reached may or may not apply to the use of commercial humates as amendments for lawns, turf, or certain problem soils, such as sands.

Origin and Nature of Commercial Humates

For the purpose of this discussion, commercial humates are defined as oxidized lignites, or products derived from them. Oxidized lignite is an earthy, medium-brown, coal-like substance associated with lignitic outcrops. The material typically occurs at shallow depths, overlying or grading into the harder and more compact lignite, a type of soft coal. A unique feature of oxidized lignites is their unusually high content of humic acids, of the order of 30 to 60% of the material as mined.

Extensive deposits of oxidized lignites occur in North Dakota, Texas, New Mexico, Idaho, and elsewhere. The deposit in North Dakota, often referred to as Leonardite, is about 1.8 m (6 ft) thick and lies 2.4 to 10.7 m (8 to 35 ft) beneath the soil surface.

Oxidized lignites are undesirable as fuel because of a low heating content. While normally discarded during mining, they have been utilized in a limited way in industry, such as for viscosity control in oil-well drilling muds. The possible use of "coal humates" as soil additives dates back at least four decades.

Chemical Properties of Lignite Humic Acids

Mined lignites have low water solubilities, the insolubility being due to the tie-up of humic acids with mineral matter. Some oxidized lignites are essentially salts of humic acids mixed with mineral matter, such as gypsum, silica, and clay.

Two main types of commercial humate preparations have been marketed for agricultural purposes.

Mined lignites   The original mined product is crushed, pulverized, and usually fortified with commercial fertilizer. The material itself does not contain significant amounts of N, P, or K but is assumed to have a favorable effect on soil properties. Inclusion of mined lignite in the fertilizer mixture is sometimes said to increase the efficiency with which plants utilize the applied nutrients.

Ammonium humate fertilizer   This is a soluble product containing available N, P, and K. In contrast to oxidized lignites, which have low solubility, NH_4^+-humates are easily dispersed in water.

Because of manufacturing and transportation costs, commercial humates are relatively expensive. Their use can only be justified if they result in extra yields from the *humates* they contain.

Promoters of coal-derived humates often give the impression that humates from oxidized lignites have biological and chemical properties similar to the humus in soil. For reasons outlined below, this is probably not the case.

Coal humates can be considered an advanced stage in the transformation of humus to fossil forms. The major changes include losses in carbohydrates, proteins, and other biochemical constituents and an increase in the oxidation state of the humic material. Thus, unlike soil humus, coal humates are essentially free of such biologically important compounds as the proteins and polysaccharides; furthermore, they contain little if any fulvic acids. In comparison to soil humic acids, lignite humic acids have higher C contents, which indicate that they will be less soluble (see Chapter 1).

The virtual absence of proteins and other nitrogenous biochemicals indicate that commercial humates per se are not good sources of N for plant growth. Mucopolysaccharide gums, the constituents in soil humus that are the most effective in forming stable aggregates, are absent from commercial humates. This suggests that the latter will be relatively ineffective as soil conditioners. Also, because of their high C contents and condensed structures, lignite humic acids will be easily immobilized in soil, even when applied in soluble forms. This means that they will have little opportunity to influence plant growth directly by acting as growth hormones.

Value for Increasing Soil Organic Matter Content

The only way commercial humates will increase the cation exchange and buffering capacities of the soil permanently is if they build up the soil humus. Because commercial humates resist microbial attack, some increase in organic matter content is to be expected. However, at the rates at which commercial humates are normally applied (200 to 400 kg/ha or less), the increases will be small.

A light-colored soil with an organic matter content of 3% will contain 67,400 kg of organic matter per hectare to plow depth. Applying 500 kg of commercial humate containing 60% humic matter (a high value) will increase organic content from 3.00 to only 3.01%, assuming no decomposition. For darker-colored soils, the relative increase will be even less.

Thus prolonged and costly applications of commercial humates will be required to bring about a noticeable increase in organic matter content. Even so, the cation exchange and buffering capacities of the soil may not be significantly increased. The potential cation exchange capacity of commercial humates (value obtained in the laboratory) is undoubtedly much higher than will be attained in the soil. To a large extent, mined humates are inert owing to their combination with mineral matter (blockage of reactive sites).

If increasing the organic matter content of the soil is of major importance, the farmer could probably achieve the same effect at a much lower cost by adding other types of organic materials, such as plant residues or farmyard manure. Thus the use of commercial humates can only be recommended when the cost is in line with other approaches.

Lignites as Controlled-Release N Fertilizers

Although commercial humates per se are known to have little fertilizer value, claims persist that they provide a continuous supply of N for plant growth in subsequent seasons. While this claim may be true, the residual value for any one year would be negligible.

The N content of oxidized lignites is on the order of 1.2 to 1.5%. Assuming that availability approaches that of native humus, only 2 to 3% of the N will become available to plants during any given growing season. From a single application of 500 kg of mined lignite per hectare, only 0.225 kg of N would become available the second season ($500 \times 0.015 \times 0.03 = 0.225$), which is equivalent to 0.3 kg of anhydrous NH_3 or 0.6 kg of NH_4NO_3. A buildup of 20,000 kg of oxidized lignite per hectare would result in the annual release of only 9 kg of N per hectare.

Growth-Promoting Effects

Numerous studies have shown that under certain circumstances, humic acids can stimulate plant growth, the usual explanation being that they act as growth hormones (auxins) by serving as activators of O_2 during photosynthesis.

For humic acids to be taken up by plants, they must be in water-soluble forms. Since mined humates are water-insoluble, they would not exhibit a direct regulatory effect. Most mineral soils have the ability to inactivate humic acids; thus the application of humates as their soluble Na^+, K^+, or NH_4^+ salts will not ensure uptake by plants. It should be emphasized that most studies on the growth-promoting properties of humic acids have been carried out in nutrient solutions or in a sand culture. In such rooting mediums, the inactivation of humic acids that takes place in mineral soils will not occur.

The slow decomposition of coal humates by soil microorganisms is sometimes said to result in the continued release or synthesis of soluble growth-promoting substances. However, the amount produced on a per hectare basis would be very small. Assume, for example, that 500 kg of commercial humate containing 300 kg of active ingredient were applied to the soil and that the material was transformed at the same rate as the native humus (2 to 3% decomposition per year). There would, therefore, be a gross synthesis of only 9 kg of new humus during the entire growing season, only a trace of which would occur as low-molecular-weight bioactive compounds. Thus the possibility seems remote that sufficient amounts of bioactive compounds would be present in the soil solution at any one time to directly affect plant growth.

Another way to visualize the problem is as follows. To such a soil, let us also incorporate the residues from a 150 bushel per acre corn crop, equivalent to about 7,500 lb/acre of plant tissue, or 8,400 kg/ha. Numerous studies using radioactive-labeled C substrates have shown that one-third of the C of crop residues will remain behind in the soil as newly formed humic substances after the first growing season, the remainder being lost as CO_2 during decomposition (see Chapter 1). During this same period, some of the stable humus will be mineralized, so there will be little if any net change in organic matter content. Ignoring resynthesis of "active humates" by decomposition of soil humus, the total amount of newly formed material derived from the corn stalk residues would be about 2,500 lb/acre (7,500 × 0.33), or 2,800 kg/ha. Obviously, the quantity of "active" humus produced through microbial transformations of the applied commercial humate (9 kg) will be negligible as compared to the amount synthesized during decay of the corn stalks (2,800 kg).

Influence on Soil Physical Properties

As suggested earlier, commercial humates may be relatively ineffective in producing stable aggregates in soil. The major points in this rationale are itemized as follows.

1 Mined humates per se are inert because of their association with mineral matter. Accordingly, they would be expected to have little if any direct effect on soil physical properties. Humates may slightly increase the water-holding capacity of the soil, but this does not necessarily mean that more water will be available to the plant, because the permanent wilting percentage may also be increased.
   While a variety of organic substances are believed to be responsible for producing stable aggregates in soil, polysaccharides and other gelatinous substances, either acting alone or in conjunction with humic substances to which they are attached, play a major role. Polysaccharides are essentially absent in oxidized humates and their humic acids.
2 Polysaccharides and other binding agents may be produced by microorganisms during decay of applied commercial humates. However, as noted in the previous section, the quantities synthesized will be negligible when considered on a per hectare basis. Plant residues would, therefore, appear superior to commercial humates as an energy source for microorganisms.
3 Fungal hyphae, as well as microscopic plant roots, are of considerable importance in forming stable soil aggregates. They serve to hold soil particles together through physical entanglement. This mechanism will only be operative to the extent that commercial humates enhance microbial life and plant growth.

These and other considerations indicate that heavy applications of commercial humates would be required to significantly modify the physical condition of most agricultural soils. Other soil builders (crop residues, animal manures) would appear to be more effective at a lower cost.

General Observations

Commercial humates applied to normally productive agricultural soils at rates recommended by their promoters would not appear to contain sufficient quantities of the necessary ingredients to produce the desired, and sometimes claimed, beneficial effects. Yield increases, if any, from the use of such products would appear to be insufficient to offset increased production costs to the farmer.

Promoters of humates often recommend a general management program that involves inclusion of a soil-building crop in the rotation, use of animal manures or a green manure crop to maintain and conserve soil organic matter, and adoption of tillage and cultural practices that conserve moisture and reduce soil compaction. In any case, any increase in yields brought about by application of the commercial humate will be due to the fertilizer nutrients (N, P, K) included therein, rather than to the "humates" per se. There is little evidence to indicate that commercial humates increase the efficiency with which plants utilize applied fertilizer nutrients.

SUMMARY

Soils vary widely in their organic matter (C) contents. In undisturbed (uncultivated) soils, the amount present is governed by the soil-forming factors of age (time), parent material, topography, vegetation, and climate. During soil development, considerable migration of organic matter occurs into the lower soil horizons, often in association with clay and polyvalent metal cations.

Organic matter is usually lost when soils are first placed under cultivation, and a new equilibrium is reached which is characteristic of cultural practices and soil type. Increases in plant yield brought about by improved varieties, more widespread use of fertilizers, or adoption of better management practices would be expected to have a positive effect on equilibrium levels of organic matter in soil through return of larger quantities of plant residues. However, the increase will generally be slight. For most soils, organic matter can only be maintained at high levels by inclusion of a sod crop in the rotation or by frequent additions of large quantities of organic residues (e.g., animal manure).

The mean residence time of C in soil is initially short but approaches that of the native humus C within a relatively short time. The mean residence time of stable humus varies from several hundred to somewhat over 1,000 years.

Humus serves as a reservoir of N, P, and S for higher plants, improves structure, drainage, and aeration; increases water-holding, buffering, and exchange capacity; and serves as a source of energy for the growth and development of microorganisms. Commercial humates rich in humic acids have been marked for agricultural use, but, as applied to normally productive agricultural soils, their cost would appear to be prohibitive when considered in terms of expected yield increases.

REFERENCES

1 H. Jenny, *Missouri Agric. Exp. Sta. Res. Bull.* **152**, 1 (1930).
2 H. Jenny, *Soil Sci.* **31**, 247 (1931).

3   H. Jenny, *Soil Sci.* **69**, 63 (1950).

4   H. Jenny, *Missouri Agric. Exp. Sta. Res. Bull.* **765**, 1 (1960).

5   H. Jenny and S. P. Raychaudhuri, *Effect of Climate and Cultivation on Nitrogen and Organic Matter Reserves in Indian Soils*, Indian Council Agric. Res., New Delhi, 1960.

6   J. W. Parsons and J. Tinsley, "Nitrogenous Substances," in J. E. Gieseking, Ed., *Soil Components, Vol 1, Organic Components*, Springer-Verlag, New York, 1975, pp. 263–304.

7   P. A. Sanchez, M. P. Gichuru, and L. B. Katz, *Trans. 12th. Intern. Congr. Soil Sci., New Delhi, Symposia Papers* **1**, 99 (1982).

8   F. J. Stevenson, "Origin and Distribution of Nitrogen in Soil," in W. V. Bartholomew and F. E. Clark, Eds., *Soil Nitrogen*, American Society of Agronomy, Madison, Wis., 1965, pp. 1–42.

9   F. J. Stevenson, "Origin and Distribution of Nitrogen in Soil," in F. J. Stevenson, Ed., *Nitrogen in Agricultural Soils*, American Society of Agronomy, Madison, Wis., 1982, pp. 1–42.

10   E. A. Paul and J. A. van Veen, *Trans. 11th Intern. Congr. Soil Sci., Edmonton* **3**, 61 (1978).

11   F. E. Allison, *Soil Organic Matter and Its Role in Crop Production*, Elsevier, New York, 1973.

12   F. J. Stevenson, *BioScience* **22**, 643 (1972).

13   F. J. Stevenson, *Trans. 12th Intern. Congr. Soil Sci., New Delhi, Symposia Papers* **1**, 137 (1982).

14   N. C. Brady, *The Nature and Properties of Soils*, Macmillan, New York, 1974.

15   R. L. Crocker and B. A. Dickson, *J. Ecol.* **45**, 169 (1957).

16   G. R. Hallberg, N. C. Wollenhaupt, and G. A. Miller, *Soil Sci. Soc. Amer. J.* **42**, 339 (1978).

17   R. M. Smith, E. H. Tyron, and E. H. Tyner, *West Virginia Agric. Exp. Sta. Bull.* **604T**, 1 (1971).

18   J. K. Syers, J. A. Adams, and T. W. Walker, *J. Soil Sci.* **21**, 146 (1970).

19   S. W. Buol and F. D. Hole, *Soil Sci. Soc. Amer. Proc.* **23**, 239 (1959).

20   M. W. Senstius, *Amer. Scientist* **46**, 355 (1958).

21   C. Enders, *Biochem. Z.* **315**, 259, 352 (1943).

22   G. W. Harmsen, *Plant and Soil* **3**, 110 (1951).

23   H. L. Shantz, *Ann. Assoc. Amer. Geog.* **13**, 81 (1923).

24   J. B. Reid and M. J. Goss, *Soil Biol. Biochem.* **15**, 687 (1983).

25   H. J. Haas, C. E. Evans, and M. L. Miles, *USDA Tech. Bull.* **1164**, 1 (1957).

26   A. R. Aandahl, *Soil Sci. Soc. Amer. Proc.* **13**, 449 (1948).

27   R. E. Terry, *Soil Sci. Soc. Amer. J.* **44**, 747 (1980).

28   C. A. Campbell, "Soil Organic Carbon, Nitrogen and Fertility," in M. Schnitzer and S. U. Khan, Eds., *Soil Organic Matter*, Elsevier, New York, 1978, pp. 173–271.

29   R. M. Salter and T. C. Green, *J. Amer. Soc. Agron.* **25**, 622 (1933).

30   H. Jenny, *Factors of Soil Formation*, McGraw-Hill, New York, 1941.

31   W. V. Bartholomew and D. Kirkham, *Trans. 7th Intern. Congr. Soil Sci.* **2**, 471 (1960).

32   J. A. Hobbs and P. L. Brown, *Kansas Agric. Exp. Sta. Tech. Bull.* **144**, 1 (1965).

33   J. S. Russell, *Soil Sci.* **120**, 37 (1975).

34   H. F. Birch, *Plant and Soil* **10**, 9 (1958).

35   H. F. Birch, *Plant and Soil* **11**, 262 (1959).

36   C. R. Clement and T. E. Williams, *J. Agr. Sci.* **69**, 133 (1967).

37   J. Giddens, W. E. Adams, and R. N. Dawson, *Agron. J.* **63**, 451 (1971).

38 R. H. Jackman, *New Zealand J. Agric. Res.* **7**, 445 (1964).

39 E. M. White, C. R. Krueger, and R. A. Moore, *Agron. J.* **68**, 581 (1976).

40 A. L. Azevedo, *Ann. Inst. Super. Agron. Univ. Tec. Lisboa*, **34**, 63 (1973).

41 A. L. Azevedo and M. L. V. Fernandes, *Ann. Inst. Super. Agron. Univ. Tec. Lisboa* **34**, 115 (1973).

42 H. Fleige and K. Baeumer, *Agro-Ecosystems* **1**, 19 (1974).

43 P. E. Rasmussen, R. R. Allmaras, C. R. Rohde, and N. C. Roager, Jr., *Soil Sci. Soc. Amer. J.* **44**, 596 (1980).

44 D. S. Jenkinson and A. E. Johnson, "Soil Organic Matter in the Hoosfield Continuous Barley Experiment," in *Rothamsted Exp. Sta. Report for 1976, Part 2*, Harpenden, Herts, England, 1977, pp. 81–101.

45 F. N. Anderson and G. A. Peterson, *Agron. J.* **65**, 697 (1973).

46 W. E. Larson, C. E. Clapp, W. H. Pierre, and Y. B. Morachan, *Agron. J.* **64**, 204 (1972).

47 A. L. Black, *Soil Sci. Soc. Amer. Proc.* **37**, 943 (1973).

48 K. Sivapalan, *Soil Biol. Biochem.* **14**, 309 (1982).

49 L. F. Elliott and F. J. Stevenson, Eds., *Soils for Management of Organic Wastes and Waste Waters*, American Society of Agronomy, Madison, Wis., 1977.

50 F. J. Stevenson, *Soil Sci.* **107**, 470 (1969).

51 R. W. Simonson, *Amer. J. Sci.* **252**, 705 (1954).

52 R. W. Simonson, *Soil Sci. Soc. Amer. Proc.* **6**, 373 (1941).

53 J. D. Hogan and M. T. Beatty, *Soil Sci. Soc. Amer. Proc.* **27**, 345 (1963).

54 R. W. Ruhe, W. Rubin, and W. H. Scholtes, *Amer. J. Sci.* **255**, 671 (1957).

55 W. Libby, *Radio Carbon Dating*, University of Chicago Press, Chicago, 1955.

56 B. J. O'Brien and J. D. Stout, *Soil Biol. Biochem.* **10**, 309 (1978).

57 E. A. Paul, C. A. Campbell, C. A. Rennie, and K. J. McCallum, *Trans. 8th Intern. Congr. Soil Sci.* **3**, 201 (1964).

58 C. A. Campbell, E. A. Paul, D. A. Rennie, and K. J. McCallum, *Soil Sci.* **104**, 81, 217 (1967).

59 Y. A. Martel and E. A. Paul, *Soil Sci. Soc. Amer. Proc.* **38**, 501 (1974).

60 R. V. Ruhe, *Soil Sci.* **107**, 398 (1969).

61 K. M. Goh and B. P. J. Molloy, *J. Soil Sci.* **29**, 567 (1978).

62 International Atomic Energy Agency (IAEA), *Isotopes and Radiation in Soil Organic Matter Studies*, FAO/IAEA, Vienna, 1968.

63 K. M. Goh, *J. Soil Sci.* **29**, 340 (1978).

64 G. Dungworth, *Chem. Geol.* **17**, 135 (1976).

65 F. J. Stevenson, *Crops and Soils* **31**, 14 (1979).

THREE

ENVIRONMENTAL ASPECTS OF THE SOIL CARBON CYCLE

Soil is often the receptacle for organic products of various types, some of which pose a threat to human and animal health. Interest in organic substances in the soil environment is centered on five main areas of concern: (1) purposeful use of land for disposal of organic wastes and waste products, (2) possible synthesis of carcinogenic compounds through the activities of soil microorganisms, (3) fate of synthetic organic chemicals applied to soil for the purpose of controlling pests and weeds, (4) accidental contamination of soil with toxic or hazardous industrial organics, and (5) contribution of soil C to the CO_2 in the atmosphere.

Because of the magnitude of the task, no one item can be covered in detail. Emphasis will be given here to the problem of land disposal of organic wastes (item 1), with somewhat lesser attention being given to the synthesis of carcinogens (item 2), persistence of synthetic organic chemicals (items 3 and 4), and organic matter as a source of CO_2 in the atmosphere (item 5).

DISPOSAL OF ORGANIC WASTE PRODUCTS IN SOIL

A major problem exists in coping with the vast quantities of organic wastes that are produced each year from domestic and municipal sewage, animal and poultry enterprises, food-processing industries (e.g., canneries), pulp and paper mills, garbage, and miscellaneous industrial organics.[1-5] A common practice in the past has been to discharge these unwanted wastes into streams, lakes, and the open sea, but this method is no longer considered acceptable.

Problems of disposing of municipal sewage and garbage—publicized in the popular press—are not restricted to the urban community because a similar situation exists regarding disposal of farm wastes, such as those generated by a feedlot operation. For example, farm animals in the United States produce 10 times more organic wastes than people, the amount being equivalent to a population of 1.9 billion. A cow will generate over 16 times more wastes than a human while the wastes from 7 chickens equals those of 1 man.[5]

Table 3.1
Estimated Solid Wastes Generated by Major Farm Animals in the United States[a]

Animal	Number of Farms	Manure Produced, tons/yr
Cattle	108,862,000	1,088,620,000
Hogs	47,414,000	379,312,000
Sheep	21,456,000	64,368,000
Poultry		
Broilers	2,568,338,000	11,557,000
Turkeys	115,507,000	2,888,000
Layers	339,921,000	15,976,000

[a] From McCalla et al.[6] The amount of sewage sludge produced is of the order of 7,278,000 metric tons/yr.[2]

The total amount of organic wastes produced in the United States from livestock and poultry production has been estimated at over 1.5 billion tons per year, about one-half of which is produced in concentrated production zones. Solid wastes generated by major farm animals are recorded in Table 3.1.

Another point of interest is that enormous quantities of organic wastes are produced by the food industry. The wastes from the cottage cheese whey industry alone are equivalent to the domestic waste from 83 million people.

Utilization of waste organics as soil amendments offers a number of advantages to society in that the fertility of the soil can be enhanced and water quality will improve as a result of decreased disposal into rivers, lakes, and streams.

Specific Problems

The application of organic wastes to the soil, such as crop residues, composts, and manures, is a historical and accepted practice and one that is being used extensively today for maintaining soil fertility. Manures, when properly utilized, have a favorable effect on soil productivity in that they are good sources of nutrients for plant growth, including N, P, S, and certain trace elements. In soils with poor tilth or water-holding capacity, organic additions have a favorable effect on physical properties; other benefits include maintenance of soil organic matter and enhancement in biological life.

The monetary value attributed to organic wastes in earlier days can be illustrated by the following statement from the 1938 *Yearbook of Agriculture* on soils.[7]

One billion tons of manure, the annual product of livestock on American farms, is capable of producing $3,000,000,000 worth of increase in crops. The potential value of this agricultural resource is three times that of the Nation's wheat crop and equivalent to $440 for each of the country's 6,800,000 farm operators.

The crop nutrients it contains would cost more than six times as much as was expended for commercial fertilizers in 1936. Its organic matter content is double the amount of soil humus annually destroyed in growing the nation's grain and cotton crops.

Because of technological and economic considerations, the monetary value of manures and other wastes has been greatly reduced in recent years, with the result that they are now regarded as a liability rather than an asset to the total farm operation. Changes in perspective regarding the value of farmyard manure have occurred because:

1 Expansion in facilities for chemical conversion of molecular N_2 to NH_3 and NO_3^- has led to the production of large quantities of fertilizer N at reasonable cost. At the same time, high costs associated with handling animal wastes have made them no longer competitive in price with chemical fertilizers as sources of plant nutrients.
2 Farming has become highly specialized and livestock and poultry production has become concentrated in large-scale confinement-type enterprises, which means that large volumes of wastes have to be disposed of in a relatively small area.

Limitations to the use of farmland for disposal of animal manures and sewage sludges can be summarized as follows:[1-4]

1 Animal manures and sewage sludges, as sources of plant nutrients, are bulky, low-grade fertilizers of variable composition. Accordingly, they cannot be transported over any great distance before transportation costs exceed the fertilizer value. Many animal confinement facilities and many large cities lack agricultural land within economic transportation distances from the production point.
2 Concentrations of nutrients, soluble salts, trace elements, and water vary tremendously and are seldom known. Optimum rates of application are thus difficult to predict.
3 Both animal manures and sewage sludges contain soluble salts that can cause problems in their use as fertilizers, particularly in irrigated soils of arid regions where soluble salts are already present in irrigation waters. In many soils, leaching of nutrients, especially NO_3^-, to groundwaters may place a limit on the application rate.
4 Sewage sludges contain heavy metals that are retained in soils and may accumulate to levels that are toxic to some plants and thus restrict the type of crop that can be grown. As a result of uptake by plants, they may be toxic to animals and humans, and this may reduce the value of the harvested crop and its use.
5 Sewage sludges contain pathogenic bacteria, viruses, and parasites that may represent a public health risk to farm workers and the public via

the food chain. The degree of risk depends on the method for processing the sludge. Experiences with sludge application in a number of locations suggest that the risk is low.

6   On-farm management problems are created by the physical properties of these materials. Application techniques are inefficient and time-consuming. When liquid sludges are applied to the soil surface, a drying interval is required, and this loss of time can result in delays, for example, in seedbed preparation. Production of animal manures and sewage sludges is continuous, while the need for fertilizers is seasonal.

7   Odors and associated nuisances, both real and imagined, create conflicts between urban residents and the farmer who could advantageously use the wastes. Once resistance and fear have been generated in a community, the people have difficulty in accepting the fact that the risks will be extremely low in a properly managed system of utilizing the sludges or manures on land.

8   An obstacle to the use of sewage sludges for crop production is the requirement by environmental protection agencies that both crop and water quality be monitored. Farmers who would otherwise use the material cannot or are unwilling to pay the additional cost of monitoring for possible environmental side effects.

The following were suggested by CAST[1] as ways of promoting the beneficial uses of animal manures and sewage sludges on farmland.

1   An increase in quality of the products and adequate quality control would make sewage sludges and animal manures more competitive with chemical fertilizers. Conservation of the N that usually volatilizes as NH_3 would greatly increase the value of many products. Decreases in the salt and trace element contents of feed rations and improvements in systems of collecting and treating waste waters or control of trace elements at the source before industrial waste waters are put into sewage systems would also be helpful. Improvements in the physical condition of many products would allow better control of placement and application rates.

2   Development of new management systems that would not delay other farm operations would create a better image for these materials in the minds of farmers.

3   Appropriate guidelines for applying sewage sludges and animal manures to croplands would promote or enhance beneficial use. The guidelines should be based on facts and acceptable risks rather than on fears or unsubstantiated claims of environmental damage.

There are many unanswered questions regarding the application of animal wastes to farmland at *high rates*, such as effect on crop quality and quantity and on the pollution of surface and ground waters.

Contamination of Lakes and Streams with Organics

A basic environmental problem common to most organic wastes is that when they are introduced into streams or waterways, either directly or indirectly through leaching and runoff, the quality of the water is adversely affected; other problems include odor, nutrient enrichment, and introduction of pathogens.

The pollution potential of organic wastes is usually characterized as BOD or COD values:

BOD (biological oxygen demand) The oxygen consumed by microorganisms in the process of oxidizing the organic matter during a 5-day incubation period. This serves as a measure of readily oxidizable material.

COD (chemical oxygen demand) A measure of total oxidizable organic material as estimated by chemical oxidation with sulfuric acid–potassium dichromate. COD is used less frequently than BOD.

The BOD demand of organic matter results in depletion of dissolved O_2 as the material is decomposed. If the waste has a high BOD, the O_2 content of the water may be reduced to such a low value that fish and other forms of water life die off (fish require about 5 ppm of dissolved O_2 for survival).

A BOD value for water of 1 ppm (i.e., 1 ppm of O_2 consumed in a 5-day incubation period) is regarded as high-quality water, whereas a BOD value of 5 indicates water of doubtful purity. Runoff entering a stream is considered objectionable if the BOD exceeds 20 ppm.

Animal wastes have relatively high BOD values; digested sewage sludge is relative low . Wastes in runoff from barnyards and feedlots, for example, can have BOD values as high as 10,000, depending on dilution and degree of decomposition. Similarly high BOD values occur in the effluent from food processing industries. The low BOD values for digested sewage sludge can be accounted for by biological oxidation during sewage treatment.

Buildup of Toxic Heavy Metals

Long-time application of organic wastes, particularly sewage sludges, leads to the accumulation of various metals, some of which are known to be toxic to animals and man.[2,4,8–14] Manures from animals where trace elements (e.g., Cu) have been used as feed additives will also be enriched with metals.[12]

The possibility exists that repeated applications of sewage sludges (and other wastes) to soil might build up the concentration of heavy metals to levels toxic to crops, and in turn, to man and animals. The elements of most concern are Cu, Cd, Ni, and Zn. Cadmium is especially toxic to man and animals, and its entry into the food chain must be kept within acceptable limits. Many of the metals in sludge are bound to the organic matter and may be released to available forms as the sludge decomposes over a period of years.

Table 3.2
Miscellaneous Trace Elements in Sewage Sludges[a]

Constituent	n^b	Range	Mean
		μg/g	μg/g
As	10	6–230	43
Ba	60	21–8,980	576
Cd	115	4–846	101
Cr	119	17–99,000	3,280
Hg	53	1–10,600	1,077
Ni	109	10–3,515	440
Pb	116	13–19,730	1,656

[a] From McCalla et al.[9]
[b] Number of sites.

The metals found in sewage sludges vary from city to city, depending on the industrial contribution. Sewage sludge from Chicago, for example, will contain rather high amounts of Cr, Zn, Cu, Pb, Ni and Cd. Values for maximum and minimum amounts of As, Ba, Ni, Cr, Cd, Pb, and Hg in sewage sludges are given in Table 3.2. Ranges of four trace elements in sludges from industrial and nonindustrial cities are shown in Table 3.3. An annual application of 20 metric tons/ha of a typical digested sludge for 20 years would result in approximately the following increases in trace element concentration in the soil (plow depth):

Co	18 μg/g
Cu	180 μg/g
Cr	540 μg/g
Mn	90 μg/g
Pb	270 μg/g
Zn	890 μg/g

Table 3.3
Concentration Range of Four Heavy Metals in Sewage Sludges as Affected by Source[a]

Heavy Metal	Sludges from Industrial and Nonindustrial Cities	Sludges from Nonindustrial Cities
	μg/g	μg/g
Cd	5–2,000	5–10
Cu	250–17,000	250–1,000
Ni	25–8,000	25–200
Zn	500–50,000	500–2,000

[a] From CAST.[1]

Table 3.4

Average Contents of Fecal Coliforms and Fecal
Streptococci in Fecal Wastes of Seveal Farm Animals[a]

Animal	Fecal Coliform	Fecal Streptococci
	millions/g	millions/g
Cow	0.23	1.3
Pig	3.3	84.0
Sheep	16.0	38.0
Poultry	1.3	3.4
Turkey	0.3	2.8

[a] From Wadleigh.[5]

Vegetable crops are the least tolerant to excesses of heavy metals; grasses are the most tolerant. Most field crops are relatively tolerant but over a rather broad range for individual crops and species. The long-time use of soil for disposal of sewage sludge may place a limit on the type of crop that can be grown.

Water-soluble organic constituents in sludge, or formed during decomposition, may move metals through the soil and into groundwaters as soluble metal–chelate complexes.

Biological Pollutants

Biological pollutants include many pathogens that are parasitic to man and animals. When stored in a lagoon or biologically oxidized, the pathogens die off rapidly. Thus little public health hazard will normally result from application of digested sewage or lagooned animal wastes to soil.[14] A common biological pollutant is the bacterium *Escherichia coli,* which is not pathogenic per se but is a normal constituent of the intestinal flora of vertebrates. Although fecal coliform and fecal streptococci bacteria are seldom pathogenic, they serve as indicators that contamination has occurred and that infectious organisms may be present. The average contents of fecal coliform and fecal streptococci bacteria in fecal wastes of various farm animals are given in Table 3.4.

Infectious agents of animals that may pollute streams include organisms that can cause anthrax, brucellosis, foot-rot, histoplasmosis, mastitis, Newcastle disease, and salmonellosis, among others.[5]

Enrichment of Surface and Ground Waters with Nutrients

The practice of spreading large quantities of manures and other wastes, such as sewage sludge, to soil may add to the nutrient enrichment of natural waters through leaching and runoff.

Table 3.5
Nutrient Content of Feedlot Manure[a]

Nutrient	Range, %	Average, %	Amount in 10 Metric Tons, kg
N	1.16–1.96	1.34	134
P	0.32–0.85	0.53	53
K	0.75–2.35	1.50	150
Na	0.29–1.43	0.74	74
Ca	0.81–1.75	1.30	130
Mg	0.32–0.66	0.50	50
Fe	0.09–0.55	0.21	21
Zn	0.005–0.012	0.009	0.9

[a] From McCalla et al.[9] and Mathers et al.[13] Moisture contents ranged from 20.9 to 54.5%. Data are from 23 feedlots.

Data giving the nutrient content of two typical types of organic wastes, feedlot manure and municipal sewage sludge, are shown in Tables 3.5 and 3.6, respectively. From the data for the feedlot, it can be calculated that the annual application of 40 metric tons of manure per hectare (17.8 tons/acre) would add about 540 kg of N per hectare (480 lb/acre), ony a portion of which would be removed by cropping.

A 2-in. application of anaerobically digested sewage sludge (<10% solids) will supply approximately the following amounts of major plant nutrients:

$NH_4^+ - N$ 252–280 kg/ha (225–250 lb/acre), subject to rapid nitrification

Organic N   336 kg/ha (300 lb/acre), slowly released during degradation

Phosphorus   200–336 kg/ha (178–300 lb/acre), 80% in the sludge organic matter

Potassium   45–90 kg/ha (40–80 lb/acre)

Excess NO_3^- left in the soil after the growing season is subject to leaching and movement into water supplies.

The presence of NO_3^- in surface and ground waters is considered undesirable because of the possible influence of NO_3^- in promoting the unwanted growth of aquatic plants (eutrophication) and health hazards associated with the presence of NO_3^- in drinking waters (methemoglobinemia). These aspects of the soil N cycle are discussed in greater detail in Chapter 6.

It is appropriate to mention that the NO_3^- in natural waters is normally derived from several diverse sources, including municipal and rural sewage,

Table 3.6

Nutrient Composition of Fresh, Heated, Anaerobically Digested Sewage Sludge[a]

	Concentration Range, %	Typical Sludge (dry-weight basis)[b]		
		%	kg/MT	lb/MT
Organic N	2–5	3	27	60
NH_4^+—N	1–3	2	18	40
Total N	1–6	5	45	100
P	6–8	3	27	60
K	0.1–0.7	0.4	4	9
Ca	1–8	3	27	60
Mg	2–5	1	9	20
S	0.3–1.5	0.9	8	18
Fe	0.1–5	4	36	80
	μg/g	μg/g		
Na	800–4,000	2,000	2	4
Zn	50–50,000	5,000	5	10
Cu	200–17,000	1,000	1	2
Mn	100–800	500	0.5	1
B	15–1,000	100	0.1	0.2

[a] Adapted from Thorne et al.[15]

[b] MT = metric ton. Values vary according to source, treatment, and other factors. Additional data for Zn, Cu, and miscellaneous nonessential trace elements are given in Tables 3.2 and 3.3.

feedlots or barnyards, food processing wastes, septic tank effluent, agricultural land (runoff and leaching), natural NO_3^- (caliche of semiarid regions), sanitary facilities of recreational areas, landfills, and miscellaneous industrial wastes.[4] The relative contribution of each will vary widely, depending on conditions existing at the particular site under consideration. The contribution from municipal sewage can be appreciable when raw or digested sewage is discharged directly into lakes or streams. As far as disposal on farmland is concerned, pollution will become increasingly serious as the loading rate is increased.

Soluble Salts

Animal wastes and sewage sludges contain inorganic salts of K, Na, Ca, and Mg. At high application rates, more salts may be added to the soil than are leached out by irrigation water or natural precipitation.[16] The buildup of salt to unacceptable levels is most likely to occur in areas of low rainfall, such as the western United States (west of line from central North Dakota through central Texas). For soils naturally high in salts, even a small application of wastes can increase salt concentration sufficiently to reduce yields.

The salt content of organic wastes is highly variable and is dependent, in part, on salt added to the system and management of the wastes. The salinity hazard of animal wastes can be reduced to some extent by lowering the salt content of food rations for animals.

General Chemical Composition of Organic Wastes

All organic wastes have properties in common in that they are made up of C, hydrogen, and oxygen; most contain N, P, and S as well. Residues from agriculturally oriented industries are made up of many of the constituents found in crop residues, such as lignin and cellulose, although not necessarily in the same proportion. The wastes from paper and sugar mills contain high amounts of carbohydrates; those of meat packing and dairy processing industries contain relatively large amounts of fatty acids and protein. Wood residues would be expected to be relatively rich in lignin.

Animal Wastes

The feces of farm animals consist mostly of undigested food that has escaped bacterial action during passage through the body. This undigested food is mostly cellulose or lignin fibers, although some modification of the lignin to humic substances has occurred.[17] The feces also contain the cells of microorganisms.

Nitrogen in manure solids occurs largely in organic forms (undigested proteins and the bodies of microorganisms); liquid manure may also contain significant amounts of NH_3, the latter having been formed from urea through hydrolysis.

Animal wastes are more concentrated than the original feed in lignins and minerals. Lipids are present along with humiclike substances. Manures also contain a variety of trace organics, such as antibiotics and hormones.

The manure applied to cropland varies greatly in nutrient content, depending on animal type, ration fed, amount and type of bedding material, and storage condition. Both N content and availability of the N to plants decreases with losses of NH_3 through volatilization and NO_3^- through leaching. The two other major plant nutrients, P and K, are as available as fertilizer sources of these nutrients. Manures aged by cycles of wetting and drying and subjected to leaching with rainwater may have lost so much N that very little will be available to the crop in the year of application.

The gross chemical composition of animal manures is approximately as shown in Table 3.7. The trace element contents of most manures are somewhat higher than crop residues.[18,19] Swine and chicken manures will contain high amounts of Cu and Zn when these elements are used as feed additives.[12]

Sewage Sludge

As used here, the term *sewage sludge* refers to heated, anaerobically digested sludge of the type produced in a typical sewage disposal facility. A flow diagram showing a typical method for treating domestic sewage is given

Table 3.7
Gross Chemical Composition of Animal Manures[a]

Constituent	Typical Range, %
Ether-soluble compounds	1.8–2.8
Cold water-soluble compounds	3.2–19.2
Hot water-soluble compounds	2.4–5.7
"Hemicellulose"	18.5–23.5
Cellulose	18.7–27.5
"Lignin"	14.2–20.7
Total N	1.1–4.1
Ash	9.1–17.2

[a] From McCalla et al.[6]

in Fig. 3.1. A brief description of the various treatment processes is given below:

PRIMARY TREATMENT. Primary treatment is the simple settling and screening of solids. The effluent contains solids that do not settle plus organic and inorganic nutrients dissolved in the water. Pathogens and other organisms are also present in the effluent.

SECONDARY TREATMENT. Secondary treatment is a process aimed at reducing the amount of solid matter and removing O_2-demanding substances. Two different biological processes are used: aerobic and anaerobic digestion. Most of the pathogens are killed and new cells are formed. The effluent still contains high amounts of nutrients. Most sewage sludges used for application to farmland are of this type.

TERTIARY OR ADVANCED TREATMENT. Tertiary or advanced treatment is an additional treatment sometimes used for further purification of the effluent, including removal of inorganics, especially P. Among the treatments used singly or in combination are lime precipitation of P, air stripping of NH_3, filtration for removal of residual biological cells and suspended organics, and activated charcoal adsorption of organics.

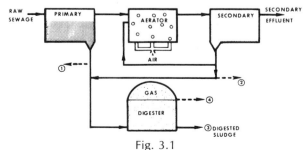

Fig. 3.1
Flow diagram of biological sewage treatment. (From Dean and Smith.[8])

Because of the volume of material produced in a sewage treatment plant, serious storage and disposal problems are being encountered by many municipalities. Utilization on agricultural land offers an attractive alternative to other disposal methods, many of which are expensive and/or damaging to the environment.

At the outset, it should be noted that little if any environmental damage is expected when sewage sludge is applied to agricultural land at rates that supply adequate but not excessive amounts of any given major nutrient or trace element. The main environmental concern is from repeated applications over a prolonged period, particularly at high rates.

The suitability of domestic sewage as a soil amendment has been greatly increased by previous biological treatment, which not only stabilizes the material but eliminates pathogens and obnoxious odors. The main purpose of treating municipal sewage has been to reduce the amount of suspended solids, to kill pathogenic bacteria, and to reduce the amount of O_2-demanding material to an acceptable level for discharge into streams or lakes. With more stringent water quality standards, greater emphasis in the future will be given to disposal on land.

The product obtained from biological treatment of domestic sewage, which is the material available for land disposal from most municipal treatment plants, is a stabilized product with an earthy odor and which does not contain raw, undigested solids. Liquid sewage sludge is blackish and contains colloidal and suspended solids. Most sludges, as produced in a sewage treatment plant, contain from 2 to 5% solids.

The solid portion of sewage sludge consists of approximately equal parts of organic and inorganic material. The latter includes numerous elements, such as N, P, K, S, Cl, Zn, Cu, Pb, Cd, Hg, Cr, Ni, Mn, B, and others. The organic component is a complex mixture consisting of: (1) digested constituents that are resistant to anaerobic decomposition, (2) dead and live microbial cells, and (3) compounds synthesized by microbes during the digestion process. The organic material is rather rich in N, P, and S. The C/N ratio of digested sludge ranges from 7 to 12, but is usually about 10.

As was the case for animal manures, the availability of N in sludges decreases as the content of NH_4^+ and NO_3^- decreases and as the organic N becomes more stable as a result of digestion during biological waste treatment. Conservation of the N that often volatilizes as NH_3 would greatly increase the value of sewage sludge as an N source.

Table 3.8 gives the range of chemical composition for digested sewage sludge collected at two sampling dates from the Sanitary District of Greater Chicago.[20] The composition of individual sludges can vary appreciably from the values shown.

Sewage sludges contain relative large quantities of minor and trace elements, as indicated earlier in Tables 3.2 and 3.6. Some of the elements are essential for plant growth, but nearly all can be toxic at some concentration. Zinc, Cu, Ni, Cd, Hg, and Pb may occur in quantities sufficient to adversely

Table 3.8
Chemical Composition of Two Samples of Digested Sewage
Sludge Collected in 1973 and 1974.[a]

	% of Organic Matter	
Constituent	1973	1974
Fast, waxes and oils	19.8	19.1
Resins	3.8	8.2
Water-soluble polysaccharides	3.2	14.4
Hemicellulose	4.0	6.0
Cellulose	3.5	3.2
Lignin-humus	16.8	14.5
Protein (N × 6.25)	24.1	39.6
Total recovered	75.1	105.0

[a] From Varanka et al.[20]

affect plants and soils. The availability of any given metal in soil will be influenced by such soil properties as pH, organic matter content, type and amount of clay, content of other metals, cation-exchange capacity, variety of crop grown, and others.

Decomposition of Organic Wastes in Soil

Organic wastes of various types would be expected to follow somewhat the same pathways of decomposition as discussed in Chapter 1. However, because of previous biological transformations as discussed above, and also the higher content of inorganic material, digested sewage sludge (and to some extent animal manures) would be expected to be more resistant than crop residues to biological decomposition in soils. This seems to have been borne out in the case of sewage sludge by the study conducted by Miller,[21] where an average retention of 80% of the C was obtained after a 6-month incubation period. The percentage of the sludge C evolved as CO_2 was highly correlated with "degree days," but in no case was over 20% of the C recovered as CO_2 during the test period. His data for an application rate of 90 metric tons/ha are reproduced in Fig. 3.2.

These data provide some insight into the quantities of C that would be expected to be retained in soil at moderate loading of organic wastes. Extent of decomposition (and subsequently C retention) would be expected to vary when applied at massive rates as compared with repeated smaller additions, and to be affected by soil texture and by natural cycles of wetting and drying and freezing and thawing. Less C would be expected to be retained in warm humid climates than in cold humid climates.

The effect of mean annual temperature on the relative rate of C (organic matter) loss at select locations for a north to south transect of the eastern

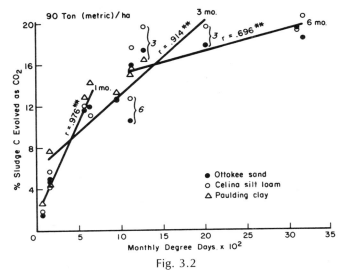

Fig. 3.2

Relationship between percent sludge C evolved as CO_2 and degree days after 1, 3, and 6 months incubation.[21] [Reproduced from *J. Environ. Qual.* **3,** 376–380 (1974) by permission of the American Society of Agronomy.]

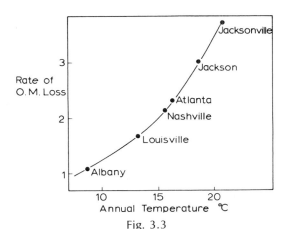

Fig. 3.3

Effect of mean annual temperature on the relative rate of organic matter loss at select locations for a north to south transect of the eastern United States.[22] See Fig. 3.4 for monthly temperatures at the various locations. [Reproduced from *Soils for Management of Organic Wastes and Waste Waters* (1977) by permission of the American Society of Agronomy.]

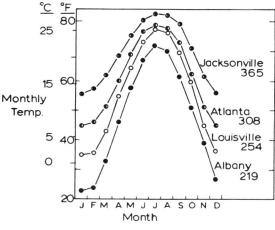

Fig. 3.4

Mean monthly temperatures for several locations along a north to south transect of the eastern United States.[22] For mean annual temperatures see Fig. 3.3. [Reproduced from *Soils for Management of Organic Wastes and Waste Waters* (1977) by permission of the American Society of Agronomy.]

United States, as visualized by Thomas,[22] is shown in Fig. 3.3. Because of temperature effects on microbial activity, greater decomposition would occur as the mean annual temperature increases. Another important aspect is that soils of the northern sections of the United States are in a frozen state part of the year and the mean annual temperature is relatively low, as can be seen from Fig. 3.4. Accordingly, decomposition rates will be lower, and C retention will be greater, than for soils of the more southern regions, where temperatures are seldom sufficiently low to inhibit microbial activity completely.

Management of Organic Wastes for N Conservation

In the future, permissible loading rates for the application of manures (and other organic wastes) to agricultural soils may be based on the amount of a given nutrient (e.g., N, P, K) that will become available to the crop during the year of application. In other cases, rates may need to be restricted on the basis of toxic element content (e.g., Cd). A discussion of alternative decision-making factors for waste-loading rates has been given elsewhere.[1–3,15]

The following is intended to furnish background information for the case where NO_3^- leaching is a problem and a need exists for restricting application rates on the basis of fertilizer N equivalents for the waste under consideration. Emphasis will be given to farmyard manure but the principles apply equally well to other nitrogenous organic wastes. The problem is extremely complex, not only because of the various forms that N assumes during trans-formations by microorganisms but because of the difficulty in predicting what these forms will be at any one time. Various aspects of the recycling of N through land application of organic wastes are discussed by Smith and Peterson.[23]

In general, it can be said that the net amount of NO_3^- produced, and subsequently available to plants or subjected to leaching and denitrification, represents the difference between the total amount of N applied and the amount that is immobilized and tied up in organic forms (e.g., amount of residual N in the soil after the growing season). The quantity of N retained is directly related to C retention, which for soils of the temperate region of the earth, is about one-third of the C applied (see Chapter 1). There is ev-idence, however, that a higher percentage (50%) will remain when manures are applied at high rates. This may be due in part to the fact that manures contain humiclike substances that are relatively resistant to decomposition.

The significance of the above-mentioned observations is that N is pre-served along with C in the ratio of 10 parts C to one part N, which is near the C/N ratio of microbial tissue. Thus for each 10 metric tons (dry weight) of organic wastes added to the soil, at least one-third (3,300 kg or 7,300 lb) will remain behind in a modified form after the first year. Assuming that 90% of the dry matter in manure is organic waste material with a C content of 50%, approximately 1,500 kg (3,300 lb) of C will be preserved, and this C will retain 150 kg (330 lb) of N. If one-half of the C is preserved, about 225 kg of N will be retained.

Tables 3.9 and 3.10 give the amounts of potential inorganic N released by application of 10 and 20 metric tons of manure (dry-weight basis) con-taining variable amounts of N (retention of one-third and one-half of the C, respectively). Feedlot manure will typically contain from 2.5 to 3.5% N, whereas fresh manure will contain 4.0% N or more. The moisture content of feedlot manure will be about 50% while fresh manure may contain 75% moisture or more.

Table 3.11 gives the approximate metric tons of manure (dry-weight basis) that would be required to provide 90 and 180 kg (200 and 400 lb) of available mineral N. It can be seen that the quantity required does not follow a direct 1:1 relationship to N content. With increasing N content, a *lower percentage* of the N will remain in the soil at the end of the first season.

It should be emphasized that the data given in Tables 3.9 to 3.11 apply to a single application of manure. For repeated applications over a period of years, these relationships do not hold, for the reason that the residual organic matter remaining in the soil after the first year, representing one-third to one-half of the manure initially applied, undergoes further decom-postion during subsequent years with the release of bound N. As a rough approximation, one-fourth to one-third of the N remaining after each year will be mineralized the succeeding year until near-complete "humification" has occurred after about 5 years. *In practical terms, this means that ap-*

Table 3.9
Nitrogen Balance for Application of 10 and 20 Metric Tons of Animal Manure
Containing Variable Amounts of N—Retention of One-Third of C (in kg)[a]

	10 Tons			20 Tons		
% N[b]	Total N	Retained in Residues	Inorganic N	Total N	Retained in Residues	Inorganic N
1.5	150	150	b	300	300	b
2.0	200	150	50	400	300	100
2.5	250	150	100	500	300	200
3.0	300	150	150	600	300	300
3.5	350	150	200	700	300	400
4.0	400	150	250	800	300	500
5.0	500	150	350	1,000	300	700

[a] Dry-weight basis. For conversion to pounds, multiply by 2.205.
[b] For N contents of 1.5% or less, there may be a net loss of mineral N from the soil through immobilization.

plication rates for an equivalent amount of inorganic N will need to be reduced each succeeding year for 5 years, after which the rate will be constant and equivalent to the addition of an equal quantity of inorganic N. In other words, after 5 years essentially all the N added in the manure must be considered to be "available" because the amount immobilized will be compensated for by mineralization of residual N from earlier applications.

Table 3.10
Nitrogen Balance for Application of 10 and 20 Metric Tons of Animal Manure
Containing Variable Amounts of N—Retention of One-half of C (in kg)[a]

	10 Tons			20 Tons		
% N[b]	Total N	Retained in Residues	Inorganic N	Total N	Retained in Residues	Inorganic N
2.0	200	225	b	400	450	b
2.5	250	225	25	500	450	50
3.0	300	225	75	600	450	150
3.5	350	225	125	700	450	250
4.0	400	225	175	800	450	350
5.0	500	225	275	1,000	450	500

[a] Dry-weight basis. For conversion to pounds multiply by 2.205.
[b] For N contents of 2% or less, there may be a net loss of mineral N from the soil through immobilization.

Table 3.11
Estimated Metric Tons of Manure per Hectare (Dry-Weight Basis)
Required to Provide 90 and 180 kg (200 and 400 lb) of Available
Mineral N

N Content of Manure, %	One-third Retention of C		One-half Retention of C	
	90 kg N	180 kg N	90 kg N	180 kg N
2.0	20	40	a	a
2.5	10	20	40	80
3.0	7	14	13	26
3.5	5	10	8	16
4.5	4	8	6	12

a Very high rates required because of net immobilization.

Successive applications for 5 or more years commonly occur when a small land area is used for disposal. For best utilization of nutrients, the manure should be spread over a broad area at a low loading rate.

SYNTHESIS OF CARCINOGENIC COMPOUNDS

Several specific types of organic compounds can be produced through the activities of microorganisms that are carcinogenic, mutagenic, or acutely toxic in very low amounts. They include methylmercury, dimethylarsine, dimethylselenide, and nitrosamines. These substances are not normal constituents of agricultural soils but can be produced under certain circumstances in aqueous sediments, and possibly in polluted soils. The role of microorganisms in synthesizing hazardous substances from innocuous precursors has been discussed by Alexander.[24]

Methylmercury

Discharge of mercury in industrial effluents has been demonstrated to lead to the formation of methylmercury (CH_3Hg^+) in waterways and streams through microbial action. This deadly poisonous compound accumulates in fish, and when they are consumed by humans this can lead to serious illness or death. Cases of death due to methylmercury have been documented in the literature.

The bottom sediments of many lakes and waterways are highly contaminated with mercury and slow conversion to the methyl form is possible for

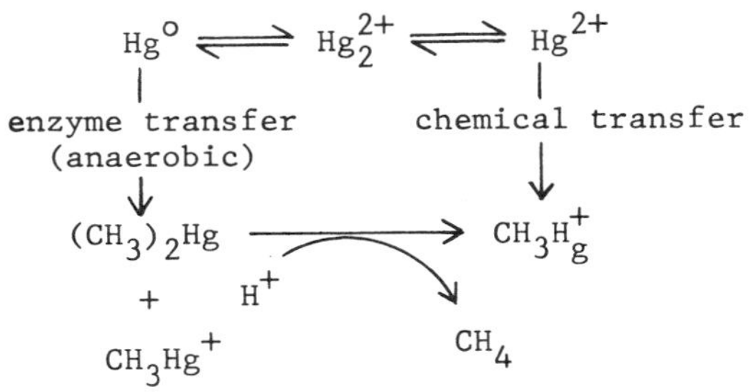

$$Hg^0 \rightleftharpoons Hg_2^{2+} \rightleftharpoons Hg^{2+}$$

enzyme transfer   chemical transfer
(anaerobic)

$(CH_3)_2Hg \longrightarrow CH_3H_g^+$

$+ \quad H^+$

$CH_3Hg^+ \qquad\qquad CH_4$

Fig. 3.5

Proposed mechanism for the formation of methylmercury by microorganisms. (From Wood.[26])

many years to come. Sources of mercury in agricultural soil include organic pesticides and organic wastes, such as municipal sewage sludges.

Many schemes have been proposed for the formation of methylmercury by microorganisms.[25,26] Pathways suggested by Wood[26] include transfer of the methyl group from methyl cobalamin (CH_3–B_{12}), which is a common coenzyme in both anaerobic and aerobic bacteria. The reaction is given in Fig. 3.5.

Dimethylarsine

Arsenic is of environmental concern because of its widespread use in pesticides and defoliants, its presence in many different ecosystems, and its toxicity to man.[27] A problem similar to that mentioned for methylmercury exists when arsenic and its derivatives are introduced into environments containing an active flora of anaerobic organisms. Like methylmercury, dimethylarsine [$(CH_3)_2AsH^{3-}$)] is deadly poisonous and can enter the food chain through accumulation in fish.

The mechanisms whereby dimethylarsine is formed appear to be similar to that involving mercury in that methyl transfer from methyl B_{12} can occur.[26] The reaction is shown in Fig. 3.6.

Dimethylselenide

Anionic forms of selenium are also subject to microbial alkylation, the product being demethylselenide.[26–28] Selenium occurs in many soils as selenate and selenite; in some regions, such high concentrations are present that plants accumulate the element to levels that are toxic to grazing animals. Francis et al.[28] suggested that the microbial methylation of selenium is potentially widespread and that losses of selenium can occur from soils as the volatile dimethylselenide.

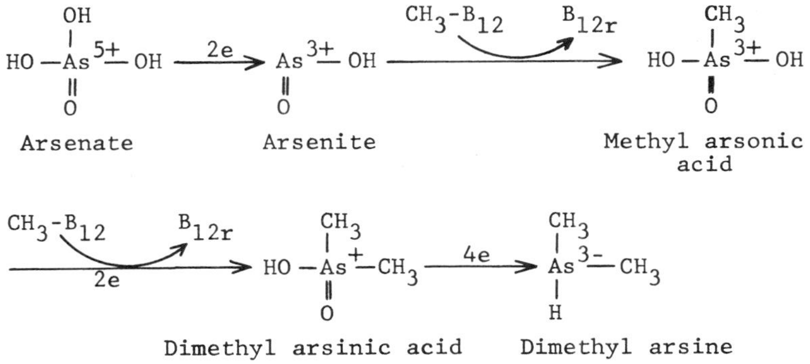

Fig. 3.6
Proposed mechanism for the formation of dimethylarsine.

Nitrosamines

Nitrosamines are formed by the chemical reaction of amines (RNH_2) with NO_2^- and both must be present concurrently in order for these carcinogenic and toxic substances to be formed. Nitrite required for the reaction is produced as an intermediate during biochemical N transformations but the compound seldom persists in soil. However, temporary accumulations are possible in microsites where NO_2^--oxidizing autotrophs (*Nitrobacter* sp.) are inhibited by free NH_3. Studies on the formation of dimethylamine and diethylamine in soils treated with pesticides that yield amines upon partial decomposition indicate that the potential exists for nitrosamine synthesis. However, it has yet to be established that these compounds are produced in this way in natural soils (see Chapter 6).

PERSISTENT SYNTHETIC ORGANIC CHEMICALS

A new dimension to the C cycle in soil has been introduced as a consequence of the widespread use of synthetic organics to control pests (insecticides, fungicides) and weeds (herbicides). In addition, a wide array of industrial organic chemicals are introduced into the soil each year as toxic pollutants. They include such synthetic organic compounds as plasticizers (e.g., phthalates), dyes of various types, flame retardants, and impurities in pesticide formulations (e.g., chlorinated dioxins). Chlorinated dioxins have become widely recognized as dangerous substances in the environment, the most notorious being 2,3,7,8-tetrachlorodibenzo-*p*-dioxin (TCDD), whose structure is shown below.

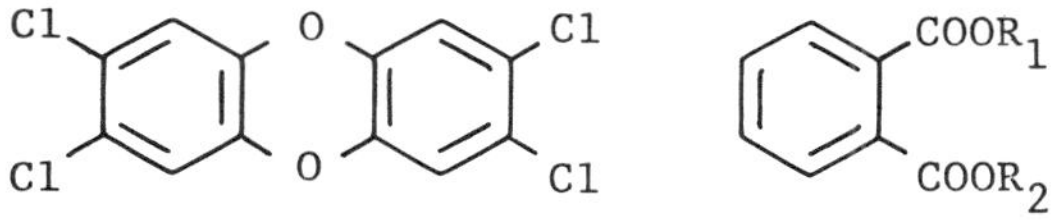

TCDD   **Phthalate Esters**

Table 3.12
Type Reactions for Transformations of Organic Chemicals of Environmental Importance[a]

Category	Reaction[b]	Example
Dehalogenation	$RCH_2Cl \rightarrow RCH_2OH$	Propachlor
	$ArCl \rightarrow ArOH$	Nitrofen
	$ArF \rightarrow ArOH$	Flamprop-methyl
	$ArCl \rightarrow ArH$	Pentachlorophenol
	$Ar_2CHCH_2Cl \rightarrow$ $Ar_2C=CH_2$	DDT
	$Ar_2CHCHCl_2 \rightarrow$ $Ar_2C=CHCl$	DDT
	$Ar_2CHCCl_3 \rightarrow$ $Ar_2CHCHCl_2$	DDT
	$Ar_2CHCCl_3 \rightarrow$ $Ar_2C=CCl_2$	DDT
	$RCCl_3 \rightarrow RCOOH$	DDT, N-Serve
Deamination	$ArNH_2 \rightarrow ArOH$	Fluchloralin
Decarboxylation	$ArCOOH \rightarrow ArH$	Bifenox
	$Ar_2CHCOOH \rightarrow Ar_2CH_2$	DDT
	$RCH(CH_3)COOH \rightarrow$ RCH_2CH_3	Dichlorfop-methyl
	$ArN(R)COOH \rightarrow$ $ArN(R)H$	DDOD
Methyl oxidation	$RCH_3 \rightarrow RCH_2OH$ and/or $\rightarrow RCHO$ and/or $\rightarrow$ $RCOOH$	Bromacil, diiso-propylnaphthalene pentachlorobenzyl alcohol
Hydroxylation	$ArH \rightarrow ArOH$	Benthiocarb, dicamba
β-oxidation	$ArO(CH_2)_nCH_2CH_2COOH$ $\rightarrow ArO(CH_2)_nCOOH$	ω-(2,4-Dichlorophenoxy)-alkanoic acids
Triple bond reduction	$RC\equiv CH \rightarrow RCH=CH_2$	Buturon
Double bond reduction	$Ar_2C=CH_2 \rightarrow Ar_2CHCH_3$	DDT
	$Ar_2C=CHCl \rightarrow$ Ar_2CHCH_2Cl	DDT
Double bond hydration	$Ar_2C=CH_2 \rightarrow$ Ar_2CHCH_2OH	DDT

[a] From Alexander.[24]
[b] R = organic moiety; Ar = aromatic.

Two classes of organic chemicals that are of particular interest from the standpoint of pollution are the phthalic acid esters and detergents. Phthalates have been under close scrutiny because large quantities are produced each year for use in construction and housing, in the medical sciences, and as carriers in pesticide formulations. It is now realized that these phthalates (polyesters of the benzene dicarboxylic acids) are general environmental

Table 3.13
Type Reactions for Cleavage of Chemicals of Environmental Importance[a]

Substrate	Reaction[b]	Example
Ester	RC(O)OR' → RC(O)OH	Malathion, phthalates
Ether	ArOR → ArOH	Chlomethoxynil, 2,4-D
	ROCH$_2$R' → ROH	Dichlorfop-methyl
C—N bond	R(R')NR" → R(R')NH and/or → RNH$_2$	Alachlor, trimethylamine
	RN(Alk)$_2$ → RNHAlk	Chlorotoluron, Trifluralin
	RNH$_2$CH$_2$R' → RNH$_2$	Glyphosate
Peptide, carbamate	RNHC(O)R' → RNH$_2$ and/or HOOCR'	Benlate, dimetholate
	R(R')NC(O)R" → R(R')NH + HOOCR"	Benzoylprop-ethyl
=NOC(O)R	RCH=NOC(O)R → RCH=NOH	Aldicarb
C—S bond	RSR' → ROH and/or HSR'	Benthiocarb, Kitazin P
C—Hg bond	RHgR' → RH and/or Hg	Ethylmercury, phenylmercuric acetate
C—Sn bond	R$_3$SnOH → R$_2$SnO → RSnO$_2$H	Tricyclohexyltin hydroxide
C—O—P[c]	(AlkO)$_2$P(S)R → AlkO(HO)P(S)R and/or → (HO)$_2$P(S)R	Gardona, malathion
P—S	RSP(O)(R')OAlk → HOP(O)(R')OAlk	Hinosan, Kitazin
Sulfate ester	RCH$_2$OS(O$_2$)OH → RCH$_2$OH and/or HOS(O$_2$)OH	Sesone
S—N	ArS(O$_2$)NH$_2$ → ArS(O$_2$)OH	Oryzalin
S—S	RSSR → RSH	Thiram

[a] From Alexander.[24]
[b] R = organic moiety; Alk = alkyl.
[c] For the reaction, S = sulfur or oxygen.

contaminants. Many scientists believe that they do not pose an eminent threat to human health, although it is recognized that certain individuals may be more affected than others. Questions have also arisen concerning possible subtle effects by persistent exposure to very low concentrations of phthalates or of derivatives obtained therefrom by partial degradation.

The amount of synthetic organic chemicals produced in the United States in 1980 alone was of the order of 100 billion lb.[29] Over the past 50 years, over 600,000 metric tons have been produced, about 15% of which is estimated to have entered the mobile environmental reserve.[30]

Synthetic organic chemicals can be divided into two general types as far as their persistence in soil is concerned, namely, those that are biodegradable and those that are recalcitrant (i.e., decompose slowly if at all). The division

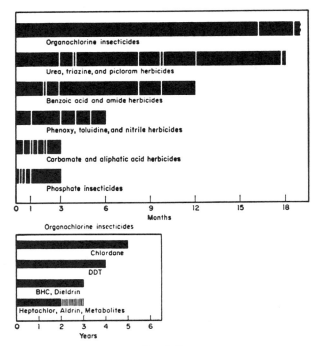

Fig. 3.7

Persistence of several groups of pesticides (A) and specific organochlorine insecticides (B). (Adapted from Kearney et al.[31] and Sethunathan et al.[32])

between the two is somewhat arbitrary in that certain organic chemicals are biodegradable under some soil conditions but not others. For example, most synthetic organics are destroyed at a much slower rate under anaerobic soil conditions than in well-aerated soils. The persistence of synthetic organic compounds in poorly drained soils and wet sediments can be attributed to several causes, the most important being the need for O_2 in key enzymatic reactions. Wet sediments contain a less varied micro- and macroflora than soils that are well drained; furthermore, there is less modification of the environment through various processes associated with crop production (mixing of the soil by plowing, incorporation of crop residues, etc.).

For the most part, biodegradable organic chemicals undergo the same type of cleavage reactions as biogenic molecules (e.g., decarboxylation, deamination, hydroxylation, β-oxidation, hydrolysis of ester linkages, and others).[24] Typical type reactions are given in Table 3.12; examples of cleavage reactions are listed in Table 3.13. The fate and biodegradability of a synthetic organic chemical can be predicted to some extent from its chemical structure.

A popular concept in the past was that essentially every organic chemical

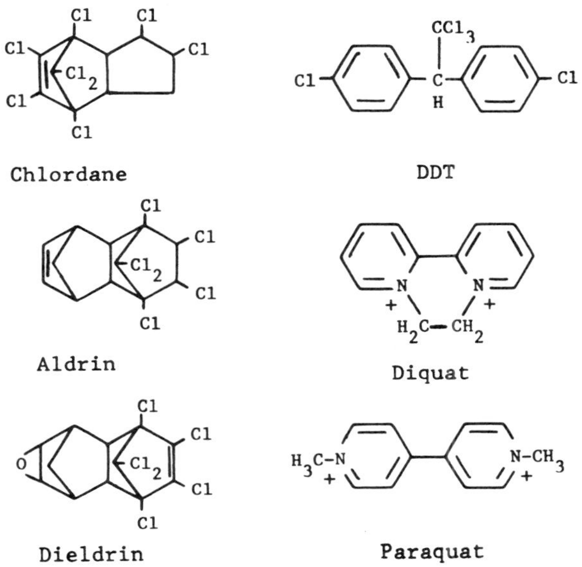

Fig. 3.8
Chemical structures of some of the more persistent pesticides in soil.

introduced into the biosphere would be metabolized within a reasonably short time by microorganisms. This concept has now been found to be incorrect. Substantial evidence indicates that many synthetic organic chemicals are recalcitrant and that they can persist in soil for long periods. They include chlorinated aromatic compounds of various types (including polychlorinated biphenyls, PCBs), certain pesticides (particularly insecticides), plastics and related synthetic polymers, and an array of industrial organic chemicals.

Comparison of the persistence of individual pesticides in Fig. 3.7 indicates the long-time persistence of organochlorine insecticides, notably DDT. The herbicides Diquat and Paraquat (data not shown) also persist in soils for long periods. Chemical formulas for some of the more persistent pesticides are given in Fig. 3.8.

A major factor affecting the persistence of synthetic organics in soils is adsorption to clay and organic matter surfaces. Adsorption depends on the chemical structure of the organic compounds and properties of the soil system (kind and amount of clay, organic matter content, pH, and type of exchange cations).

Adsorption by soil organic matter has been shown to be a key factor in the behavior of synthetic organics in soil.[32–34] It has been well established, for example, that the rate at which an adsorbable pesticide must be applied to the soil in order to achieve adequate pest control can vary as much as

20-fold, depending on the nature of the soil and the amount of organic matter it contains. Soils that are black (e.g., most Mollisols) have higher organic matter contents than those which are light in color (e.g., Alfisols), and pesticide application rates must often be adjusted upward on the darker soils to achieve the desired result.

Adsorption by soil colloids (humus and clay) poses severe problems in ascertaining the long-time fate of synthetic organics in the environment. For example, adsorbed organics can be transported during soil erosion to lakes and reservoirs where conditions may be unfavorable for microbial detoxification. The adsorption phenomena must also be taken into account in assessing the impact of industrial processes that lead to widespread dispersal of organics into the ecosystem, such as conversion of coal to natural gas.

A full accounting of the fate of synthetic organics in soil is beyond the scope of the present chapter. Numerous volumes have been prepared on both pesticide degradation and pollution of soils with industrial organics. Some key reviews and books are listed in references 30 to 35.

CONTRIBUTION OF SOIL C TO CO_2 OF THE ATMOSPHERE

It is well known that the CO_2 content of the atmosphere has been increasing at a steady rate over the past century.[36-41] This is a matter of grave concern because of the possible warming of the earths surface by the so-called greenhouse effect. Global temperature increases of 2 to 3°C are predicted for early in the next century when the content of CO_2 in the atmosphere is expected to have doubled. The fear exists that the increased climatic warming will have a calamitous effect and should be suppressed.[36-40] An opposing view has been expressed by Idso,[41] who has conluded that an increase in CO_2 will have a beneficial effect on agricultural productivity.

Current increases in the CO_2 content of the atmosphere cannot be attributed to losses of soil C, for the reason that the decline in humus content for the vast majority of the world's cultivated soils occurred prior to the mid-nineteenth century (see Chapter 2). Under steady-state conditions, the net contribution of soil organic matter to CO_2 in the atmosphere will be negligible. Bolin[36] estimated that from 10 to 40 $\times$ 10^9 tons of C have been added to the atmosphere since the early nineteenth century through changes in soil C reserves.

Recent increases in atmospheric CO_2 content appear to be due mainly to the burning of fossil fuels (e.g., coal), although agricultural practices of various types also play a role. They include the harvest of forests and accompanying oxidation of humus on the forest floor, slash-and-burn agriculture of the tropics, drainage of wetlands with accelerated decomposition of organic matter, and utilization of peat for agricultural purposes.

OTHER EFFECTS

Humic substances in leaching waters can pose a public health problem when introduced into water supplies. Chlorine (used for chlorination) has been shown to react with humic and fulvic acids to produce a family of compounds known as halomethanes,[42-44] one being chloroform, a cancer-causing agent.

Hydrogen peroxide (H_2O_2) is ubiquitous in the hydrosphere, including surface and groundwaters. Recent evidence indicates that the H_2O_2 may be generated through the photochemical action of humic substances in reducing elemental O_2.[45] Since sunlight is required, the reaction is not expected to take place to any extent in most agricultural soils, but may be of importance in natural waters containing dissolved humic matter.

SUMMARY

Large quantities of organic wastes are generated each year from domestic and municipal sewage, animal and poultry enterprises, food processing industries, pulp and paper mills, municipal garbage, and others. It is imperative to have safe and economic practices for the utilization and disposal of these wastes. With increased fertilizer and energy costs, some organic wastes (e.g., animal manures and selected municipal organic wastes) are again being considered as beneficial soil amendments. Use of some products (e.g., municipal and domestic organic wastes) incurs special problems, such as heavy metals, pathogens, nutrient imbalances, soluble salts, odors, and social acceptance.

Organic wastes must be managed so that they do not pollute air or soil–water resources, nor should they be used in any way that might introduce a toxic or pathological component into the food chain. There is a challenge to find more effective ways to manage organic wastes in soils so as to conserve soil resources and protect the environment.

An array of synthetic organic chemicals is deliberately or inadvertently introduced into soil, and many of them are recalcitrant and can persist for long periods. Persistence of a given organic chemical is also affected by adsorption to clay and organic colloids.

The organic matter content of most agricultural soils is at steady state. Accordingly, current increases in the CO_2 content of the atmosphere cannot be attributed to losses of soil C. The main causes are the burning of fossil fuels and deforestration.

REFERENCES

1 CAST, *Utilization of Animal Manures and Sewage Sludges in Food and Fiber Production*, Report 41, Council for Agricultural Science and Technology, Ames, Iowa, 1975.

2 CAST, *Application of Sewage Sludge to Cropland: Appraisal of Potential Hazards of the Heavy Metals to Plants and Animals*, Report 64, Council for Agricultural Science and Technology, Ames, Iowa, 1976.

3 L. F. Elliott and F. J. Stevenson, Eds., *Soils for Management of Organic Wastes and Waste Waters*, American Society of Agronomy, Madison, Wis., 1977.

4 T. L. Willrich and G. E. Smith, Eds., *Agricultural Practices and Water Quality*, Iowa State University Press, Ames, 1970.

5 C. H. Wadleigh, *Wastes in Relation to Agriculture and Forestry*. USDA Misc. Publ. No. 1065, 1968, pp. 1–112.

6 T. M. McCalla, J. R. Peterson, and C. Lue-Hing, "Properties of Agricultural and Municipal Wastes," in L. F. Elliott and F. J. Stevenson, Eds., *Soils for Management of Organic Wastes and Waste Waters*, American Society of Agronomy, Madison, Wis., 1977, p. 9–43.

7 R. M. Salter and C. J. Schollenberger, "Farm Manure," in *Soils and Men: Yearbook of Agriculture*, U.S. Government Printing Office, Washington, D.C., 1938.

8 R. B. Dean and J. E. Smith, Jr., "The Properties of Sludges," in *Recycling Municipal Sludges and Effluents on Land*," *National Association of State Universities and Land-Grant Colleges, Washington D.C., 1973, pp. 39–47.*

9 T. M. McCalla, L. R. Frederick, and G. L. Palmer, "Manure Decomposition and Fate of Breakdown Products in Soil," in T. L. Willrich and G. E. Smith, Eds., *Agricultural Practices and Water Quality*, Iowa State University Press, Ames, 1970, pp. 241–255.

10 A. L. Page, "Fate and Effects of Trace Elements in Sewage Sludge when Applied to Agricultural Lands," *Environmental Protection Agency Technology Series, Program Element 1B2043*, EPA Cincinnati, Ohio, 1974.

11 G. W. Leeper, *Managing the Heavy Metals on the Land*, Sheaffer and Roland, Inc., Chicago, 1978.

12 A. L. Sutton, D. W. Nelson, V. B. Mayrose, and D. T. Kelly, *J. Environ. Qual.* **12**, 198 (1983).

13 A. C. Mathers, B. A. Stewart, J. D. Thomas, and B. J. Blair, "Effect of Cattle Feedlot Manure on Crop Yields and Soil Conditions," *Proc. Symp. Animal Waste Management, Texas Tech. Rep. 11*, 1973, pp. 1–13.

14 R. H. Miller, "Soil Microbiological Aspects of Recycling Sewage Sludges and Waste Effluents on Land," in *Recycling Municipal Sludges and Effluents on Land*, National Association of State Universities and Land-Grant Colleges, Washington D.C., 1973, pp. 79–90.

15 M. D. Thorne, T. D. Hinesly, and R. L. Jones, "Utilization of Sewage Sludge on Agricultural Land," *Illinois Agricultural Experiment Station Agronomy Fact Sheet SM-29*, 1975, pp. 1–8.

16 B. A. Stewart and B. D. Meek, "Soluble Salt Considerations with Waste Application," in L. F. Elliott and F. J. Stevenson, Eds., *Soils for Management of Organic Wastes and Waste Waters*, American Society of Agronomy, Madison, Wis., 1977, pp. 218–232.

17 S. L. Jansson, *Lantbruks. Högskol Ann.* **36**, 51, 135 (1960).

18 H. J. Atkinson, G. R. Giles, and J. G. Desjardins, *Can. J. Agric. Sci.* **34**, 76 (1954).

19 D. C. Whitehead, "Nutrient Minerals in Grassland Herbage," *Commonwealth Bureau of Pastures and Field Crops Mimeo. Publ. No. 1*, 1966, pp. 1–83.

20 M. W. Varanka, Z. M. Zablocki, and T. D. Hinesly, *J. Water Pollut. Control Fed.* **48**, 1728 (1976).

21 R. H. Miller, *J. Environ. Qual.* **3**, 376 (1974).

22 G. W. Thomas, "Land Utilization and Disposal of Organic Wastes in Hot, Humid Regions," in L. F. Elliott and F. J. Stevenson, Eds., *Soils for Management of Organic Wastes and Waste Waters*, American Society of Agronomy, Madison, Wis., 1977, pp. 492–507.

23   J. H. Smith and J. R. Peterson, "Recycling of Nitrogen Through Land Application of Agricultural, Food Processing, and Municipal Wastes," in F. J. Stevenson, Ed., *Nitrogen in Agricultural Soils,* American Society of Agronomy, Madison, Wis., 1982, pp. 791–831.

24   M. Alexander, *Science* **211,** 132 (1981).

25   A. Jernelöv and A. L. Martin, *Ann. Rev. Microbiol.* **29,** 61 (1975).

26   J. M. Wood, *Science* **183,** 1049 (1974).

27   D. P. Cox and M. Alexander, *J. Microb. Ecol.* **1,** 136 (1974).

28   A. J. Francis, J. M. Duxbury, and M. Alexander, *Appl. Microb.* **28,** 248 (1974).

29   G. F. White, *Science* **209,** 183 (1980).

30   National Research Council (NCR), *Polychlorinated Biphenyls,* National Academy of Sciences, Washington, D.C., 1979.

31   P. C. Kerney, E. A. Woolson, J. R. Plimmer, and A. R. Isensee, *Residue Rev.* **28,** 137 (1969).

32   N. Sethunathan et al., *Residue Rev.* **68,** 91 (1977).

33   W. D. Guenzi, Ed., *Pesticides in Soil and Water,* American Society of Agronomy, Madison, Wis., 1974.

34   L. G. Morrill, B. Mahilium, and S. H. Modiuddin, *Organic Compounds in Soils: Sorption, Degradation, and Persistence,* Ann Arbor Science Publishers, Ann Arbor, Mich., 1982.

35   M. R. Overcash, *Decomposition of Toxic and Non-toxic Organic Compounds in Soils,* Ann Arbor Science Publishers, Ann Arbor, Mich., 1981.

36   B. Bolin, *Science* **196,** 613 (1977).

37   W. S. Broecker, T. Takahashi, H. J. Simpson, and T.-H. Peng, *Science* **206,** 409 (1979).

38   J. Hansen et al., *Science* **213,** 957 (1981).

39   M. Stuiver, *Science* **199,** 253 (1978).

40   G. M. Woodwell et al., *Science* **199,** 141 (1978).

41   S. B. Idso, *Carbon Dioxide: Friend or Foe?* IBR Press, Institute for Biospheric Research, Tempe, Ariz., 1982.

42   J. J. Rook, *Environ. Sci. Tech.* **11,** 1478 (1977).

43   A. A. Stevens, C. J. Slocum, D. R. Seeger, and G. G. Robeck, *Amer. Water Works Assoc. J.* **68,** 615 (1976).

44   N. Wade, *Science* **196,** 1421 (1977).

45   W. J. Cooper and R. G. Zika, *Science* **220,** 711 (1983).

FOUR

THE NITROGEN CYCLE IN SOIL: GLOBAL AND ECOLOGICAL ASPECTS

The N cycle in soil (Fig. 4.1) is an integral part of the overall cycle of N in nature.[1] The source of the soil N is the atmosphere, where the strongly bonded gaseous molecule (N_2) is the predominant gas (79.08% by volume of the gases). The significance of N arises from the fact that, after C, hydrogen, and oxygen, no other element is so intimately associated with the reactions carried out by living organisms. The cycling of other nutrients, notably P and S, is closely associated with biochemical N transformations.

Although considered as a sequence, an "N cycle" as such does not exist in nature. Rather, any given N atom moves from one form to another in an irregular or random fashion. Also, the soil contains an *internal cycle* that is distinct from the overall cycle of N but that interfaces with it (see Chapter 5). A key feature of the internal cycle is the turnover of N through mineralization–immobilization.

Gains in soil N occur through fixation of molecular N_2 by microorganisms and from the return of ammonia (NH_3) and nitrate (NO_3^-) in rainwater; losses occur through crop removal, leaching, and volatilization. The conversion of molecular N_2 to combined forms occurs through *biological N_2 fixation*. Organic forms of N, in turn, are converted to NH_3 and NO_3^- by a process called *mineralization*. The conversion to NH_3 is termed *ammonification*; the oxidation of this compound to NO_3^- is termed *nitrification*. The utilization of NH_3 and NO_3^- by plants and soil organisms constitutes *assimilation* and *immobilization*, respectively. Combined N is ultimately returned to the atmosphere as molecular N_2, such as through biological *denitrification*, thereby completing the cycle.

Not all transformations of N in soil are mediated by microorganisms. Ammonia and nitrite (NO_2^-), produced as products of the microbial decomposition of nitrogenous organic materials, are capable of undergoing chemical reactions with organic substances, in some cases leading to the evolution of N gases. Through the association of humic materials with mineral matter, organo–clay complexes are formed whereby the N compounds are protected against attack by microorganisms. The positively charged ammonium ion (NH_4^+) undergoes substitution reactions with other cations of the exchange complex, and it can be fixed by clay minerals.

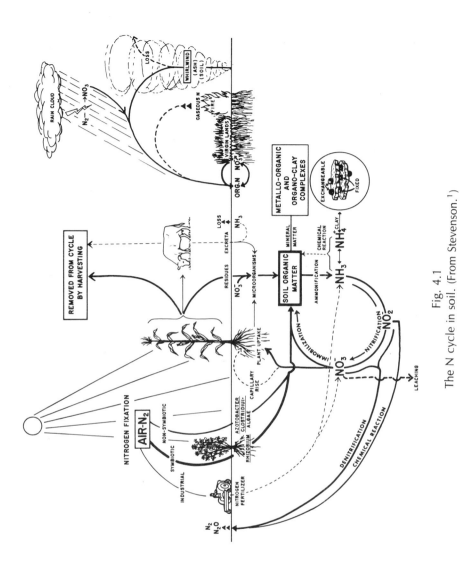

Fig. 4.1

The N cycle in soil. (From Stevenson.[1])

The basic feature of biological N transformations centers on oxidation and reduction reactions. In the oxidized ($+$) state, the outer electrons of N serve to complete the electron shells of other atoms; in the reduced ($-$) state, the three electrons required to fill the outer shell are supplied by other atoms. Typical oxidation states are -3 for NH_3, $+3$ for NO_2^-, and $+5$ for NO_3^-. Other compounds have intermediate values ($N_2{=}O$).

Topics covered in this chapter include global aspects of the soil N cycle, gains and losses of N from the soil–plant system, and the flux of soil N with other ecosystems. The next two chapters deal specifically with the "internal" cycle (Chapter 5) and environmental aspects (Chapter 6). Universal aspects of the N cycle have been covered in numerous articles and reviews.[1-12]

GEOCHEMICAL DISTRIBUTION OF N

Nitrogen is an important constituent of the four recognized spheres of the earth, namely, the lithosphere, atmosphere, hydrosphere, and biosphere. The inventory of N in the four spheres is given in Table 4.1.

A striking characteristic of the distribution pattern is the enormous size of the inert reservoir (e.g., N contained in primary rocks). The amount of N held by igneous rocks of the crust and mantle is about 50 times that present

Table 4.1
Inventory of N in the Four Spheres of the Earth[a]

Sphere	N Content, $\times\ 10^{16}$ kg
Lithosphere	16,360
Igneous rocks	
Of the crust	100
Of the mantle	16,200
Core of the earth	13
Sediments (fossil N)	35–55
Coal	0.007
Sea-bottom organic	
compounds	0.054
Terrestrial soils	
Organic matter	0.022
Clay-fixed NH_4	0.002
Atmosphere	386
Hydrosphere	
Dissolved N_2	2.19
Combined N	0.11
Biosphere	0.028–0.065

[a] Most estimated are from Burns and Hardy[3] and Söderlund and Svensson.[9] The values for terrestrial soils are from Stevenson.[1]

in the atmosphere. Relatively small amounts of N are found in the hydrosphere and biosphere.

Origin of Soil N

The original source of combined N in soils (and sediments) was the atmosphere. This N, in turn, is believed to have originated from fundamental rocks of the earth's crust and mantle.[11] One popular theory is that the earth was formed by the accretion of small solid particles called planetismals and that the atmosphere arose through the gradual evolution of gases from the interior as the newly formed earth warmed up from the heat generated by compression, by decay of radioactive elements, and possibly by other exothermic processes. Vapors and gases were driven from the interior because of their evaporation with the rising temperature. Later, as the earth cooled, the vapors condensed to form the oceans. The N, which is believed by many geochemists to have consisted mostly of NH_3, was ejected largely during the early stages of the existence of the earth, small quantities have been liberated during the course of geological times, and the process is continuing today. Free oxygen of the atmosphere is believed to have formed through photosynthesis by green plants, as well as by photochemical dissociation of water vapor in the atmosphere. As the atmosphere became enriched with oxygen, reduced N (NH_3) became oxidized to molecular N_2. Small additions of N have been made to the atmosphere over geologic times by volatilization of N compounds from meteorites during entry into the earth's atmosphere.[11]

Nitrogen in the Hydrosphere

Nitrogen in aquatic systems occurs as molecular N_2, NH_4^+, NO_2^-, NO_3^-, and dissolved and particulate organic matter.[8] Most of this N ($\sim$95%) occurs as dissolved N_2 (see Table 4.1). The remaining N occurs in various organic and inorganic forms, but mostly the former. Only the N present as NH_3, NO_3^-, NO_2^-, and organic matter belongs to what might be called the active N fraction. Molecular N_2 occurs as a dissolved gas and is virtually unaffected by chemical or biological activity in the water.

For all practical purposes, the N reserve of the ocean can be considered to be in a state of quasi-equilibrium. Variations in abundance of the different forms of N occur with depth, season, biological activity, and other factors. However, in the long run, the amount of each form remains relatively constant.

New sources of N to the ocean are the land and atmosphere, from which combined N is carried by rivers and rain. The total amount of N added each year is believed to be about 78×10^9 kg, of which 19×10^9 kg, or about one-fourth of the total, is transported by rivers.

The loss of N by deposition of organically bound N to bottom sediments

Table 4.2

Total Amounts of N (kg) in Various Pools of the Soil–Plant–Animal System

Component	Burns and Hardy[3]	Söderlund and Svensson[9]
Plant biomass	1.0×10^{14}	$1.1–1.4 \times 10^{13}$
Animal biomass	1.0×10^{12}	2.0×10^{11}
Litter	–	$1.9–3.3 \times 10^{12}$
Soil organic matter	5.5×10^{14}	3.0×10^{14}
Soil biomass	–	5.0×10^{11}
Fixed NH_4^+	–	1.6×10^{13}
Soluble inorganic	1.0×10^{12}	?

is estimated to be 8.6×10^9 kg/yr. The unaccounted-for N, amounting to about 70×10^9 kg, is believed to be lost by bacterial denitrification.

Pools of N for the Plant–Soil System

A breakdown of N in the soil–plant–animal system is given in Table 4.2. The organic N fraction constitutes the largest reservoir of potentially available N in soils. The amount of N contained in soils in organic forms, or as clay fixed NH_4^+, far exceeds that which is present in soluble inorganic forms (NO_3^- and exchangeable NH_4^+). Somewhat more N resides in the plant biomass than in the animal biomass.

The values given in Table 4.2 for the amounts of N in the organic matter of terrestrial soils ($3.0–5.5 \times 10^{14}$ kg) are of the same order as those given in Table 4.1 (2.2×10^{14} kg). The former are from the work of Burns and Hardy[3] and Söderlund and Svensson[9] while those of Table 4.1 were estimated from published data for total organic C in soil associations of the world.[1] For Table 4.1, the assumption was made that the average C/N ratio was 10 for the mineral soils and 30 for the organic soils (Histosols). An average of 10% of the N in the mineral soils was assumed to occur as clay-fixed NH_4^+.

A further breakdown of total, organic, and clay-fixed NH_4^+ in soil associations of the world is given in Table 4.3. Because of their high organic matter and N contents, Histosols are major contributors to the total soil N (28.7×10^{12} kg, or more than 10% of the total). It should be noted that Histosols are often considerably deeper than the 1 m depth upon which the calculations were made. The relatively low amounts of N in the mineral soils of South America are due to the fact that most of them are tropical soils low in organic matter.

Estimated amounts of N to depths of 15 and 100 cm for the major soil associations of the United States are recorded in Table 4.4. Most of the N resides in those soils classified as Mollisols (listed as Prairie, Chernozem, and Chestnut soils).

Table 4.3

Nitrogen in Some Soil Associations of North America, South America, and Other Areas (Values Represent Amounts to a Depth of 1 m)[a]

| Association | Area, 10^5 km² | N, 10^{12} kg | | |
		Organic	Fixed NH$_4^+$	Total
North America				
Histosols	13.3	8.9	–	8.9
Podzols	31.5	5.9	0.7	6.6
Cambisols	9.4	5.0	0.6	5.6
Haplic Kastanozems	32.0	4.6	0.5	5.1
Eutric Gleysols	10.9	3.0	0.3	3.3
Gelic Regosols	15.9	2.9	0.3	3.2
Dystric Cambisols	3.6	1.9	0.2	2.1
Phaeozems	10.4	1.9	0.2	2.1
All others	76.0	10.7	1.2	11.9
	203.0	44.8	4.0	48.8
South America				
Ferrasols	89.8	9.7	1.1	10.8
Dystric Histosols	3.8	2.5	–	2.5
Cambisol–Andisols	6.2	3.1	0.4	3.5
Cambisols	4.2	2.3	0.3	2.6
Acrisol–Xerosol– Kastanozems	16.6	1.2	0.1	1.3
All others	61.4	3.9	0.4	4.3
	182.0	22.7	2.3	25.0
Asia, Africa, Europe, Oceania				
Histosols	26.0	17.3	–	17.3
Cambisols	47.0	25.2	2.8	28.0
Podzols	130.0	24.3	2.7	27.0
Kastanozems	131.0	18.9	2.1	21.0
Chernozems	51.0	18.0	2.0	20.0
Ferrasols	141.0	15.3	1.7	17.0
Cambisol–Vertisols	28.0	9.9	1.1	11.0
All others	282.0	19.8	2.2	22.0
	836.0	148.7	14.6	163.3
Totals		216.2	20.9	237.1

[a] From Stevenson.[1]

Table 4.4

Estimated Amounts of N to Depths of 15 cm and 1 m for the Major Soil Associations of the United States[a]

Soil Association	Approximate Area	Average Amount of N per ha		Total N in Association	
		to 15 cm	to 1 m	to 15 cm	to 1 m
	$(\times\ 10^6$ ha)	$(\times\ 10^3$ kg)	$(\times\ 10^3$ kg)	$(\times\ 10^9$ kg)	$(\times\ 10^9$ kg)
Brown forest	72.8	2.8	7.5	203.8	546.0
Red and Yellow	60.7	2.2	4.5	133.5	273.2
Prairie	45.7	3.9	17.9	178.2	818.0
Chernozem and Chernozem-like	49.8	5.0	17.9	249.0	891.4
Chestnut	41.3	3.3	12.0	136.3	495.6
Brown	21.0	2.8	9.0	58.8	189.0
Totals				959.6	3213.2

[a] From Stevenson.[1]

According to Rosswall,[7] about 95% of the N that cycles annually within the pedosphere interacts solely within the soil–microbial–plant system. On this basis only 5% of the total flow is concerned with exchanges to and from the atmosphere and hydrosphere. As will be shown later, losses of fertilizer N through leaching and denitrification are much higher than the 5% value might suggest.

The average mean residence time for N in soils has been estimated to be of the order of 175 years.[7] Some components will have much longer ages, perhaps 1,000 years or more.

The Biosphere

Many difficulties are encountered in determining the distribution of N in living matter (the biosphere). Unlike other spheres, the biosphere is in a constant state of change. Also, living organisms are not uniformly distributed, and the N contents of different organisms vary widely. The estimate given in Table 4.1 for total N in the biosphere (0.028 to 0.065 $\times\ 10^{16}$ kg) is at best an approximation.

NITROGEN AS A PLANT NUTRIENT

Nitrogen occupies a unique position among the elements essential for plant growth because of the rather large amounts required by most agricultural crops. A deficiency of N is shown by yellowing of the leaves and by slow and stunted growth. Other factors being favorable, an adequate supply of N in the soil promotes rapid plant growth and the development of dark-green color in the leaves. Major roles of N in plant nutrition include: (1) component of chlorophyll; (2) component of amino acids, the building blocks of proteins; (3) essential for carbohydrate utilization; (4) component of enzymes, vitamins, and hormones; (5) stimulative of root development and activity; and (6) supportive to uptake of other nutrients.[13]

Nitrate is the main form of N taken up by most crop plants, the most notable exception being lowland rice. The first step in NO_3^- utilization by plants is reduction to the NH_3 form:

$$NO_3^- + 2e^- \xrightarrow[\text{reductase}]{\text{nitrate}} NO_2^-$$

$$NO_2^- + 6e^- \xrightarrow[\text{reductase}]{\text{nitrite}} NH_3$$

Reduction of NO_3^- to NO_2^-—the rate-limiting step in the transformation—is catalyzed by nitrate reductase, a metaloprotein containing Fe and Mo as a constituent.[13,14] Considerable energy is required; thus reduction is closely linked to photosynthesis, as illustrated in Fig. 4.2. In the presence of light and CO_2, sugars and starches, along with phosphorylated intermediates of carbohydrate metabolism, are produced in the chloroplasts and move to the cytoplasm, where reduced NADH (nicotinamide adenine dinucleotide) is generated from the utilization of 3-phosphoglyceraldehyde. The NADH is the source of energy whereby NO_3^- is reduced to NO_2^-.

The final step (reduction of NO_2^- to NH_3) is catalyzed by nitrite reductase in the leaf tissue. Although the reaction involves the transfer of six electrons, no other free intermediates (e.g., hydroxylamine or hyponitrite) are known. In some systems the electrons required for reduction are provided by ferredoxin, an Fe–S protein. Nitrate may be protonated to HNO_2 before transport across the chloroplast membrane. The overall sequence is:

$$NO_2^- + H^+ \rightleftharpoons HNO_2 \quad \text{cytoplasm}$$

Since the reduction of NO_3^- to NO_2^- is the step that limits the rate of synthesis of amino acids and proteins, the activity of nitrate reductase is sometimes regarded as a good indicator of growth rate.[14] The level of nitrate reductase in plants varies during the day, over the course of the growing season, and during periods of moisture or heat stress.

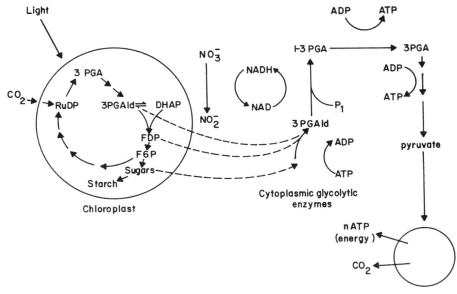

Fig. 4.2

A simplified metabolic scheme illustrating the interrelationship between photosynthate, glycolytic metabolism, and NO_3^- reduction in plants. Key abbreviations are 3PGA, 3-phosphoglyceric acid; 3PGAld, 3-phosphoglyceraldehyde; ADP and ATP, adenosine di- and triphosphate; NAD and NADH, oxidized and reduced nicotinamide adenine dinucleotides, respectively. (From Viets and Hageman.[14])

The NH_3 produced by nitrite reductase seldom accumulates in plants but is rapidly metabolized and incorporated into glutamic and aspartic acids, the two main compounds from which other amino acids and N-containing biochemicals are formed. The overall pathway of N metabolism in plants is shown in Fig. 4.3.

The amount of N consumed by plants varies greatly from one species to another, and, for any given species, the amount varies with genotype and the environment.[14] Also, considerable variation exists in the relative amount of the N contained in the different plant parts (grain, stems, leaves, roots, etc.). Examples of N removals in the harvested portion and residues of some important crops under conditions of good yields are given in Fig. 4.4. Substantial variation from the reported values can occur depending on soil N status, fertilization practice, and climate. In general, more N is contained in the harvested portion than in the stover, vines, straw, or roots. Nitrogen uptake by plants is very rapid during the period of rapid vegetative growth, as illustrated for three crops in Fig. 4.5.

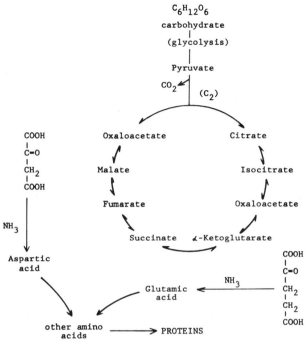

Fig. 4.3
Simplified pathway for N metabolism in plants.

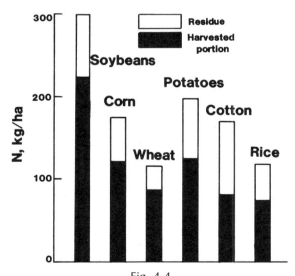

Fig. 4.4
Nitrogen contained in the harvested portion and residue of good yields of some major agricultural crops. (Adapted from Olson and Kurtz.[13])

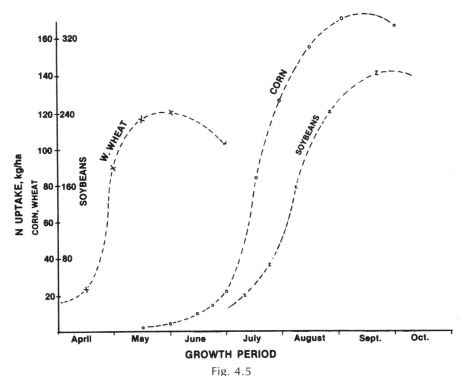

Fig. 4.5

Average rates of N accumulation in the above-ground crop of nonirrigated wheat and irrigated corn and soybeans in Nebraska.[13] (Reproduced from *Nitrogen in Agricultural Soils* (1982) by permission of the American Society of Agronomy.)

GAINS IN SOIL N

Except for irrigation and flooding, practically all the N that enters the soil by natural processes is due to biological N_2 fixation and atmospheric deposition of NH_3, NH_4^+, and NO_3^-. Nitrogen is also added in crop residues and animal manures, but these represent recycling within the soil–plant system.

Biological N_2 Fixation

Although a vast supply of N occurs in the earth's atmosphere (386×10^{16} kg), it is present as an inert gas and cannot be used by higher forms of plant and animal life. The covalent triple bond of the N_2 molecule ($N\equiv N$) is highly stable and can be broken chemically only at elevated temperatures and pressures. Nitrogen-fixing microorganisms, on the other hand, perform this seemingly impossible task at ordinary temperatures and pressures. Nitrogen

Table 4.5
Estimated Average Rates of Biological N_2 Fixation for Specific Organisms and Associations[a]

Organism or System	N_2 Fixed, kg/ha · year
Blue-green algae	25
Free-living microorganisms	
Azotobacter	0.3
Clostridium pasteurianum	0.1–0.5
Plant–algal associations	
Gunnera	12–21
Azollas	313
Lichens	39–84
Legumes	
Soybeans (*Glycine max* L. Merr.)	57–94
Cowpeas (*Vigna, Lespedeza, Phaseolus,* and others)	84
Clover (*Trifolium hybridium* L.)	104–160
Alfalfa (*Medicago sativa* L.)	128–600
Lupines (*Lupinus* sp.)	150–169
Nodulated nonlegumes	
Alnus	40–300
Hippophae	2–179
Ceanothus	60
Coriaria	150

[a] From Evans and Barber.[16]

fixation in prehistoric times created the combined N that is currently present in many commercially important natural deposits, such as coal, petroleum, and the caliche of the Chilean desert.

Because of the extensive research on biological N_2 fixation, only a brief summary will be given here. For detailed information, the reader is directed to several recent books and reviews on the subject.[3,15–21]

Data given later on global N fluxes will show that the total amount of N returned to the earth each year through biological N_2 fixation is of the order of 139×10^9 kg, of which about 65% (89×10^9 kg) is contributed by nodulated legumes grown for grain, pasture, hay, and other agricultural purposes.[3,4,8] Estimated average rates of biological N_2 fixation for some specific organisms and associations are given in Table 4.5.

The ability of a few bacteria, actinomycetes and blue-green algae to fix molecular N_2 can be regarded as being second in importance only to photosynthesis for the maintenance of life on this earth. The two basic biochemical processes in nature are often considered to be *photosynthesis* and *respiration*; to this list should be added *biological N_2 fixation* and possibly *denitrification*.

Numerous approaches have been used to determine the extent of biological N_2 fixation in natural ecosystems. They include the following.[17,22]

1 From the increase in total combined N in the system under study. The Kjeldahl method is usually used for this purpose.
2 By using nodulating and nonnodulating soybean isolines. In this case the amount fixed is equal to the N content of the nodulating strain minus that of the nonnodulating strain.
3 From the amount of ^{15}N taken up by the test plant when grown in a closed chamber containing ^{15}N-enriched N_2.
4 By growing legumes in soil where the organic matter has been enriched with ^{15}N. A lower ^{15}N content of the plant tissue suggests an alternate N source (i.e., the atmosphere).
5 From estimates of available soil N (A-values) as measured from the amounts of applied ^{15}N taken up by a nonleguminous plant and the legume.
6 From $\Delta^{15}N$ values for nonleguminous and leguminous plants grown on the same soil. This approach is based on the observation that the ^{15}N content of the soil N is higher than for N_2 of the atmosphere.
7 By acetylene reduction. In the presence of the enzyme nitrogenase, acetylene ($CH{\equiv}CH$) is reduced to ethylene ($CH_2{=}CH_2$). Estimates of the latter by gas-liquid chromatography provide a sensitive test for biological N_2 fixation.

Each method has its advantages and disadvantages, as discussed elsewhere.[17,22] The most direct approach is through use of ^{15}N-labeled N_2 but a closed system is required. A popular technique used at the present time for demonstrating N_2 fixation under field conditions is by acetylene reduction.[22] A system designed for in situ N_2 (C_2H_2) fixation is illustrated in Fig. 4.6. An open cylinder is buried in the soil, plants are grown, the soil surface is sealed, and an acetylene–air mixture is allowed to flow through the nodulated root zone. The gas leaving the exit port is analyzed for ethylene, from which N_2 fixation rates are estimated.

The route whereby molecular N_2 is converted to a reduced form (NH_3) has been worked out in broad outline. The overall equation is:

$$N_2 + 6e^- + 6H^+ \xrightarrow{\quad \text{nitrogenase} \quad} 2NH_3$$

The nitrogenase system consists of two protein complexes. The first has a molecular weight of about 180,000 and contains both Mo and Fe; the second has a molecular weight of about 51,000 and contains nonheme Fe. The larger Mo–Fe protein is the N_2 reductase enzyme while the smaller Fe protein provides the electrons for reduction. Other electron-transporting agents include ferredoxin, an Fe–S protein, or flavodoxin. The system also requires

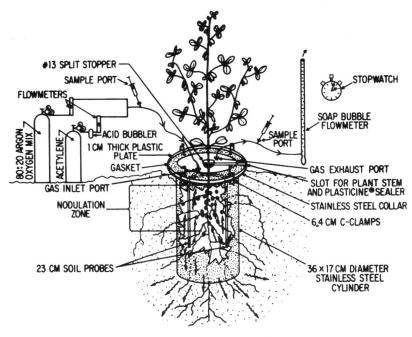

Fig. 4.6

Continuous flow technique for the in situ measurement of biological N_2 fixation in the field.[17] (Reproduced from *Nitrogen in Agricultural Soils* (1982) by permission of the American Society of Agronomy.)

energy in the form of ATP (adenosine triphosphate), with about 10 molecules being required per molecule of NH_3 formed. A simplified reaction scheme is as follows.

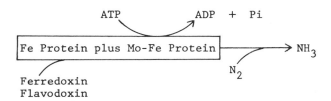

A postulated reaction sequence of biological N_2 fixation is illustrated in Fig. 4.7. The H_2 produced from pyruvate serves as a hydrogen donor for N_2 reduction; ATP furnishes energy for various steps of the reduction process. The NH_3 thus formed is used for the synthesis of amino acids and other N-containing biochemicals. The pathways involved are identical to those shown earlier in Fig. 4.3.

The organisms that fix N_2 are conveniently placed into two groups: (1)

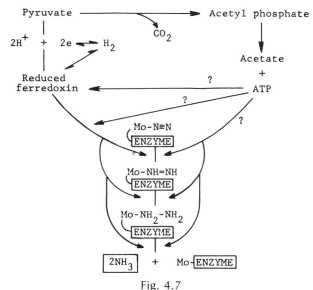

Fig. 4.7
General scheme for biological N_2 fixation. (Adapted from Burris.[21])

the nonsymbiotic fixers, or those that fix molecular N_2 apart from the specific host; and (2) the symbiotic fixers, or those that fix N_2 in association with higher plants, including some nonlegumes. The nonsymbiotic fixers include a number of blue-green algae of the family Nostocaceae, various photosynthetic bacteria (e.g., *Rhodospirillum*), several aerobic bacteria (e.g., *Azotobacter, Beijerinckia, Derxia*), and certain anaerobic bacteria of the genus *Clostridium*. A variety of other organisms, including some actinomycetes and fungi, have been reported to fix molecular N_2, but these claims have not all been verified and, in any event, the amounts of N_2 fixed would appear to be too small to be of practical significance in soils. The major exception is the actinomycete *Frankia*, the endophyte that fixes N_2 in association with nonleguminous plants.

Blue-Green Algae

There is now satisfactory evidence for N_2 fixation by over a dozen species of blue-green algae (Cyanobacteria).[23,24] The organisms for which fixation has been firmly established include species of *Anabaena, Nostoc, Aulosira, Cylindrospermum*, and *Calothrix*, among others. Nitrogen-fixing blue-green algae function as microsymbiants in various genera of the gymnosperm family Cycadacea and in the angiosperm genus *Gunnera* (family Haloragaceae).

The blue-green algae are an archaic group of organisms that have persisted during long epochs of the earth's history. They can be found in practically every environmental situation where sunlight is available for photosynthesis, such as uninhabited wastelands and barren rock surfaces. Their ability to colonize virgin landscapes can be accounted for by the fact that they are completely autotrophic and thereby able to synthesize all their cellular material from CO_2, molecular N_2, water, and mineral salts. It is of additional significance that they form symbiotiotic relationships with a variety of other organisms, such as the lichen fungi.

Geographically, lichens are widely distributed over land masses of the earth. They make up a considerable portion of the vegetation on the arctic continent. Besides being the pioneering plants on virgin landscapes, they bring about the disintegration of rocks to which they are attached, thereby forming soil on which higher plants can grow. In desert areas of the southwestern United States, they form surface crusts of varying density and cling to surface stones. Crusts of blue-green algae have been found in semiarid soils of eastern Australia and the Great Plains of the United States, where their favorite habitat is the undersurface of translucent pebbles.

The importance of blue-green algae in supplying fixed N to most agricultural soils is probably limited to the initial stages of soil formation. These organisms fix N_2 only in the presence of sunlight; consequently, their activity is confined almost exclusively to superficial layers of the earth's crust. There is abundant evidence to indicate that blue-green algae are important agents in fixing N_2 in rice paddy fields, as well as eutrophic lakes.

Azolla, a freshwater fern, is used extensively in parts of southeast Asia as a green manure crop and substitute for N fertilizers in rice paddy culture.[24,25] This plant forms a symbiotic relationship with the blue-green algae *Anabaena azollae*. The alga inhabits a cavity on the ventral surface of *Azolla* and lives symbiotically with the fern while fixing N_2. The practical importance of the relationship to rice production in many parts of Asia cannot be overemphasized. The fern grows readily over the paddy water, often covering the surface completely as an *Azolla* "bloom." *Azolla* is the principal source of N for over 1 million hectares of rice in China.[26]

Free-Living Bacteria

Classical examples of N_2 fixation by free-living bacteria are by species of the photosynthetic *Rhodospirillum*, the anaerobic heterotroph *Clostridium*, and the aerobic heterotroph *Azotobacter*. To this list can be added *Beijerinckia*, *Derxia*, and *Azospirillum*. The latter forms an associative symbiosis with the roots of certain cereal crops.[27,28]

The requirement of photosynthetic N_2-fixing bacteria for both irradiation and anaerobiosis restricts their activities to shallow, muddy ponds or estuarine muds. They generally are found as a layer overlying the mud and covered by a layer of algae; fixation of N_2 is possible because pigments of the photosynthetic bacteria absorb light rays in the region of the spectra not absorbed by pigments of the overlying algae.

The anaerobic fixer *Clostridium* is universally present in soils, including those too acid for *Azotobacter*. The normal condition of *Clostridium* is the spore form, vegetative growth occurring only during brief anaerobic periods following rains.

Azotobacter is also widely distributed in soils. The most common species, *A. chroococcum,* is found worldwide but mostly in near-neutral or slightly alkaline soils. In contrast, *Beijerinckia* and *Derxia* are typical inhabitants of acid soils. *Beijerinckia* has been found in soils of India, Southeast Asia, tropical Africa, South America, and northern Australia but is essentially absent from temperate zone soils.

Under natural soil conditions the N_2-fixing capabilities of free-living bacteria are greatly restricted. These organisms require a source of available energy, a factor that limits their activities to environments with relatively high organic matter contents. In an early discussion of the subject, Jensen[29] suggested that many of the estimates for N_2 fixation by nonsymbiotic N_2-fixing bacteria—frequently as high as 20 to 50 kg N/ha·yr (18 to 45 lb N/acre·yr)—were much too high. He concluded that the level of available organic matter in most soils was too low to support fixation of this magnitude. The consensus of many soil scientists is that no more than about 6 kg N/ha·yr is added to soils of the United States by the combined activities of nonsymbiotic N_2 fixing microorganisms; in semiarid soils, no more than 3 kg N/ha·year may be fixed. On this basis, the amount of N_2 fixed by nonsymbiotic N_2 fixers in soils under intensive cultivation would appear to be too low to have much practical impact.

Conditions for optimum N_2 fixation by free-living microorganisms include the presence of adequate energy substrates (such as organic residues), low levels of available soil N, adequate mineral nutrients, near-neutral pH, and suitable moisture. In view of the large number of microorganisms that have been reported to fix molecular N_2, it would appear that N gains under field conditions result from the cumulative action of numerous organisms fixing rather small amounts of N rather than through fixation by only one or two organisms. The subject of gains in soil N through the activities of free-living N_2-fixing microorganisms have been reviewed elsewhere.[30,31]

Arguments in support of the view that free-living bacteria do not provide large amounts of combined N for plants in most cultivated soils include the following:

1 Free-living N_2 fixers are heterotrophic and inefficient users of carbohydrates, with from only 1 to 10 mg of N being fixed per gram of carbohydrate used.

2 Heterotrophic N_2 fixers are in severe competition with other bacteria, actinomycetes, and fungi for available organic matter, which is normally in short supply in most soils.

3 Fixation of N_2 is greatly reduced in the presence of readily available combined N. In productive agricultural soils, levels of available N are often sufficiently high to seriously inhibit fixation of N_2.

Substantial gains in N, frequently of the order of 66 to 112 kg/ha·year, or 50 to 100 lb/acre·year, have been reported in many soils in the apparent absence of legumes or other plants known to form a symbiotic relationship with N_2-fixing microorganisms. Some investigators believe that these increases cannot be attributed entirely to errors inherent in measuring N gains under field conditions, but that, under certain circumstances, significant amounts of N can be added to the soil through the combined activities of free-living, N_2-fixing organisms. For example, the incorporation of organic residues in soil can enhance fixation when used as an energy source by nonsymbiotic N_2 fixers.

It is appropriate to mention that considerable difficulty is encountered in evaluating reports for gains in soil N under field conditions. In addition to faulty experimental techniques, lack of statistical control, and errors inherent in measuring small increases in soil N by the Kjeldahl method, the possibility exists that N has accumulated through other means, such as by upward movement of NO_3^- in solution, recycling via plant roots, and accretion from the atmosphere (discussed later).

Evidence that nonsymbiotic N_2 fixers often make a significant contribution to the soil N is as follows:

1 More kinds and numbers of N_2-fixing bacteria are present in soil than formerly thought. The number of described species of nonsymbiotic fixers has increased greatly in recent years, now totaling about 100. While the amount of N_2 fixed by any given species may be small, the combined total for all organisms could be appreciable.

2 The discovery of new N_2 fixers in tropical soils, such as *Beijerinckia* and *Derxia gummosa,* points to the importance of these organisms in tropical agriculture.

3 Appreciable gains in soil N observed for legume-free grass sods suggest extensive fixation in the rhizosphere of crop plants.

4 The ability to grow crops continuously on the same land for years without N fertilizers, and without growing legumes, is well known and may be due in part to nonsymbiotic N_2 fixation.

At one time, Russian scientists claimed that inoculation of soils and seeds with *Azotobacter* and *Clostridium* resulted in improved growth of wheat and cotton. Extensive programs of inoculation were carried out in Russia and elsewhere from 1930 to recent years. Attempts have been made from time to time to confirm the early Russian claims, but, in general, the findings have failed to show increases in growth that could be ascribed positively to N_2 fixation.[32]

Data summarizing N gains in soil as reported by Moore[31] are given in Table 4.6. In some but not all cases, nonsymbiotic organisms may have been responsible for fixation of N_2. For example, some research was conducted

Table 4.6
Gains in Soil N Through Biological N$_2$ Fixation[a]

No. of References Cited	Conditions	N Gains, kg/ha · year[b]
5	Soil amended with crop residues	15–78 (13–70)
4	Field plots under sodlike crops	14–56 (12–50)
4	Lysimeter studies	25–67 (22–60)
4	Stands of *Pinus* sp. or other monoculture, nonnodulated trees	36–67 (22–60)

[a] As recorded by Moore.[31]
[b] Values in brackets are in lb/(acre year).

on grass plots kept legume-free to assess the exact contribution of nonsymbiotic fixers.

The values in Table 4.6 represent the typical or usual range reported by Moore.[31] Both lower and higher N gains were reported, with some of the higher values exceeding 224 kg/ha·yr (200 lb/acre·yr). The actual amount of N$_2$ fixed under most agricultural systems is probably much less than suggested by Table 4.6.

Many of the unexplained increases in soil N have been associated with the growth of grass, indicating that a specialized association may exist between the grasses and certain N$_2$-fixing microorganisms. An associative symbiosis has been observed between certain rhizosphere bacteria and the root surfaces of corn, wheat, and tropical grasses.[17,27,28] Species of *Azospirillum* are believed to be responsible for N$_2$ fixation.[17]

The rhizosphere, or that part of the soil near plant roots, would appear to be a particularly favorable site for N$_2$ fixation because of organic material excreted or sloughed off by roots. In natural plant communities, a relatively low rate of fixation (<10 kg/ha·yr) may be adequate, whereas under intensive farming conditions fixation of rather large amounts (often over 50 kg/ha·) would be required for optimum yields.

Symbiotic N$_2$ Fixation by Leguminous Plants

The symbiotic partnership between bacteria of the genus *Rhizobium* and leguminous plants has had a long and comprehensive development. The importance of this relationship is emphasized by the fact that, even with the tremendous expansion in facilities for producing fertilizer N over the past three decades, legumes are still the main source of fixed N for a large portion of the world's soils. On the conservative estimate of an average fixation of

Table 4.7
Distribution of the Leguminosae[a]

Subfamily	Number of Genera and Species	Genera and Species in Tropics and Subtropics	Genera and Species in Temperate Regions	Genera Occurring in Both Tropic and Temperate Zones
Mimosoideae	31–1,341	31–1,200	1–141	1
Caesalpinioideae	95–1,032	89–988	7–44	1
Papilionateae	305–6,514	176–2,430	141–3,084	12
Total	431–8,887	296–4,618	149–3,269	14

[a] From Norris.[34]

55 kg of N for the 84 million hectares of legumes planted each year in the United States, a total of over 4×10^9 kg of N are fixed annually. Legumes of agricultural significance can be broadly divided into grain and forage legumes, the distinction being that seeds of the former are harvested for food (e.g., peas and beans).

According to LaRue and Patterson,[33] the role of legumes as sources of N is certain to increase in importance. Factors leading to more extensive use of legumes or other N-fixing plants include the need to (1) make marginal lands more productive, (2) control erosion and desertification, (3) reduce fertilizer costs, especially as fertilizer N becomes more expensive, and (4) reclaim drastically disturbed land, such as from strip mining.

The Leguminosae family contains from 10,000 to 12,000 plant species, most of which are indigenous to the tropics. Thus far, only about 1,200 species have been examined for nodulation, of which about 90% have been found to bear nodules. Less than 100 species are used in commercial food production. Whereas greatest attention has been given to the cultivated legumes, wild species are of considerable importance for fixation of N_2 in natural ecosystems.

Norris[34] summarized available information regarding the global distribution of the Leguminosae, from which he compiled the tribal and species distribution. A summary is given in Table 4.7. A smaller number of genera and species are indigenous to the temperate regions of the earth than to the tropics and subtropics. The data are undoubtedly obsolete, particularly with respect to the number of species. Nevertheless, the material is adequate for arriving at some generalizations regarding the distribution of the Leguminosae. With the exception of one genus (141 species), the subfamily Mimosoideae is entirely tropical and subtropical, while 89 of the 95 genera of plants in the subfamily Caesalpinioideae (over 95% of the species) are confined to the tropics and subtropics. In the subfamily Papilionateae there are 141 genera of plants (3,084 species) that are located in the temperate regions, while 176 genera (2,430 species) occur in the tropics and subtropics.

Table 4.8
Classification Scheme of *Rhizobium*–Legume Associations

Rhizobium	Cross-inoculation Group	Host Genera	Legumes Included
R. meliloti	Alfalfa	*Medicago*	Alfalfa
		Melilotus	Sweet clover
		Trigonella	Fenugreek
R. trifolii	Clover	*Trifolium*	Clovers
R. leguminosarum	Pea	*Pisum*	Pea
		Vicia	Vetch
		Lathyrus	Sweetpea
		Lens	Lentil
R. phaseoli	Bean	*Phaseolus*	Beans
R. lupini	Lupine	*Lupinus*	Lupine
		Ornithopus	Serradella
R. japonicum	Soybean	*Glycine*	Soybean
	Cowpea[a]	*Vigna*	Cowpea
		Lespedeza	Lespedeza
		Crotalaria	Crotalaria
		Pueraria	Kudzu
		Arachis	Peanut
		Phaseolus	Lima bean

[a] This group has not been accorded species status.

The bacterial symbionts, all members of the genus *Rhizobium*, are gram-negative, nonspore-forming rods that measure 0.5 to 0.9 μm by 1.2 to 3.0 μm. The following species are generally recognized: *R. meliloti, R. leguminosarum, R. phaseoli, R. japonicum, R. lupini,* and *R. trifolii.* A collection of leguminous plants that exhibit specificity for a common *Rhizobium* species is referred to as a *cross-inoculation group.* The validity of these bacterial–plant associations has often been challenged because the boundaries between the groups overlap, and because some strains of rhizobia form nodules on plants occurring in several different groups. A list of the important legumes with which the above-mentioned species of *Rhizobium* are associated is given in Table 4.8.

Estimates for the amounts of N fixed by various legumes are tabulated in Table 4.9. Much of the work has been carried out using lysimeters or controlled experimental plots, and extrapolation of the data to agricultural soils in general is speculative. LaRue and Patterson[33] arrived at the following conclusions regarding N₂ fixation by legumes:

1 There is not a single crop for which valid estimates are available for N fixed by agricultural legumes.

2 There is no evidence that any legume crop satisfies all its N requirements

Table 4.9
Estimates of N Fixation by Some Typical Legumes[a]

Forage Crops		Pulses	
Species	kg N/ha	Species	kg N/ha
Alfalfa (*Medicago sativa*)	148–290	*Phaseolus vulgaris*	10
White clover (*Trifolium repens*)	128–268	*Pisum sativum*	17–69
Ladino clover (*Trifolium repens*)	165–189	*Vicia faba*	121–171
Red clover (*Medicago pratense*)	17–154	Lupine	121–157
Subclover (*Trifolium subterraneum*)	21–207	Chick-pea	67–141
Egyptian clover (*Trifolium alexandrinum*)	62–235	Lentil	62–103
Vetch (*Vicia villosa*)	184	*Arachis hypogea*	87–122

[a] Adapted from LaRue and Patterson,[33] from which specific references can be obtained. A wide range was reported for the amounts fixed by soybeans (15 to over 200 kg N/ha).

by fixation, the highest percentages (80%) being typical of low-fertility soils or soils artificially made N-poor by amendment with carbonaceous organic residues.

3 Some legumes (e.g., soybeans) may actually deplete soil N.

For the most part, the rhizobia are capable of prolonged independent existence in the soil; however, N_2 fixation takes place only when symbiosis is established with the plant. Maintenance of a satisfactory population of any given *Rhizobium* species in the soil depends largely on the previous occurrence of the appropriate leguminous plant. High acidity, lack of necessary nutrients, poor physical condition of the soil, and attack by bacteriophages contribute to their disappearance. The desirability of legume inoculation to insure nodulation with host-specific effective rhizobial strains is well known. A discussion of the potential for increasing protein production by legume inoculation has been given by Dawson.[35] The subject of legume seed inoculants has been covered by Roughley.[36]

Factors that affect symbiotic N_2 fixation include light, temperature, water relations, and soil pH, among others. Maximum fixation is obtained only when the supply of available mineral N in the soil is low. However, during the early stages of plant growth, small amounts of fertilizer N may improve nodulation and N_2 fixation, especially on N-poor soils. Presumably, the added N alleviates the N starvation period that can occur between the exhaustion of seed N and the onset of N_2 fixation. The optimum amount of N required varies with the leguminous species.

The presence of high amounts of available N in the soil tends to depress nodule formation, a result that appears to be associated with a low carbohydrate–N ratio in the plant and, consequently, an inadequate supply of carbohydrates to the roots. Numerous studies have shown that the quantity of N_2 fixed by rhizobia decreases as the ability of the soil to provide mineral N increases. However, the amount of N released from the soil organic matter is rarely adequate to suppress N_2 fixation entirely.

In many places in the world an abundance of fertilizer N at reasonable cost has prompted a reevaluation of the role of legumes in crop production. It now seems certain that under certain circumstances, legumes in a rotation can be replaced effectively by nonlegumes, provided that chemically fixed N is applied. Provided crop residues are returned to the soil following harvest, the increased production of plant material brought about through adequate fertilization may allow nonlegumes to assume some of the functions historically assigned to legumes, namely, to improve soil tilth, to prevent erosion, to increase the storehouse of soil N, and to enhance the activities of desirable microorganisms in such a way that soil structure is improved, and thereby plant growth.

The extent to which fertilizer N will replace legumes in crop rotations will depend on the availability of inexpensive N fertilizers, the need for increased acreage of nonleguminous crops, and the ability of legume-free cropping systems to maintain soil fertility and prevent erosion.

Microorganisms Living in Symbiosis with Nonleguminous Plants

Nitrogen fixation of a nature similar to the symbiotic relationship between the *Rhizobium* and leguminous plants has been demonstrated for many angiosperms, including plants belonging to the families Betulaceae, Casuarinaceae, Coriariaceae, Elaegnaceae, Myricaceae, Rhamnaceae, and Rosaceae. Contrary to popular belief, nodulated nonlegumes are not freak plants of limited distribution but they are important sources of fixed N for plants in general. Many are shrubs or trees located on poor soils, where they function as "pioneering plants" and become established because of their ability to obtain combined N through biological N_2 fixation.

The geographical distribution of the nonleguminous families for which N_2 fixation has been confirmed is outlined in Table 4.10. Nodulated nonlegumes occur in 12 genera or 7 families of dicotyledon plants. The genera of nodulated nonlegumes comprise about 300 plant species, of which about one-third have been reported to bear nodules. Very little information is available concerning the organisms responsible for N_2 fixation. It has been established, however, that the endophyte is an actinomycete (*Frankia*). Docu-

Table 4.10
Distribution of Nodulated Nonlegumes

Family and Genera	Incidence of Nodulating Species[a]	Geographical Distribution
Betulaceae		
Alnus	25/35	Cool regions of the Northern Hemisphere
Casuarinaceae		
Casuarina	14/45	Tropics and subtropics, extending from East Africa to the Indian Archipelago, Pacific Islands, and Australia
Coriariaceae		
Coriaria	12/15	Widely separated regions, chiefly Japan, New Zealand, Central and South America, the Mediterranean region
Elaeagnaceae		
Elaeagnus	9/45	Asia, Europe, North America
Hippophae	1/1	Asia and Europe, from the Himalayas to the Arctic Circle
Shepherdia	2/3	Confined to North America
Myricaceae		
Myrica	12/35	Temperate regions of both hemispheres
Rhamnaceae		
Ceanothus	30/55	Confined to North America
Discaria	1/10	Temperate, subtropical and tropical regions
Rosaceae		
Cerocarpus	1/20	Cool regions of temperate zone
Dryas	3/4	Cool regions of temperate zone
Purshia	2/2	Cool regions of temperate zone

[a] Incidence refers to ratio of species bearing nodules to total number of species as reported by Silver.[38] See also Becking.[37]

mentary evidence for N_2 fixation by microorganisms living in association with nonleguminous plants has been given in several reviews.[17,37,38]

The family Betulaceae, consisting of the birches and alders, is found almost entirely in the cool temperate and arctic zones of the northern hemisphere. Thus far, only the alder (*Alnus*) has been found to bear nodules. Crocker and Major[39] estimated an annual gain of 62 kg/ha (55 lb/acre) by *A. crispa* during colonization of the recessional moraines of Alaskan glaciers.

Nitrogen fixation has been reported for 14 of the 45 species of Casuarinaceae, the main nonleguminous angiosperm family of nodulating plants occurring in tropical and subtropical areas. Plants of this family are of great ecological significance in the Australian environment.[40]

Twelve of the 15 species of the family Coriariaceae (genus *Coriaria*) have been found to bear nodules. Bond and Montserrat[41] suggested that the discontinuous distribution of this family indicates that, in ancient times, it made a far greater contribution to the supply of fixed N than at present.

The family Elaeagnaceae, consisting of the genera *Elaeagnus*, *Hippophae*, and *Shepherdia*, is distributed widely in the temperate regions of both hemispheres. The genus *Elaeagnus*, with 45 species (9 of which bear nodules), occurs in Asia, Europe, and North America. *Sphepherdia*, a plant confined to North America, consists of three species, two of which nodulate. Crocker and Major[39] reported that *Shepherdia*, in company with *Alnus*, colonized the moraines of receding glaciers in Alaska.

The family Myricaceae (the galeworts) is distributed widely in the temperate regions of both hemispheres. The family consists of about 35 species of *Myrica*, of which 12 species have been found to nodulate. A few species occur in the tropics. Bog myrtle (*M. gale*) may be involved in the fixation of N_2 in acid peats.

The family Rhamnaceae contains 40 genera of trees and shrubs that are spread over most of the globe. However, 30 of the 31 species that have been reported to nodulate occur in *Ceanothus*, a genus of about 55 species confined to North America. More than half of the species are found in the southwestern part of the United States. The accumulation of N during soil development on the Mt. Shasta mudflows in California has been attributed to N_2-fixing microorganisms living in association with species of *Ceanothus*.[42]

The family Rosaceae occurs typically in cool regions of the temperate zone. Species occurring in three genera have been reported to nodulate.

Future Trends in Biological N_2 Fixation

Considerable attention is now being given to ways of maximizing biological N_2 fixation as a source of combined N for plants. Interest in this subject has developed from the urgent need to solve practical problems related to energy, the environment, and world food requirements. Some goals of research on biological N_2 fixation are as follows.[16]

1   Transfer N_2-fixing genes from bacteria to higher plant cells, thus endowing the plant with the capability for utilizing molecular N_2.
2   Transfer N_2-fixing genes into a beneficial bacterium capable of invading plant cells and establishing an effective N_2-fixing system, such as a nodule.
3   Use protoplast fusion methods to create new symbiotic associations between microorganisms and higher plants.
4   Select or develop by genetic means N_2-fixing bacteria capable of living

on the roots of such cereal crops as corn and wheat and providing adequate fixed N for optimum plant growth.

5 Develop by genetic manipulation *Rhizobium* strains that are insensitive to soil NH_4^+ and NO_3^- concentrations that normally inhibit nodulation and N_2 fixation.

6 Develop by use of plant-breeding methods legumes that have increased photosynthetic capabilities and, therefore, greater capacities for providing energy to N_2-fixing bacteria in the nodules.

Scientists have long yearned for the day when N_2-fixing microorganisms could be induced to live in or on the roots of cereal crops, such as corn and wheat. The expectation is that this would permit the farmer to decrease reliance on N fertilizers without reduction in yields. As noted in item 4, research may ultimately provide a solution to this problem.

Atmospheric Deposition of Combined N

Combined N, consisting of NH_3, NO_2^-, NO_3^-, and organically bound N, is a common constituent of atmospheric precipitation.[43–45] Nitrite occurs in trace amounts and is usually ignored or included with the NO_3^- determination. The organically bound N is probably associated with cosmic dust and does not represent a new addition to land masses of the world.

The amount of N added to the soil each year in atmospheric precipitation (per hectare basis) is normally too small to be of significance in crop production. However, this N may be of considerable importance to the N economy of mature ecosystems, such as undisturbed natural forests and native grasslands. Natural plant communities, unlike domesticated crops, are not subject to continued large losses of N through cropping and grazing, and the N in precipitation serves to restore the small quantities that are lost by leaching and denitrification.

Eriksson[43] summarized earlier measurements (to 1952) for combined N in atmospheric precipitation. For the United States and Europe, the estimates for NH_4^+- plus NO_3^--N ranged from 0.78 to 22.0 kg/ha (0.7 to 19.6 lb/acre) per year. Many high values recorded in the literature may represent analytical or sampling errors, although higher than normal amounts would be expected near highly industrial areas because of burning of fossil fuels.

The concentration of N in precipitation decreases with increasing latitude. Tropical air contains 10 to 30% more mineral N than polar air and nearly twice as much as arctic air.[44] In temperate regions of the earth, the mineral N in precipitation appears to be highest during the warmer periods, and, for any given rainfall, concentrations decrease progressively with the duration of precipitation. Rain contains higher quantities of NH_4^+ and NO_3^- than snow, a result that may be due to their greater adsorption in the liquid phase.

Hutchinson[45] gives the following sources of combined N in atmospheric precipitation:

1 From soil and the ocean
2 From fixation of atmospheric N
    a. electrically
    b. photochemically
    c. in the trail of meteorites
3 From industrial contamination

Important sources of NH_3 in the atmosphere include volatilization from land surfaces, combustion of fossil fuel, and natural fires. The quantity of N fixed in the trail of meteorites is negligible.

The origin of NO_3^- in atmospheric precipitation is not known with certainty but formation by electrical discharge during thunderstorm activity has long been a favored theory. However, NO_3^- distribution patterns have seldom correlated with thunderstorm activity. According to Hutchinson,[45] only 10 to 20% of the NO_3^- in precipitation can be accounted for by electrical discharge.

Soils have the ability to absorb atmospheric NH_3 and earlier investigators (notably Liebig[46]) placed considerable emphasis on this process as a means of providing N to plants. The opinion of most soil scientists is that the process is of little practical significance except under special circumstances. High values for NH_3 sorption have been observed for soils near industrial areas, as well as downwind from cattle feedlots.[47,48] A rapid circulation of NH_3 within a crop canopy has been reported by Denmead et al.,[49] who observed that NH_3 produced near the ground level was almost completely absorbed within the plant cover.

In conclusion, N gains from precipitation plus gaseous adsorption can vary from insignificant amounts to 50 kg N/ha·yr or more, depending on geographical location and proximity of an NH_3 source.

NITROGEN LOSSES FROM SOIL

Of all the nutrients required for plant growth, N is by far the most mobile and subject to greatest loss from the soil–plant system. Even under the best circumstances, no more than two thirds of the N added as fertilizer can be accounted for by crop removal or in the soil at the end of the growing season; losses of as much as one-half are not uncommon. Numerous attempts have been made to account for the low recoveries, and it is now known that available mineral forms of N, whether added as fertilizer or produced through decay of organic matter, will not remain very long in most soils.

Five main channels of N loss are discussed in this section, including bacterial denitrification, chemodenitrification, NH_3 volatilization, leaching, and erosion. Brief mention will be made of volatile losses from plants. The possible formation of N_2O during nitrification will be discussed in Chapter 5.

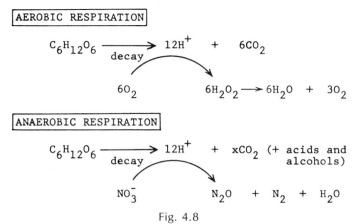

Fig. 4.8

General course of the breakdown of organic matter (e.g., carbohydrates) under aerobic and anaerobic condition. In the absence of oxygen, NO_3^- serves as a terminal electron acceptor, with formation of N_2 and N_2O. The lower equation is not balanced.

At the outset, it should be noted that none or all of the factors may operate in any given soil. In general, N is lost by several avenues, such as by a combination of leaching and denitrification from medium to heavy-textured soils of humid and semihumid zones.

Bacterial Denitrification

Under suitable conditions, NO_3^- is lost rapidly from soil through denitrification.[50–57] The ability to convert NO_3^- to N_2 and N_2O is limited to a few organisms that are able to utilize the oxygen of NO_3^- (and NO_2^-) as a substitute for O_2 in conventional metabolism. The process of denitrification is depicted in Fig. 4.8.

The geochemical significance of denitrification arises from the fact that the process acts as a balance on biochemical N_2 fixation. It is analogous to the relation between photosynthesis and respiration in the C cycle. Just as organically bound C is returned to the atmosphere (as CO_2) through respiration, combined N is returned through denitrification.

Some scientists believe that the reason N_2 is the principal constituent of the earth's atmosphere is because of the continued activity of denitrifying microorganisms throughout geological history. In any event, it is likely that most of the atmospheric N_2 has passed at least once through the denitrification cycle. The annual exchange of N between the atmosphere and the biosphere has been reported to range from 0.017 to 0.034 mg/cm$^2 \cdot$yr.[45] This corresponds to a cycle length between 44 and 220 million years, or from one-tenth to one-half of the time span from the Cambrian to the present.

The general requirements for denitrification are (1) presence of bacteria

Table 4.11
Genera of Bacteria Capable of Denitrification[a]

Genus	Comments
Alcaligenes	Commonly found in soils
Agrobacterium	Commonly found in soils
Azospirillum	Capable of N_2 fixation, commonly associated with grasses
Bacillus	Thermophilic denitrifiers reported
Flavobacterium	Denitrifying species recently isolated
Halobacterium	Requires high salt concentrations for growth
Hyphomicrobium	Grows on one-carbon substrates
Paracoccus	Capable of both lithotrophic and heterotrophic growth
Propionibacterium	Fermentors capable of denitrification
Pseudomonas	Commonly found in soils
Rhodopseudomonas	Photosynthetic bacteria
Thiobacillus	Generally grow as chemoautotrophs

[a] Adapted from Firestone.[52]

possessing the metabolic capacity to do so, (2) suitable electron donors such as organic C compounds, reduced S compounds, or molecular hydrogen, (3) anaerobic conditions or restricted O_2 availability, and (4) a supply of NO_2^- or NO_3^- to serve as terminal electron acceptors.[52]

The capacity to denitrify has been reported for about 33 genera of bacteria. Some of the more common ones, and for which denitrification has been firmly established, are listed in Table 4.11. The organisms primarily involved are heterotrophic (living off organic matter) and belong to the genera *Alcaligenes, Agrobacterium, Bacillus,* and *Pseudomonas.* Several chemoautotrophs (i.e., species of *Thiobacillus*) are also capable of utilizing NO_3^-, with production of N gases. However, they are of little importance in most agricultural soils.

The probable sequence of bacterial denitrification is as follows:

$$NO_3^- \longrightarrow NO_2^- \longrightarrow NO \longrightarrow N_2O \longrightarrow N_2$$
$$+5 \qquad +3 \qquad +2 \qquad -1 \qquad 0$$

The reduction of NO_3 in bacteria serves two different physiological purposes. In the one case, NO_3^- is reduced to NH_3 and is utilized for cell synthesis (*assimilatory NO_3^- reduction*). In the second case, NO_3^- is used as a terminal electron acceptor during respiration (*dissimilatory NO_3^- reduction*) and is referred to as denitrification.

The initial step in denitrification (reduction of NO_3^- to NO_2^-) is catalyzed by the enzyme *dissimilatory nitrate reductase*. This enzyme contains labile sulfide groups, Mo, and Fe in both heme and nonheme forms. Two types of enzymes are believed to be involved in the reduction of NO_2^- to NO:

cytochrome cd and a *Cu-containing protein.* The enzymes responsible for NO and N_2O reduction have not been adequately characterized. The highly reactive nature of NO has made the isolation of NO reductase difficult. The enzyme responsible for N_2O reduction is labile and cannot be easily isolated from the host cell.

Not all denitrifers have the ability to reduce N_2O to N_2. Several bacteria (e.g., *Corynebacterium nephridii, Pseudomonas aurefacieus,* and *Pseudomonas chlororaphis*) produce N_2O as the terminal product of NO_3^- reduction. However, these organisms are rare and it appears that most soil denitrifyers are capable of reducing NO_3^- completely to N_2.

Nitrous oxide is a gas and can escape from the soil before being reduced further to N_2. The ratio of N_2 to N_2O in the gases evolved from soil depends on such factors as soil pH, moisture content, E_h, temperature, NO_3^- concentration, and content of available organic C.

The following approaches have been used in attempts to estimate N losses from soils through bacterial denitrification.[22,52,55]

1 By mass-balance calculations. The amount of N lost from the soil–plant system is determined from the difference between N inputs and outputs, including crop removal and leaching. Systematic errors for all analyses are incorporated in the estimate for gaseous loss.[58]

2 By following the disappearance of NO_3^- from the system under study. This approach is generally unsatisfactory because some of the NO_3^- may be reduced to NH_3 and organic N.[59,60]

3 By estimates of N_2 and N_2O in the evolved gas. The experiments are usually carried out in closed chambers under an N_2-free atmosphere (i.e., He or Ar gas). The evolved gases have been detected by gas chromatography, although the ^{15}N tracer technique has also been applied.[22]

4 Through use of ^{15}N-labeled NO_3^- in field studies. The soil is covered with a chamber that allows for sampling and detection of ^{15}N-labeled N_2 in the evolved gases.[61,62] Nitrate highly enriched with ^{15}N is required and the approach is both time-consuming and expensive.

5 By measurement of residual ^{15}N-labeled NO_3^- in field plots. A complication of this approach is the possibility that some NO_3^- may be lost by leaching below the sampling depth. Also, sampling errors can arise because of nonuniform distribution of the applied [15]N.

6 By the acetylene inhibition method. This approach is based on the observation that acetylene inhibits the reduction of N_2O to N_2 by denitrifying microorganisms. The procedure has been used in both laboratory and field studies to estimate gaseous losses of N by measuring the amounts of N_2O produced by denitrification of NO_3^- in closed chambers.[63–66] Problems to be considered in application of the acetylene inhibition method have been discussed by Bremner and Hauck.[22]

Concerns regarding the potential adverse effect of nitrogenous fertilizers in contributing to the concentration of N_2O in the atmosphere has stimulated the development of chamber techniques for measuring the evolution of N_2O from field soils. However, the detection of N_2O does not necessarily prove that denitrification has occurred because there is evidence to indicate that this gas may be produced by nitrifying bacteria during the oxidation of NH_4^+ to NO_3^-.[57]

Finally, it should be stated that most laboratory studies have been carried out under optimum conditions for denitrification. In general, results of such studies cannot be directly related to conditions existing in the field. As noted in the following discussion, the method of sample preparation, such as air drying, can profoundly affect denitrification rates.

Nitrogen losses through denitrification vary greatly, depending on NO_3^- levels, available organic matter, temperature, and moisture status of the soil. Optimum conditions for denitrification are as follows:

Poor drainage   Moisture is of importance from the standpoint of its effect on aeration. Denitrification is negligible at moisture levels below about two-thirds of the water-holding capacity but is appreciable in flooded soils. The process may occur in anaerobic microenvironments of well-drained soils, such as pores filled with water, the rhizosphere of plant roots, and the immediate vicinity of decomposing plant and animal residues. Nitrate, introduced into aqueous sediments in surface or ground waters, applied as fertilizer, or formed at the oxidizing surface layer, is particularly subject to gaseous loss through denitrification.

Temperature of 25°C and above   Denitrification proceeds at a progressively slower rate at temperatures below 25°C (77°F) and practically ceases at 2°C (36°F).

Soil reaction near neutral   Denitrifying bacteria are sensitive to high hydrogen ion concentrations. Soils low in pH contain such a sparse population of denitrifying bacteria that NO_3^- losses by denitrification may be negligible.

Good supply of readily decomposable organic matter.   The amount of organic matter available to denitrifying microorganisms is generally appreciable in the surface horizon but negligible in the subsoil. Significant amounts of soluble organic matter may be found under cattle feedlots, as well as in the lower horizons of soils amended with large quantities of organic wastes.

A prerequisite for denitrification in soil is the presence of inorganic N in the NO_3^- form. When the soil becomes saturated or partially inundated with water, such as following a heavy rain, any NO_3^- present in the soil is subject to conversion to N_2 and N_2O through denitrification. Subsequent drainage and reestablishment of aerobic conditions can lead to further nitrification of NH_4^+, thereby providing additional substrate for the denitrifying bacteria. Losses of soil and fertilizer N through denitrification would be expected to be especially severe during prolonged wet periods brought about by intermittent heavy rains.

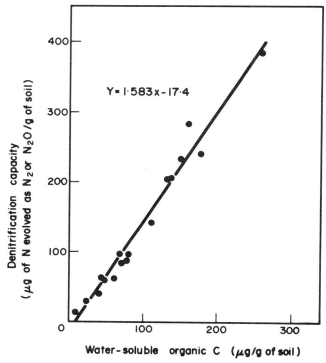

Fig. 4.9

Correlation between water-soluble organic C and the denitrification capacity of the soil. (Reprinted with permission from *Soil Biology and Biochemistry*, Vol. 7, J. R. Burford and J. M. Bremner, "Denitrifying Capacities of Soils," © 1975, Pergamon Press, Ltd.)

A second requirement for denitrification is the presence of decomposable organic matter. Only a general relationship has been noted between the total amount of organic C in soil and denitrification, which is to be expected because the bulk of the organic matter in most soils is biologically stable. Better correlations have been obtained with water-soluble or readily decomposable C.[68-70] A typical result is given in Fig. 4.9. An increase in denitrification rate occurs when the soil is dried out or frozen prior to the denitrification measurement,[71-72] which can be explained in part by release of organic compounds to soluble forms.

Nitrogen losses through denitrification are affected by mineralization–immobilization and will be especially severe in soils that contain high amounts of decomposable organic matter with a low C/N ratio. Under such circumstances, *net mineralization* occurs and NO_3^- can accumulate. Additions of carbonaceous plant residues (i.e., those with a high C/N ratio) will conserve N through *net immobilization,* thereby reducing denitrification losses. A potential beneficial effect of returning crop residues to the soil following harvest is that N losses through denitrification (and leaching) may be reduced.

There is considerable controversy as to the effect that plant roots have on denitrification, both positive and negative effects having been recorded. Plant roots can influence denitrification in several ways, as follows:[52]

1 Provide organic matter, which serves to support a population of denitrifying bacteria and acts as an electron donor for NO_3^- reduction when O_2 availability is low.
2 Create anaerobic zones by depletion of O_2 supply.
3 Create a dryer soil, thereby increasing the rate of O_2 diffusion to the root zone.
4 Deplete NO_3^- supply through plant uptake.
5 Increase O_2 availability near roots of aquatic plants in rice soils or sediments.

The net effect of the previously mentioned factors may be positive or negative, thereby accounting for the variable results reported in the literature. Other factors affecting denitrification in the root zone include O_2 diffusion rate and supply of NO_3^- in the rhizosphere.

Denitrification can be considered desirable when it occurs below the rooting zone, because of reduction in the NO_3^- content of ground water. Denitrifying microorganisms are known to be present at considerable depths in soil, and it is possible that some of the NO_3^- leached into the subsoil may be volatilized as N_2 and N_2O before reaching the water table. Meek et al.[73] concluded that much of the NO_3^- leached into the subsoils of some irrigated California soils was lost through denitrification.

Chemodenitrification

In most soils the oxidation of NO_2^- to NO_3^- by *Nitrobacter* proceeds at a faster rate than the conversion of NH_4^+ to NO_2^- by *Nitrosomonas*; consequently, NO_2^- seldom persists in detectable amounts. High levels are sometimes found, however, when NH_3 or NH_4^+-type fertilizers are applied to soil at high application rates.[74-78] The high NO_2^- accumulations have been attributed to inhibition of the second step of the nitrification process, presumably due to NH_3 toxicity to *Nitrobacter*. The buildup of NO_2^- is undesirable because of phytotoxicity[76] and because NO_2^- is relatively unstable and can undergo a series of reactions leading to the formation of N gases. The possibility that gaseous loss of N may accompany temporary NO_2^- accumulations has been noted by several investigators.[74,79-81]

Accumulations of NO_2^- have been observed following additions of anhydrous NH_3, NH_4^+ salts, and urea to soils.[75-79] According to Hauck and

Stephenson,[77] large fertilizer granules, high application rates, and an alkaline pH in the immediate vicinity of the granule are particularly favorable for NO_2^- accumulations. The importance of soil–fertilizer geometry on NO_2^- accumulations has been emphasized by Bezdicek et al.[82]

Several investigators have suggested that blockage of nitrification at the NO_2^- stage occurs under certain field conditions (discussed previously), and that nonenzymatic loss of N results from chemodenitrification. The term *side-tracking of nitrification* has sometimes been used to designate N loss by this mechanism.[79]

A factor of some importance in determining the activity of NO_2^- is the change in pH accompanying nitrification. Nitrite is particularly reactive at low pHs, a condition that may be attained in localized zones in soil following application of NH_3 or NH_4^+-type fertilizers. Nitrification of applied anhydrous NH_3, for example, starts in peripheral zones of moderately high NH_3 concentrations and proceeds inward toward the center of the retention zone. As a result of biological conversion of NH_3 to NO_2^- and NO_3^-, the pH of the soil is lowered; in peripheral zones, values as low as 4.2 to 4.8 have been observed. The pH of the soil immediately outside the retention zone also may be lowered because of migration of NO_2^- and NO_3^-. In these regions of low pHs, further oxidation of NO_2^- may be hindered by sensitivity of *Nitrobacter* to high concentrations of H^+. A similar sequence may occur at the soil–particle interface of an individual urea or $(NH_4)_2SO_4$ granule.

Reactions of Nitrite with Humic Substances

Components of the organic matter have been shown to react chemically with NO_2^- to form N gases. The role of organic matter in chemodenitrification is depicted by the diagram shown in Fig. 4.10. The possibility that losses of N occur through this mechanism is of considerable interest because there

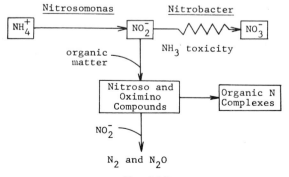

Fig. 4.10

Possible role of organic matter in promoting the decomposition of NO_2^- with production of N_2 and N_2O.

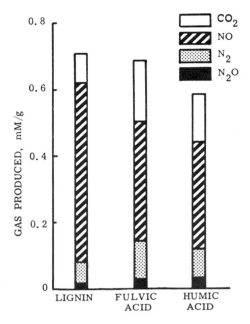

Fig. 4.11
Composition of the gasses obtained by the reaction of NO_2^- with lignin, fulvic acid (FA), and humic acid (HA) at pH 6. (From Stevenson et al.[85])

is as yet no fully suitable explanation for the losses of fertilizer N from normally aerobic soils.

Evidence for a NO_2^-–organic matter interaction has been obtained in studies where ^{15}N-labeled NO_2^- has been applied to soil,[83,84] in which case part of the NO_2^--N was fixed by the organic matter and part converted to N gases. The quantity of NO_2^--N converted to organic forms increases with increasing NO_2^- concentration and decreasing pH and is related to the C content of the soil.[83]

The gases that have been obtained through the reaction of NO_2^- with lignins and soil humus preparations in buffer solutions at pH 6 are shown in Fig. 4.11; results of a similar study using lignin building blocks have given a similar array of N gases. The main N gas identified in these studies was NO; other gases included N_2, N_2O, and CO_2.[85] Still another gas (methyl nitrite, CH_3ONO) has been observed in the gases produced from lignins under highly acidic conditions,[86] but there is no evidence that this gas is formed under conditions existing in the natural soil.

Mechanisms leading to the evolution of N gases by nitrosation of lignins and humic substances are not fully understood but the initial reaction may involve the formation of nitroso and oximino derivatives, which subsequently react with excess NO_2^- to give N gases. The sequence is shown in Fig. 4.12. An excellent discussion of mechanisms involved in nitrosation has been given by Austin.[87]

Nitric oxide can be formed in several ways, such as by reaction of HNO_2 with enolic compounds, as illustrated.

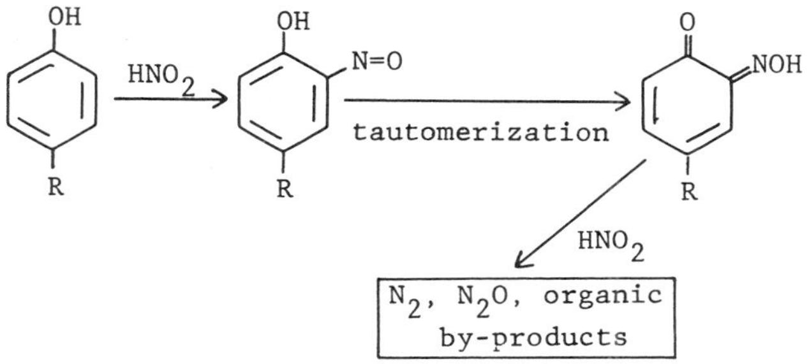

Fig. 4.12
Sequence for the formation of N_2 by nitrosation of phenolic compounds.

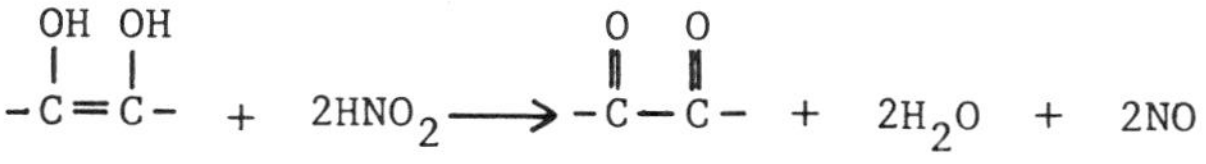

Classical Nitrite Reactions

Nitric acid (or NO_2^-) reacts with amino acids and other reduced forms of soil N, such as NH_3, to form N gases. The reactions are

$$NH_3 + HNO_2 \rightarrow NH_4NO_2 \rightarrow 2H_2O + N_2 \tag{1}$$

$$RNH_2 + HNO_2 \rightarrow ROH + H_2O + N_2 \tag{2}$$

The reaction of HNO_2 with amino acids (reaction 2) occurs much more readily than the reaction with NH_3 (reaction 1). Rather low pH values are required, and since free amino acids are normally present in soil in only trace quantities, serious losses of N by either of the preceding processes would appear unlikely under most soil conditions. It should be noted, however, that the pH at the surface of clay particles may be somewhat lower than that of the soil proper. Another factor to consider is that the tendency of NO_2^- to convert to NO_3^- and NO by chemical dismuation is considerably stronger than its tendency to react with NH_3 or amino compounds. This reaction is as follows:

$$3HNO_2 \rightarrow HNO_3 + H_2O + 2NO \tag{3}$$

The NO formed in reaction 3 will not escape from the soil in an unmodified form because of its strong tendency to react with O_2 to form nitrogen dioxide (NO_2).

$$2NO + O_2 \rightarrow 2NO_2 \tag{4}$$

The NO_2 can react further with water to form NO_2^- and NO_3^-. The cyclic nature of NO_2^- decomposition can thus be depicted as follows (equations not balanced):

$$NO_2^- \underset{OH^-}{\overset{H^+}{\rightleftharpoons}} HNO_2 \longleftarrow \longrightarrow NO_2 + NO$$
$$\downarrow H_2O$$
$$NO_2^-$$
$$+$$
$$NO_3^-$$

Arguments both for and against appreciable loss of N as NO_2 or NO can be given. Desiccation of the soil following partial nitrification of NH_4^+-type fertilizers would be particularly favorable for conversion of NO_2^- to gaseous products and for escape of NO and NO_2 into the atmosphere. An imperceptible slow evolution of N gases could lead to significant N losses on a per hectare year basis.

Ammonia Volatilization

Under suitable conditions, NH_3 can be lost from soils by volatilization.[13,74,88-90] Ammonia entering the atmosphere from worldwide terrestrial sources has been estimated by Söderlund and Svensson[9] to be as follows: 2 to 6 $\times$ 10^9 kg/yr from wild animals, 20 to 35 $\times$ 10^9 kg/yr from domestic animals, and 4 to 12 $\times$ 10^9 kg/yr from combustion of fossil fuel, giving a total of 26 to 53 $\times$ 10^9 kg/yr.

Ammonia losses from field soils have been estimated in several ways. In one approach a chamber is placed over the soil surface and the ambient NH_3 is trapped in an acid solution or acid-impregnated glass wool or filter paper. Two types of chamber techniques are used. In one, the NH_3 is trapped under static conditions; in the second, NH_3-free air is passed over the soil surface. A criticism of these methods is that artificial conditions are created over and around the area being sampled.[91] A modification of the approach has been to allow the soil to remain exposed most of the time to natural conditions but to close the lid of the volatilization chamber only for short intervals while the NH_3 loss measurements are being made.

In recent times a micrometeorological technique has been used.[49,92-94] The approach is similar to one used extensively in meteorological research to measure rates of gas exchange above natural surfaces, and it does not impose unnatural conditions on the study area. In this method volatilized NH_3 is collected in acid traps placed at various heights above the soil surface and the amount of NH_3 lost on a per hectare basis is estimated from meteorological data for wind speed and direction.

A summary of results from field studies of NH_3 losses from surface-applied NH_4^+ or urea is given in Table 4.12. Losses have ranged from as little as 3% to as much as 50% of the applied N, depending on such factors as soil texture, pH, and the amount of fertilizer applied. In general, losses of N through NH_3 volatilization are small when the fertilizer is incorporated

Table 4.12
Summary of NH_3 Losses as Measured in the Field[a]

Fertilizer Type and System	Added N Evolved as NH_3, %
Urea	
Silt loam soil, pH 6.3	19
Fine sandy loam soil, pH 5.6–5.8	9–40
Loamy sand, pH 7.7	22
Grass sod	20
Forest litter	3.5–25
Flooded rice soils	6–8
$(NH_4)_2SO_4$	
Silt loam soil, pH 6.3	4
Clay soil, pH 7.6	35
Surface of grass sod	50
Flooded rice soils	3–7
NH_4NO_3	
Loamy sand, pH 7.7	17

[a] Adapted from Nelson,[74] where specific references can be obtained. For other summaries see Mikkelsen and DeDatta.[89]

into the soil, especially those that are acidic or neutral. Large amounts of NH_3 may be lost when NH_3-forming fertilizers are applied to the surface of alkaline or calcareous soils. Losses are accentuated when weather conditions favor drying of the soil.

Conditions in rice paddy fields are conducive to NH_3 volatilization.[89] Submergence causes the pH of most soils to converge near neutrality, which along with turbulence at the water–air interface, leads to NH_3 loss.

A factor of considerable importance in reducing or eliminating losses of soil and fertilizer N through NH_3 volatilization is the ability of NH_4^+ to form electrostatic bonds with clay minerals and organic colloids. The basic reaction involves protonation of NH_3 to form NH_4^+, which subsequently undergoes exchange reactions with other cations on exchange sites of clay and humus particles. The sequence is

$$NH_3 + H_2O \rightarrow NH_4^+ + OH^- \tag{6}$$

$$NH_4^+ + CX \rightarrow NH_4X + C^+ \tag{7}$$

where C is an exchangeable cation on either the clay or organic matter.

In acid soils, NH_3 can react directly with H^+ on the exchange complex.

$$NH_3 + HX \rightarrow NH_4X \tag{8}$$

Ammonium ions that occur in the aqueous phase of the soil enter into an equilibrium reaction with NH_3.

$$NH_4^+ \rightleftharpoons NH_3(aq) + H^+ \tag{9}$$

The aqueous NH_3, in turn, is subject to gaseous loss through the reaction:

$$NH_3(aq) \rightarrow NH_3(air) \tag{10}$$

The pK_a value for reaction 9 is 9.5. At pH values of 6, 7, 8, and 9, approximately 0.036, 0.36, 3.6, and 36% of the total reduced N in the soil solution will be present as $NH_3(aq)$, respectively. It can thus be seen that loss of NH_3 to the atmosphere will be related to both pH and the concentration of NH_4^+ in the solution phase. Losses are also affected by temperature and wind speed over the soil surface.

Considerable attention has been given in recent years to loss of NH_3 by application of NH_4^+-containing fertilizers to the surface of agricultural soils.[95-98] As one might expect, losses have been substantially higher from calcareous soils as compared to noncalcareous ones. The overall reaction of $(NH_4)_2SO_4$ with $CaCO_3$ of calcareous soils is as follows:

$$(NH_4)_2SO_4 + CaCO_3 \rightarrow 2NH_3 + CO_2 + H_2O + CaSO_4 \tag{11}$$

Nitrogen is a critical nutrient in most grazing ecosystems. Volatilization of NH_3 from urine and dung usually accounts for the major part of the N loss from grazed pastures.[99] Appreciable losses of N through NH_3 volatilization also occur when farmyard manure is spread directly on the surface of cultivated soils, even those that are acid, because of localized high pH resulting from generation of NH_4OH (reaction 6). Recent research indicates that NH_3 volatilized from cattle feedlots can contribute to the pollution of lakes and streams.[47]

The popularity of anhydrous NH_3 as a fertilizer is well known. In the usual practice, anhydrous NH_3 is injected in bands in the soil to depths of 10 to 15 cm. The liquid volatilizes and some NH_3 invariably escapes to the atmosphere in injection slits and soil cracks.

The facts concerning NH_3 volatilization can be summarized as follows:

1 Losses are of greatest importance on calcareous soils, especially when NH_4^+-containing fertilizers are applied on the soil surface. Only slight losses occur in soils of pH 6 to 7, but losses increase markedly as the pH of the soil increases.

2 Losses increase with temperature, and they can be appreciable when neutral or alkaline soils containing NH_4^+ near the surface are dried out.

3 Losses are greatest in soils of low cation-exchange capacities, such as sands. Clay and humus absorb NH_4^+ and prevent its volatilization. In

soil with an alkaline reaction, little NH_3 will be lost provided adequate moisture is present.

4   Losses can be high when nitrogenous organic wastes, such as farmyard manures, are permitted to decompose on the soil surface. Gaseous loss of NH_3 accounts for a large part of the N turnover in a grazing ecosystem.[99]

5   Appreciable amounts of NH_3 are lost when urea is applied to soils under grass or pasture, a result that has been attributed to hydrolysis of urea by the enzyme urease, with subsequent volatilization of NH_3.

6   Losses of soil and fertilizer N through NH_3 volatilization are reduced in the presence of growing plants. Not only are NH_4^+ levels reduced through plant uptake, but some of the evolved NH_3 may be readsorbed by the plant canopy.[49]

Leaching

Transfer of N from soil to lakes and streams occurs through leaching. The amount of N lost from the total land area of the United States through leaching has been estimated at 2 to 3.7×10^9 kg/yr.

Nitrogen is leached mainly as NO_3^-, although NH_4^+ may be lost from sandy soils. In intensively cropped soils where fertilizer has not been applied, loss through leaching is greatly reduced, for the reason that the NO_3^- content of the soil is lower and less water passes through the soil.

Leaching losses occur when two prerequisits are met: (1) soil NO_3^- levels are high, and (2) downward movement of water is sufficient to move NO_3^- below the rooting depth.[58] These conditions are met in soils of the humid and subhumid zones, to a lesser extent in soils of the semiarid zone, and only frequently if at all in arid zone soils. Nitrates seldom accumulate in grassland soils, even those of the humid region, thereby restricting N losses through leaching.

In humid and subhumid regions, any NO_3^- remaining in the soil after the end of the growing season is subject to leaching, denitrification, or both. In some cases, NO_3^- can accumulate in the subsoil and move downward into the ground water, depending on the soil, climate, fertilizer, and management practices.[58] In soils of arid and semiarid regions, residual NO_3^- within the rooting depth represents a source of available N for the following crop and is used as a basis for assessment of soil N availability.[100]

The leaching of NO_3^- in irrigation agriculture has received considerable attention in recent years because of possible pollution of ground waters. McNeal and Pratt[101] found that leaching losses accounted for from 13 to 100% of the fertilizer N added to some irrigated California soils and commonly averaged 25 to 50% of the N applied in most cropping situations. Aspects of the management of N for maximum efficiency and minimal pollution have been discussed by Keeney.[6]

Erosion and Runoff

In addition to leaching, considerable N may be lost from the soil as a result of erosion and surface runoff. Sheet erosion is highly selective in that the eroded fraction contains several times more N than the original soil.

According to current estimates, 5×10^{12} kg of soil are lost annually through erosion, with about 80% begin lost as waterborne sediments and the remainder by wind erosion. From one-half to three-fourths of the eroded soil is from agricultural land. Assuming a loss of 3×10^{12} kg of agricultural soil and an average N content of 0.15%, an estimated 4.5×10^9 kg of N would be lost annually.[58] Most of the N lost by soil erosion is in organic forms and will eventually be deposited in streams, lakes, and the oceans with little opportunity of being recycled into agricultural systems. Bottom lands are often enriched with nutrients, including N, by periodic flooding.

Loss of N from Plants

Measurements for total N in many crops (e.g., wheat, corn, soybeans) show a rapid accumulation during vegetative growth, followed by a slight decline after flowering.[13,102] Various explanations have been given for the loss, one being volatilization of amines, NH_3, or N oxides from the plant following senescence. Whereas accurate quantitative data for N losses from plants are lacking, the magnitude of any such losses would appear to be small when compared to leaching or denitrification. There may be major exceptions, however, and the reader is directed to recent reviews[12,13,102] for additonal details.

FLUX OF SOIL N WITH OTHER ECOSYSTEMS

The soil N cycle (see Fig. 4.1) is connected to the universal N cycle through several pathways, the main ones being biological N_2 fixation and denitrification. Other transfer mechanisms include leaching of NO_3^-, volatilization of NH_3, and accession of organic and mineral forms of N in atmospheric precipitation (as noted in the previous section).

Considerable difficulty is encountered in obtaining accurate values for N transfer between soil and other ecosystems. Nitrogen fluxes are not only highly dependent on the type of ecosystem but on environmental conditions at any specific location. Accordingly, extrapolations based on N fluxes at the local level are of doubtful validity when expanded on a regional or global basis. For any given ecosystem, inputs through biological N_2 fixation are relatively accurate; those for soil N losses through denitrification are the least reliable. The subjects of N fluxes and N transfers has been covered in several reviews.[4-10]

Estimates for N transfers and fluxes for the global terrestrial system are

Table 4.13
Estimates of Global N Fluxes per Year for the Terrestrial System (10^9 kg N/year)

Fluxes	Burns and Hardy[3]	Söderlund and Svensson[9]
	Inputs	
Biological N_2 fixation		
Agricultural land	89	–
Forests	40	–
Others (e.g., the ocean)	10	–
Total biological fixation	139	139
Abiological N fixation		
Industrial (e.g., fertilizer N)	30	36
Combustion (e.g., fossil fuels)	20	19
Total abiological fixation	50	55
NO_3^- atmospheric deposition	32	
	Outputs	
Denitrification	140	107–161
	Transfers	
NH_3 volatilization		
Animals and coal burning	–	26–53
Others (assumed)	–	87–191
Total NH_3 volatilization	160	113–244
NH_3/NH_4^+ atmospheric deposition	73	91–186
NO_x atmospheric deposition	200	32–83
Organic N deposition	–	10–100
Land to river, ocean discharge		
$NH_4^+ + NO_x$	–	5–11
Organic N	–	8–13
Subtotal of $NH_4^+/NO_x/$ organic N	–	13–24
NO_3^-	–	5–11
Subtotal inorganic N	15	–
Total river discharge	–	18–33

recorded in Table 4.13. Total biological N_2 fixation for the overall terrestrial system amounts to 139×10^9 kg/yr, most of which (89×10^9 kg) is fixed in soils used for agricultural crops. Additional sources of combined N include industrial fixation (36×10^9 kg in 1976) and burning of fossil fuels (20×10^9 kg).

Losses of N from the terrestrial ecosystem occur through bacterial denitrification (107 to 161×10^9 kg/yr) and river discharge (18 to 33×10^9 kg/yr). Transfers of N to and from the global terrestrial system occur through

Table 4.14

Inputs and Outputs of N in Harvested Croplands and Total Land Area in the United States[a]

Estimated Fluxes	1930	1947	1967	1970
	10^9 kg of N per year for croplands			10^9 kg of N per year for total area
Inputs of N				
Fertilizer N	0.3	0.7	6.8	7.5
Symbiotic N_2 fixation	1.5	1.7	2.0	3.6
Nonsymbiotic N_2 fixation	1.0	1.0	1.0	1.2
Barnyard manure	0.9	1.3	1.0	–
Rainfall	0.8	1.0	1.5	5.6
Irrigation	<0.1	–	–	–
Roots and unharvested portions	1.0	1.5	2.5	–
Total inputs	5.5	6.2	14.8	17.9[b]
Outputs of N				
Harvested crops	4.2	6.5	9.5	16.8
Erosion	4.5	4.0	3.0	–
Leaching of soil N	3.7	3.0	2.0	–
Denitrification	–	?	?	8.9
Volatilization	–	–	–	5.6
Total outputs	12.4	13.5	14.5	19.5

[a] From Hauck and Tanji[5] as tabulated from data of Lipman and Corybeare,[103] and Stanford et al.[104]
[b] An additional 3.1×10^9 kg was assumed to be derived from the soil organic matter, giving a total of 21.0×10^9 kg.

NH_3 volatilization (113 to 244 $\times$ 10^9 kg/yr) and atmospheric deposition of NH_3, NH_4^+, and NO_x (121 to 279 $\times$ 10^9 kg/yr).

The global terrestrial model can be further subdivided on a regional or subregional basis. Estimated inputs and outputs of N for croplands of the United States (1930–1970 period) are given in Table 4.14. The main trend for this period was an increase in N removed by harvested crops; a slight increase is shown for symbiotic N_2 fixation. No estimates are given for gaseous N loss through denitrification but the trend would be expected to be significant and upward because of increased fertilizer N use during this period.

Nitrogen fluxes for select agroecosystems in the United States, as recorded for 1978, are given in Table 4.15. As expected, total N inputs varied greatly, depending on the amount of N added as fertilizer or fixed biologically (soybean ecosystem). Two of the five agroecosystems are shown to have negative N balances. Paul[105] concluded that the flow of N through any given

Table 4.15
Nitrogen Fluxes in Selected Agroecosystems in the United States for 1978
(kg of N/ha)[a]

	Maize for Grain, Northern Indiana	Soybeans for Grain, Northeast Arkansas	Wheat, Central Kansas	Potatoes, Maine	Cotton, California
N inputs					
Fertilizer	112	–	34	168	179
N_2 fixation	t	123	t	t	t
Irrigation water and flooding	10	–	–	–	50
Atmospheric deposition	–	10	6	6	3
Crop residue	41	30	20	65	48
Total inputs	163	163	60	239	280
N outputs from soil					
Net plant uptake	126	120	56	145	127
Denitrification	15	15	5	15	20
Volatilization	t	t	t	t	t
Leaching	15	10	4	64	83
Runoff (inorganic N)	6	3	1	5	50
Runoff (organic N)	10	13	4	10	t
Wind erosion (dust)	–	t	t	t	t
Total outputs	172	161	70	239	280
N inputs–N outputs	−9	2	− 10	0	0

[a] From Thomas and Gilliam.[10] t = trace amounts.

ecosystem is very dependent on the flow of C, a subject discussed in Chapter 1.

As noted from Table 4.15, the net amount of N taken up by plants varies from one species to another. For any given species, the amount varies with genotype and the environment. Among the factors affecting yield, and thereby N removal, are climate, fertilization practice, and the productive capacity of the soil.[13]

SUMMARY

Nitrogen is a transient nutrient and an understanding of the factors affecting gains and losses by natural processes is of paramount importance in developing management practices for its efficient use by plants. Practically all soils gain N in one way or another, albeit a small gain in some cases. By the same token, all soils lose N, frequently in appreciable amounts, such as when fertilizer N is applied in excess of crop need.

Greater emphasis in the future needs to be given to biological N_2 fixation by microorganisms, both by free-living forms and the rhizobia. Legumes have long been used to provide N for crop production, but the process is rather inefficient, particularly in productive soils where soil N levels are naturally high. Also, the amount of N_2 fixed is inversely related to the quantity of available N in the soil. Superior strains of rhizobia need to be developed that not only will be highly efficient in fixing N_2 but that will continue to fix N_2 in the presence of available forms of combined N, such as NH_4^+ and NO_3^-. Ultimately, this ability will be transferred to other crop plants, possibly even corn.

In addition to the symbiotic system, many microorganisms that live free in the soil are able to fix atmospheric N_2. It is often thought that the amounts fixed are too low to be of practical importance. However, the list of organisms that have this ability is increasing. In any event, there is the exciting prospect that superior types of free-living, N_2-fixing organisms might be found or developed that could benefit mankind by providing a source of combined N to crops.

The main avenue of N loss from soils of humid and semihumid regions is through leaching and denitrification. Losses through NH_3 volatilization are of greatest importance in grazing ecosystems, when nitrogenous organic wastes are deposited on the soil surface, and when NH_4^+ containing fertilizers are applied to the surface of calcareous soils. Losses of N through chemodenitrification are associated with temporary NO_2^- accumulations and may occur when anhydrous NH_3 or NH_4^+-containing fertilizers are applied to the soil at high rates, particularly in a band.

REFERENCES

1 F. J. Stevenson, "Origin and Distribution of Nitrogen in Soils," in F. J. Stevenson, Ed., *Nitrogen in Agricultural Soils,* American Society of Agronomy, Madison, Wis., 1982, pp. 1–42.

2 W. H. Baur and F. Wlotzka, "Nitrogen," in K. H. Wedepole, Ed., *Handbook of Geochemistry,* Vol. II/1, Springer-Verlag, New York, 1972, pp. 7-A-1–7-O-11.

3 R. C. Burns and R. W. F. Hardy, *Nitrogen Fixation in Bacteria and Higher Plants,* Springer-Verlag, New York, 1975.

4 F. E. Clark and T. Rosswell, Eds., *Terrestrial Nitrogen Cycles: Processes, Ecosystem Strategies, and Management Impacts,* Ecol. Bull. 33, Stockholm, 1981.

5 R. D. Hauck and K. K. Tanji, "Nitrogen Transfers and Mass Balances," in F. J. Stevenson, Ed., *Nitrogen in Agricultural Soils,* American Society of Agronomy, Madison, Wis. 1982, pp. 891–925.

6 D. R. Keeney, Nitrogen Management for Maximum Efficiency and Minimum Pollution, in F. J. Stevenson, Ed., *Nitrogen in Agricultural Soils,* American Society of Agronomy, Madison, Wis., 1982, pp. 605–649.

7 T. Rosswall, "The Internal Nitrogen Cycle between Microorganisms, Vegetation, and Soil," in B. H. Svensson and R. Söderlund, Eds., *Nitrogen, Phosphorus, and Sulphur— Global Cycles.* SCOPE Report 7, Ecol. Bull. 22, Stockholm, 1976, pp. 157–167.

8 R. Söderlund and T. Rosswall, "The Nitrogen Cycle," in O. Hutzinger, Ed., *The Handbook of Environmental Chemistry,* Vol. 1, Springer-Verlag, New York, 1982, pp. 61–81.

9 R. Söderlund and B. H. Svensson, "The Global Nitrogen Cycle," in B. H. Svensson and R. Söderlund, Eds., *Nitrogen, Phosphorus, and Sulphur—Global Cycles.* SCOPE Report 7, Ecol. Bull. 22, Stockholm, 1976, pp. 23–73.

10 G. W. Thomas and J. W. Gilliam. "Agro-ecosystems in the U.S.A.," in M. J. Frissel, Ed., *Cycling of Mineral Nutrients in Agricultural Ecosystems,* Elsevier, New York, 1978, pp. 182–243.

11 R. W. Fairbridge, Ed., *Encyclopedia of Geochemistry and Environmental Sciences,* Vol. IVA, Van Nostrand Reinhold, New York, 1972, pp. 795–801, 836–837, 849.

12 J. R. Freney and J. R. Simpson, Eds., *Gaseous Loss of Nitrogen from Plant–Soil Systems: Developments in Soil Science,* Vol. 9, Martinus-Nijhoff, The Hague, 1983.

13 R. A. Olson and L. T. Kurtz, "Crop Nitrogen Requirements, Utilization and Fertilization," in F. J. Stevenson, Ed., *Nitrogen in Agricultural Soils,* American Society of Agronomy, Madison, Wis., 1982, pp. 567–604.

14 F. G. Viets, Jr., and R. H. Hageman, *Factors Affecting the Accumulation of Nitrate in Soil, Water, and Plants,* Agricultural Handbook No. 413, U.S. Department of Agriculture, Washington, D.C., 1971.

15 W. J. Brill, *Microbiol. Rev.* **44,** 449 (1980).

16 H. J. Evans and L. E. Barber, *Science* **197,** 332 (1977).

17 U. D. Havelka, M. G. Boyle, and R. W. F. Hardy, "Biological Nitrogen Fixation," in F. J. Stevenson, Ed., *Nitrogen in Agricultural Soils,* American Society of Agronomy, Madison, Wis., 1982, pp. 365–422.

18 J. M. Vincent, *Nitrogen Fixation in Legumes,* Academic Press, Australia, Sydney, 1982.

19 R. W. F. Hardy and A. H. Gibson, Eds., *A Treatise on Dinitrogen Fixation,* Vols. III and IV, Wiley, New York, 1977.

20 J. J. Child, "Biological Nitrogen Fixation," in E. A. Paul and J. N. Ladd, Eds., *Soil Biochemistry,* Vol. 5, Dekker, New York, 1981.

21 R. H. Burris, "Nitrogen Fixation," in J. Bonner and J. E. Varner, *Plant Biochemistry,* Academic Press, New York, 1965, pp. 961–979.

22 J. M. Bremner and R. D. Hauck, "Advances in Methodology for Research on Nitrogen Transformations in Soils," in F. J. Stevenson, Ed., *Nitrogen in Agricultural Soils,* American Society of Agronomy, Madison, Wis., 1982, pp. 467–502.

23 W. D. P. Stewart, *Plant Soil* **32,** 555 (1970).

24 W. D. P. Stewart, Ed., *Nitrogen Fixation by Free-Living Microorganisms,* Cambridge University Press, New York, 1975.

25 Y. R. Dommergues and H. G. Diem, Eds., *Microbiology of Tropical Soils and Plant Productivity: Developments in Plant and Soil Sciences,* Vol. 5, Martinus Nijhoff, The Hague, 1982.

26 L. C. Chu, "Use of *Azolla* in Rice Production in China, in W. G. Rockwood and C. Mendoza, Eds., *Nitrogen and Rice,* International Rice Research Institute, Laguna, Philippines, 1979, pp. 375–394.

27 J. G. Torrey, *BioScience* **28,** 586 (1978).

28 P. B. Vose and A. P. Ruschel, Eds., *Associative N_2-Fixation,* CRC Press, Boca Raton, Fla., 1981.

29 H. L. Jensen, *Trans. 4th Intern. Congr. Soil Sci.* **I,** 165 (1950).

30 M. F. Jurgensen and C. B. Davey, *Soils Fert.* **33,** 435 (1970).

31 A. W. Moore, *Soils Fert.* **29,** 113 (1966).

32 E. N. Mishustin, *Plant Soil* **32,** 545 (1970).

33   T. A. LaRue and T. G. Patterson, *Adv. Agron.* **34**, 15 (1981).

34   D. O. Norris, *Emp. J. Exp. Agric.* **24**, 247 (1956).

35   R. C. Dawson, *Plant Soil* **32**, 655 (1970).

36   R. J. Roughley, *Plant Soil* **32**, 675 (1970).

37   J. H. Becking, *Plant Soil* **32**, 611, (1970).

38   W. S. Silver, "Physiological Chemistry of Non-Leguminous Symbiosis," in J. R. Postgate, Ed., *The Chemistry and Biochemistry of Nitrogen Fixation,* Plenum Press, New York, 1971.

39   R. L. Crocker and J. Major, *J. Ecol.* **43**, 427 (1955).

40   D. O. Norris, "The Biology of Nitrogen Fixation," in *A Review of Nitrogen in the Tropics with Particular Reference to Pastures,* Bull. 46, Commonwealth Agriculture Bureaux, Harpenden, England, 1962, pp. 113–129.

41   G. Bond and P. Montserrat, *Nature* **182**, 474 (1958).

42   B. A. Dixon and R. L. Crocker, *J. Soil Sci.* **4**, 142 (1953).

43   E. Eriksson, *Tellus* **4**, 215 (1952).

44   A. A. Ångstrom and L. Hogberg, *Tellus* **4**, 31 (1952).

45   G. E. Hutchinson, *Amer. Scient.* **32**, 178 (1944).

46   Justus von Liebig, *Organic Chemistry and Its Application to Agriculture and Physiology, 4th ed.,* Taylor, London, 1847.

47   G. L. Hutchinson and F. G. Viets, Jr., *Science* **166**, 514 (1969).

48   L. F. Elliott, G. E. Schuman, and F. G. Viets, Jr., *Soil Sci. Soc. Amer. Proc.* **35**, 752 (1971).

49   O. T. Denmead, J. R. Freney, and J. R. Simpson, *Soil Biol. Biochem.* **8**, 161 (1976).

50   B. A. Bryan, "Physiology and Biochemistry of Denitrification," in C. C. Delwiche, Ed., *Denitrification, Nitrification, and Atmospheric Nitrous Oxide,* Wiley-Interscience, New York, 1981, pp. 67–84.

51   C. C. Delwiche and B. A. Bryan, *Ann. Rev. Microbiol.* **30**, 241 (1976).

52   M. K. Firestone, "Biological Denitrification," in F. J. Stevenson, Ed., *Nitrogen in Agricultural Soils,* American Society of Agronomy, Madison, Wis., 1982, pp. 289–326.

53   D. D. Focht and W. Verstraete, "Biochemical Ecology of Nitrification and Denitrification," in M. Alexander, Ed., *Advances in Microbial Ecology,* Vol. 1, Plenum Press, New York, 1977, pp. 135–214.

54   B. A. Haddock and C. W. Jones, *Bacteriol. Rev.* **41**, 47 (1977).

55   R. Knowles, "Denitrification," in E. A. Paul and J. N. Ladd, Eds., *Soil Biochemistry,* Vol. 5, Dekker, New York, 1981, pp. 323–369.

56   W. J. Payne, *Denitrification,* Wiley-Interscience, New York, 1981.

57   W. J. Payne, *Bacteriol. Rev.* **37**, 409 (1973).

58   J. O. Legg and J. J. Meisinger, "Soil Nitrogen Budgets," in F. J. Stevenson, Ed., *Nitrogen in Agricultural Soils,* American Society of Agronomy, Madison, Wis., 1982, pp. 503–566.

59   G. Stanford, J. O. Legg, and T. E. Staley, "Fate of ^{15}N-Labelled Nitrate in soils under Anaerobic Conditions," in E. R. Klein and D. P. Klein, Eds., *Proc. 2nd Intern. Conf. on Stable Isotopes,* Argonne National Laboratory, Ill., 1975, pp. 667–673.

60   G. Stanford, J. O. Legg, S. Dzienia, and C. E. Simpson, *Soil Sci.* **120**, 147 (1975).

61   D. E. Rolston, F. E. Broadbent, and D. A. Goldhamer, *Soil Sci. Soc. Amer. J.* **43**, 703 (1979).

62   D. E. Rolston, M. Fried, and D. A. Goldhamer, *Soil Sci. Soc. Amer. J.* **40**, 259 (1976).

63   M. S. Smith, M. K. Firestone, and J. M. Tiedje, *Soil Sci. Soc. Amer. J.* **42**, 611 (1978).

64   R. Knowles, *Appl. Environ. Microbiol.* **38**, 486 (1979).

65   J. C. Ryden, L. J. Lund, and D. D. Focht, *Soil Sci. Soc. Amer. J.* **43**, 104 (1979).

66   J. C. Ryden, L. J. Lund, J. Letey, and D. D. Focht, *Soil Sci. Soc. Amer. J.* **43**, 110 (1979).

67   J. M. Bremner and A. M. Blackmer, *Science* **199**, 295 (1978).

68   J. R. Burford and J. M. Bremner, *Soil Biol. Biochem.* **7**, 389 (1975).

69   R. J. K. Myers and J. W. McGarity, *Plant Soil* **35**, 145 (1971).

70   G. Stanford, R. A. Vander Pol, and S. Dzienia, *Soil Sci. Soc. Amer. Proc.* **39**, 284 (1975).

71   D. K. Patten, J. M. Bremner, and A. M. Blackmer, *Soil Sci. Soc. Amer. J.* **44**, 67 (1980).

72   E. McKenzie and L. T. Kurtz, *Soil Sci. Soc. Amer. Proc.* **40**, 534 (1976).

73   M. B. Meek, L. B. Grass, and A. J. MacKenzie, *Soil Sci. Soc. Amer. Proc.* **33**, 575, (1969).

74   D. W. Nelson, "Gaseous Losses of Nitrogen Other Than Through Denitrification," in F. J. Stevenson, Ed., *Nitrogen in Agricultural Soils*, American Society of Agronomy, Madison, Wis., 1982, pp. 327–363.

75   P. M. Chalk, D. R. Keeney, and L. M. Walsh, *Agron. J.* **67**, 33 (1975).

76   M. N. Court, R. C. Stephen, and J. S. Waid, *Nature (London)* **194**, 1263 (1974).

77   R. D. Hauck and J. M. Stephenson, *J. Agric. Food Chem.* **13**, 486 (1965).

78   R. Wetselaar, J. B. Passioura, and B. R. Singh, *Plant Soil* **36**, 159 (1972).

79   F. E. Clark, *Trans. 4th Intern. Congr. Soil Sci.* **IV-V**, 153 (1962).

80   D. W. Nelson and J. M. Bremner, *Soil Biol. Biochem.* **1**, 229 (1969).

81   F. E. Broadbent and F. J. Stevenson, "Organic Matter Interactions," in M. H. McVickar et al., Eds., *Agricultural Anhydrous Ammonia*, American Society of Agronomy, Madison, Wis., 1966, pp. 169–187.

82   D. F. Bezdicek, J. M. MacGregor, and W. P. Martin, *Soil Sci. Soc. Amer. Proc.* **35**, 997 (1971).

83   F. Führ and J. M. Bremner, *Atompraxis* **10**, 109 (1964).

84   C. J. Smith and P. M. Chalk, *Soil Sci. Soc. Amer. J.* **44**, 288 (1980).

85   F. J. Stevenson, R. M. Harrison, R. Wetselaar, and R. A. Leeper, *Soil Sci. Soc. Amer. Proc.* **34**, 430 (1970).

86   F. J. Stevenson and R. J. Swaby, *Soil Sci. Soc. Amer. Proc.* **28**, 773 (1964).

87   A. T. Austin, *Sci. Prog.* **49**, 619 (1961).

88   H. A. Mills, A. V. Barker, and D. N. Maynard, *Agron. J.* **66**, 355 (1974).

89   D. S. Mikkelsen and S. K. DeDatta, "Ammonia Volatilization from Wetland Rice Soils," in W. G. Rockwood and C. Mendoza, *Nitrogen and Rice,* International Rice Institute, Laguna, Philippines, 1982, pp. 135–155.

90   G. L. Terman, *Adv. Agron.* 31, 189 (1979).

91   O. T. Denmead, J. R. Freney, and J. R. Simpson, *Science* **185**, 609 (1974).

92   D. E. Kissel, H. L. Brewer, and G. F. Arkin, *Soil Sci. Soc. Amer. J.* **41**, 1133 (1977).

93   O. T. Denmead, J. R. Simpson, and J. R. Freney, *Soil Sci. Soc. Amer. J.* **41**, 1001 (1977).

94   O. T. Denmead, R. Nulsen, and G. W. Thurtell, *Soil Sci. Soc. Amer. J.* **42**, 840 (1978).

95   Y. Avnimelech and M. Laher, *Soil Sci. Soc. Amer. J.* **41**, 1080 (1977).

96   L. B. Fenn, *Soil Sci. Soc. Amer. Proc.* **39**, 366 (1975).

97   L. B. Fenn and D. E. Kissel, *Soil Sci. Soc. Amer. J.* **40**, 394 (1976).

98   L. B. Fenn and R. Escarzaga, *Soil Sci. Soc. Amer. J.* **41**, 358 (1977).

99   R. G. Woodmansee, *BioScience* **28**, 448 (1978).

100 G. Stanford, "Assessment of Soil Nitrogen Availability," in F. J. Stevenson, Ed., *Nitrogen in Agricultural Soils,* American Society of Agronomy, Madison, Wis., 1982, pp. 651–688.

101 B. L. McNeal and P. F. Pratt, "Leaching of Nitrate from Soils," in P. F. Pratt, Ed., *Proc. Natl. Conf. on Management of Nitrogen in Irrigated Agriculture,* University of California, Riverside, 1978, pp. 195–230.

102 R. Wetselaar and G. D. Farquhar, *Adv. Agron.* **33**, 263 (1980).

103 J. G. Lipman and A. B. Corybeare, *Preliminary Note on the Inventory and Balance Sheet of Plant Nutrients in the United States,* New Jersey Agric. Exp. Station Bull. 607, 1936.

104 G. Stanford, C. B. England, and A. W. Taylor, *Fertilizer and Water Quality*, ARS 41-168, U.S. Dept. of Agriculture, Washington, D.C., 1970.

105 E. A. Paul, "Nitrogen Cycling in Terrestrial Ecosystems," in J. O. Nriagu, Ed., *Environmental Biogeochemistry,* Vol. 1, Ann Arbor Science Publishers, Ann Arbor, Mich., 1976, pp. 225–243.

FIVE

THE INTERNAL CYCLE
OF NITROGEN IN SOIL

A useful concept that has evolved in recent years is than an internal N cycle exists in soil that is distinct from the overall cycle of N but that interfaces with it (see Fig. 4.1). A key feature of the internal cycle is the biological turnover of N through mineralization–immobilization, as follows:

$$\text{Organic N} \xrightleftharpoons[\text{immobilization}]{\text{mineralization}} NH_3 \text{ , } NO_3^-$$

Essentially, biological turnover through mineralization–immobilization leads to the interchange of inorganic forms of N with the organic N. A decrease in mineral levels with time indicates *net immobilization;* an increase suggests *net mineralization.* The fact that levels of mineral N remain unchanged does not necessarily mean that an internal cycling is not operating, but that mineralization–immobilization rates, even though vigorous, are equal.[1]

Several interrelated organic matter fractions must be taken into account when considering N–organic matter interactions in soil. As can be seen from Fig. 5.1, they include plant and animal residues, the microbial biomass, partially stabilized dead organic matter (e.g., cellular remains of microorganisms, melanins produced by fungi, and newly formed humic substances), and the stable humus fraction. As plant residues undergo decay in soil, inorganic N is incorporated into microbial tissue (biomass), a portion of which is converted to newly formed humic substances and ultimately into stable humus. The mean residence time of the N in any given pool can range from a few days or weeks for some components of the biomass to 1,000 or more years for the stable humus fraction. Under conditions where steady-state levels of organic matter have been attained, mineralization of native humus is compensated for by synthesis of new humus.[1]

Biochemical processes such as ammonification, nitrification, denitrification, and assimilation are responsible for many of the transformations that occur within the soil. Nevertheless, fixation reactions of NH_4^+ by clay minerals and of NH_3 by the soil organic matter also play a prominent role and these topics are also discussed herein.

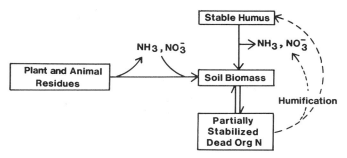

Fig. 5.1

Pools of organic N in soils and their interactions. The decay of organic residues leads to incorporation of N into the biomass, part of which is converted into partially stabilized forms and ultimately into stable humus through a process called humification (see text).

BIOCHEMISTRY OF AMMONIFICATION AND NITRIFICATION

The process whereby organic residues undergo decay in soil is complex and is brought about by the joint activities of a wide range of macro- and microfaunal organisms. The overall decay process is covered in considerable detail in Chapter 1 and only those aspects that relate directly to the release of combined N to available mineral forms will be discussed here. In this connection it should be noted that the release of C and N from residues differ, in that part of the C is lost as CO_2 (see Chapter 1) while N tends to be conserved, particularly when the residues have a high C/N ratio.

Conversion of organic N to available mineral forms (NH_4^+, NO_3^-) occurs through biochemical transformations mediated by microorganisms and is influenced by those factors that affect microbial activity (temperature, moisture, pH, etc.). The first step (ammonification) involves conversion of organic N to NH_3 and is carried out exclusively by heterotrophic microorganisms. The subsequent conversion of NH_3 to NO_3^- occurs primarily through the combined activities of two groups of autotrophic bacteria, *Nitrosomonas*, which converts NH_4^+ to NO_2^-, and *Nitrobacter*, which converts NO_2^- to NO_3^-.

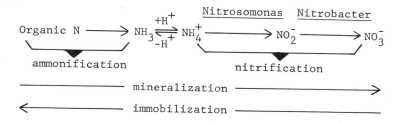

As noted earlier, mineralization is nearly always accompanied by immobilization, and for this reason results obtained for NH_4^+ and NO_3^- accumulations cannot be used to calculate a "mineralization" rate, nor can increases in NO_3^- levels be used to determine a "nitrification rate."

Both aerobic and anaerobic microorganisms are involved in the ammonification process, whereas only aerobes oxidize NH_4^+ to NO_3^-. Thus conditions that restrict the supply of O_2 permit NH_4^+ to accumulate, such as in rice paddy fields. Nitrate is the predominant available form of N in cultivated soils that are well aerated.

Ammonification

Ammonification is an enzymatic process in which the N of nitrogenous organic substances is liberated as NH_3. The review of Ladd and Jackson[2] shows that a wide array of enzymes are involved, each acting on a specific class or type of organic compounds. The initial substrate is often a macromolecule (protein, nucleic acid, aminopolysaccharide), from which simpler N-containing biochemicals are formed (e.g., amino acids, purines, pyrimidines, amino sugars). The biochemical compounds are then attacked by other enzymes, with formation of NH_3.

Brief discussions of the pathways for the enzymatic cleavage of proteins, nucleic acids, and amino sugars follow, but it should be noted that the list of enzymes is incomplete and that other organic N compounds also serve as sources of NH_3.

Breakdown of Proteins and Peptides

The decomposition of proteins and peptides in soil involves an initial cleavage of peptide bonds by proteinases and peptidases to form amino acids, from which NH_3 is released through the action of such enzymes as amino acid dehydrogenases and oxidases.

$$\begin{array}{c}
\text{Proteins} \\
\searrow \quad \overset{\text{proteinases}}{\underset{\text{peptidases}}{\longrightarrow}} \text{ Amino Acids} \xrightarrow[\text{and oxidases}]{\text{amino acid dehydrogenases}} NH_3 \\
\nearrow \\
\text{Peptides}
\end{array}$$

A typical reaction for the oxidative deamination of amino acids is as follows:

$$\underset{NH_2}{R-CH-COOH} + NAD^+ \xrightarrow{\quad NADH + H^+ \quad} \underset{NH}{R-C-COOH} \xrightarrow{\quad H_2O \quad} \underset{O}{R-C-COOH} + NH_3$$

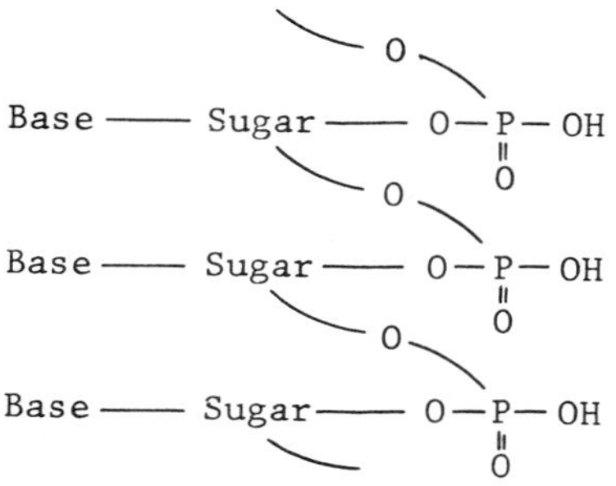

Fig. 5.2
Portion of a polynucleotide chain showing the linkage of individual mononucleotide units through a phosphate ester bond.

Decomposition of Nucleic Acids

Conversion of the N of nucleic acids to NH_3 also requires the action of a variety of enzymes. The nucleic acids, which occur in the cells of all living organisms, consist of individual mononucleotide units (base–sugar–phosphate) joined by a phosphoric acid ester linkage through the sugar (Fig. 5.2). Two types are known, ribonucleic (RNA) and deoxyribonucleic acid (DNA). The two types are identified by the nature of the pentose sugar (ribose or deoxyribose, respectively). Both contain the purine bases adenine (I) and guanine (II) and the pyrimidine base cytosine (III). In addition, RNA contains the pyrimidine uracil (IV); DNA also contains thymine (V); plant DNA contains 5-methylcytosine (VI).

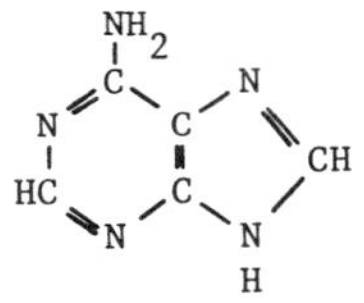

Adenine	Guanine	Cytosine
I	II	III

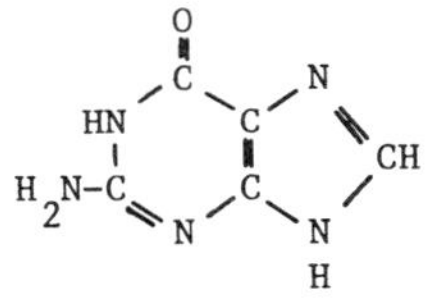

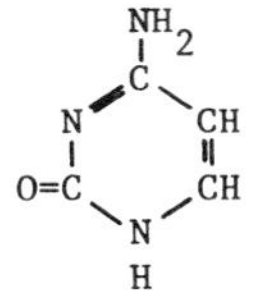

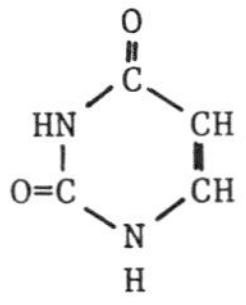

Uracil	Thymine	5-Methylcytosine
IV	V	VI

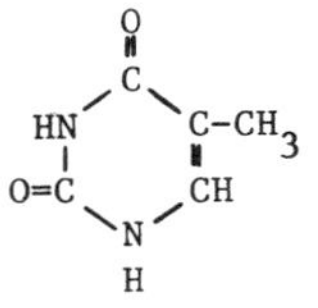

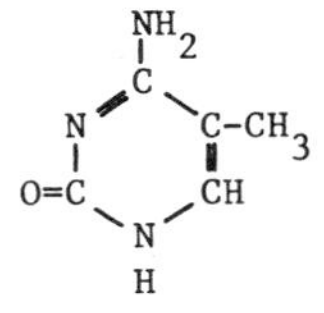

The initial reaction in the breakdown of nucleic acids involves conversion of the nucleic acid to mononucleotides by the action of nucleases, which catalyze the hydrolysis of ester bonds between phosphate groups and pentose units. The mononucleotides, in turn, are converted to nucleosides (*N*-glycosides of purines or pyrimidines) and inorganic PO_4^{3-} by nucleotidases.

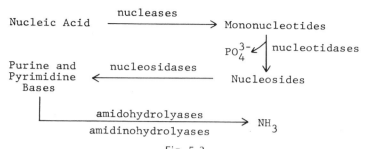

Fig. 5.3
Generalized sequence for the metabolism of nucleic acids.

The nucleosides are then hydrolyzed to the purine or pyrimidine bases and the pentose component by the nucleosidases, which are subsequently converted to NH_3 by reactions catalyzed by amidohydrolyases and amidinohydrolyases. The overall sequence is shown in Fig. 5.3.

Decomposition of Amino Sugars

The formation of NH_3 from two amino sugars (glucosamine and N-acetylglucosamine) by microorganisms is shown in Fig. 5.4. Two enzymes are involved in the breakdown of N-acetylglucosamine (N-acetyglucosamine kinase and N-acetylglucosamine deacetylase), with formation of N-acetylglucosamine-6-phosphate and glucosamine. These two compounds are converted to glucosamine-6-phosphate by the action of N-acetyglucosamine-6-

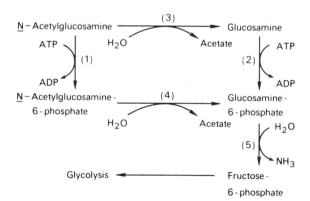

Fig. 5.4
Formation of NH_3 from glucosamine and N-acetylglucosamine in microorganisms.[2] Enzymes involved: (1) N-acetylglucosamine kinase; (2) glucosamine kinase; (3) N-acetylglucosamine deacetylase; (4) N-acetylglucosamine 6-phosphate deacetylase; (5) glucosamine 6-phosphate isomerase. [Reproduced from *Nitrogen in Agricultural Soils* (1982) by permission of the American Society of Agronomy.]

phosphate deacetylase and glucosamine kinase, respectively. Ammonia is subsequently formed from glucosamine-6-phosphate through the action of glucosamine-6-phosphate isomerase.

Urease in Soil

In addition to nitrogenous components of plant or animal tissues, soils may contain organic N as urea, which is a constituent of the urine of grazing animals and is often added to the soil as fertilizer. Decomposition of urea to yeild NH_3 (and CO_2) is brought about through the action of urease, an enzyme found in practically all soils.

$$\begin{array}{c} NH_2 \\ \diagdown \\ C=O + H_2O \xrightarrow{\text{urease}} CO_2 + 2NH_3 \\ \diagup \\ NH_2 \end{array}$$

Nitrification

Except for poorly drained or submerged soils, the NH_3 formed through ammonification (see previous section) is readily converted to NO_3^-. Energy yielded by the reactions is utilized for biosynthesis and cell maintenance.

$$NH_3 \underset{-H^+}{\overset{+H^+}{\rightleftharpoons}} NH_4^+ \xrightarrow{\text{Nitrosomonas}} NO_2^- \xrightarrow{\text{Nitrobacter}} NO_3^-$$

The biological nature of nitrification was established in 1889 by Winogradsky,[3] who related the process to the metabolism of two groups of autotrophic bacteria, one group (*Nitrosomonas*) bringing about the oxidation of NH_4^+ to NO_2^- and the other group (*Nitrobacter*) the oxidation of NO_2^- to NO_3^-. Recent reviews on nitrification include those of Belser,[4] Focht and Verstraete,[5] Painter,[6] and Schmidt.[7]

The organisms directly linked to nitrification in soil are the gram-negative chemoautotrophic (chemolithotrophic) bacteria comprising the family Nitrobacteriaceae. Five genera of NH_4^+ oxidizers and three genera of NO_2^- oxidizers are listed in the eighth edition of Bergey's *Manual of Determinative Bacteriology*.[8] A listing is given in Table 5.1. A single species of each group is considered to be primarily responsible for NH_4^+ and NO_2^- oxidation in soil—*Nitrosomonas europaea* and *Nitrobacter winogradskyi*, respectively.

Nitrosomonas is identified in pure culture as short rods that exhibit straight-line motility; *Nitrobacter* typically occurs as large cells (1.0–1.5 × 1.0–3.0 μm) with an irregular lobular shape and a tumbling motility. Contrary to popular belief, the organisms are not obligate autotrophs (use only inorganic C for cell synthesis). Notwithstanding, there is little doubt that in

Table 5.1
Listing of Chemoautotrophic Nitrifiers[a]

Genus	Species	Habitat
	Oxidize NH_4^+ to NO_2^-	
Nitroso-	europaea	Soil, water, sewage
monas	briensis	Soil
Nitrosospira	nitrosus	Marine, soil
Nitrosococ-	oceanus	Marine
cus	mobilis	Marine
Nitrosolobus	multiformis	Soil
Nitrosovibrio	tenuis	Soil
	Oxidize NO_2^- to NO_3^-	
Nitrobacter	winogradskyi[b]	Soil
	(agilis)[b]	Soil, water
Nitrospira	gracilis	Marine
Nitrococcus	mobilis	Marine

[a] As recorded by Schmidt.[7]
[b] N. winogradskyi is comprised of at least two serotypes, one of which has been referred to traditionally as N. agilis.

soil practically all of the C for cell synthesis comes from an inorganic source, namely, HCO_3^- contained in the soil water.

Pathway of Nitrification

The six-electron transfer accompanying the oxidation of NH_4^+ (oxidation state of -3) to NO_2^- (oxidation state of $+3$) by *Nitrosomonas* suggests at least two intermediates, the most likely candidates being hydroxylamine (NH_2OH) and nitroxyl (NOH).

$$NH_4^+ \xrightarrow[-H^+]{+1/2 O_2} NH_2OH \xrightarrow{-2H^+} NOH \xrightarrow[-H^+]{+1/2 O_2} NO_2^-$$

-3	-1	$+1$	$+3$
Ammonium	Hydroxylamine	Nitroxyl	Nitrite

Energy released by the reaction (65 Kcal/mole) is used by the organism for carrying out its life activities.

Several studies have indicated that N_2O is a by-product of NH_4^+ oxidation.[9,10] This gas may arise by chemical dismutation of nitroxyl (NOH) and through the action of nitrite reductase, as shown in the following illustration. Recent research suggests that some of the N_2O produced in soil, and subsequently evolved into the atmosphere, is generated during nitrification.

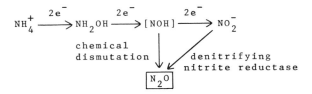

The oxidation of NO_2 to NO_3 by *Nitrobacter* involves a two-electron change in the oxidation state of N (from $+3$ to $+5$), with release of 17.8 kcal/mole of energy:

$$NO_2^- + 1/2\ O_2 \xrightarrow{\ 2e^-\ } NO_3^-$$

$$\text{Nitrite} \qquad\qquad\qquad \text{Nitrate}$$
$$+3 \qquad\qquad\qquad\qquad +5$$

No intermediates are suspect. The reaction is facilitated by an NO_2^- oxidase system with the electrons being carried to O_2 via a cytochrome system, with generation of ATP.

Isolation of Nitrifying Microorganisms

Approaches for the isolation of nitrifying bacteria in soil are limited by the complexities of the soil microhabitat and the physiology of the organisms involved. Isolation by plating methods is generally unsuitable because of problems associated with slow growth, small colony size, and overgrowth by heterotrophs. Most isolation procedures involve an enrichment step, which has distinct disadvantages in that a single isolate may achieve dominance during isolation. The most common approach for enumeration is indirect, involving some modification of the most probable number technique for statistical estimation of nitrifiers in an inoculum.

Factors Affecting Nitrification

The main factors that affect nitrification in soil are temperature, moisture, pH, and the substrates NH_4^+, O_2, and CO_2. Production of NO_3^- decreases with decreasing temperature below 30 to 35°C; below 5°C very little NO_3^- is formed. As one might expect, nitrification proceeds at a very slow rate in cold, wet soils. Farmers in the northern section of the USA are encouraged to wait until soil temperatures fall below 5°C (about 40°F) before applying ammoniacal fertilizers in the fall.

The O_2 and CO_2 (as HCO_3^-) required by the nitrifying organisms are contained in the solution phase of the soil; consequently, moisture content is of major importance. Depletion of O_2 is favored by (1) the presence of easily decomposable organic matter, which increases O_2 demand by heterotrophs; (2) excess moisture, which saturates soil pores and restricts recharge of O_2 from the gaseous phase; and (3) high soil temperatures, which reduce the solubility of O_2.

Under most soil conditions, oxidation of NO_2^- proceeds at a more rapid rate than the oxidation of NH_4^+; thus NO_2^- is seldom found in more than trace amounts. Conditions that favor the presence of free NH_3 (high pH and low CEC) restrict nitrification resulting from NH_3 toxicity. *Nitrobacter* is somewhat more sensitive than *Nitrosomonas;* consequently, conditions that promote NH_3 formation will sometimes result in the temporary accumulation of NO_2^-. As noted in Chapter 4, high levels of NO_2^- accompanying the application of anhydrous NH_3 and urea to soil can result in gaseous losses of N through chemodenitrification.

A wide variety of organic compounds, including certain amino acids and N-ring bases, is known to inhibit the growth of nitrifiers in enrichment cultures. Evidence that nitrification is inhibited by decomposition products of organic residues in soil, or by metabolites excreted by plants or microorganisms, is still only suggestive. Grassland soils consistently contain moderate amounts of exhangeable NH_4^+ but little if any NO_3^-, which has been attributed to inhibition of nitrification due to substances secreted by grass roots. Evidence for and against this hypothesis has been discussed elsewhere.[7]

The nitrifiers are among the most sensitive of the common soil bacteria to organic chemicals added to the soil as herbicides, insecticides, and fungicides. However, any reduction in numbers appears to be of short duration. Various reviews on the subject (e.g., see Goring and Laskowski[11]) indicate that most pesticides, when applied at recommended rates, are unlikely to affect nitrification adversely.

Several synthetic organic chemicals (e.g., nitrapyrin) have been proposed for the inhibition of nitrification in soil, the objective being to conserve fertilizer N and to enhance its use by plants. Presumably, by inhibiting the conversion of NH_4^+ to NO_2^- and NO_3^-, losses associated with leaching and denitrification will be mitigated and economic and environmental benefits will result.

Heterotrophic Nitrification

In addition to the autotrophic bacteria, several heterotrophs have been shown to produce NO_2^- or NO_3^- from NH_4^+ and organic N compounds in pure culture. They include a number of bacteria, actinomycetes, and fungi, as noted in Table 5.2. Several algae have also been reported to produce NO_2^- or NO_3^- from NH_4^+.

The ecological importance of heterotrophic nitrification has yet to be established with certainty. Nitrification occurs in soil under a broader range of environmental conditions than has been predicted from biochemical and physiological studies of the auotrophic nitrifiers, suggesting the participation of heterotrophs in the process. For example, nitrification proceeds in soil

Table 5.2
Examples of Hetrotrophic Nitrifying Microorganisms

Bacteria	Actinomycetes	Fungi
Arthrobacter sp.	*Streptomycetes*	*Aspergillus flavus*
Azotobacter sp.	*Nocordia*	*Neurospora crossa*
Pseudomonas fluorescens		*Penicillium* sp.
Aerobacter aerogenes		
Bacillus megaterium		
Proteus sp.		

at pH values well below the optimum observed for *Nitrosomonas* and *Nitrobacter* in pure culture.

NET MINERALIZATION VERSUS NET IMMOBILIZATION

Some, and often all, of the NH_4^+ and NO_3^- formed through ammmonification and subsequent nitrification is simultaneously consumed by the heterotrophic microflora and converted into microbial tissue, that is, the inorganic N is said to be *immobilized*. The biochemical pathways involved are the reverse of those described earlier for ammonification and nitrification in that NO_3^- is reduced to NH_3 (assimilatory reduction), which subsequently combines with a C substrate to form first simple N-containing biochemicals (amino acids, purine and pyrimidine bases, amino sugars, etc.) and then complex molecules (proteins, nucleic acids, aminopolysaccharides, etc.).

Since both mineralization and immobilization occur simultaneously in soil, the amounts of mineral N (NH_4^+ and NO_3^-) found at any one time represent the difference in the magnitude of the two opposing processes. A decrease in mineral N levels with time indicates *net immobilization;* an increase indicates *net mineralization.*

The C/N Ratio

The decay of organic residues in soil is accompanied by conversion of C and N into microbial tissue. In the process, part of the C is liberated as CO_2 (see Chapter 1). As the C/N ratio is lowered, and as microbial tissues are attacked (with synthesis of new biomass), a portion of the immobilized N is released through net mineralization.

The N content of organic residues, as reflected through the C/N ratio, is of primary importance in regulating the magnitude of the two opposing processes of mineralization and immobilization. Residues that have C/N ratios greater than about 30, equivalent to N contents of about 1.5 or less, result in lowering of mineral N reserves because of *net immobilization* by micro-

Table 5.3
Typical C/N Ratios of Some Organic Material

Material	C/N
Microbial tissues	6–12
Sewage sludge	5–14
Soil humus	10–12
Annual manures	13–25
Legume residues and green manures	13–25
Cereal residues and straw	60–80
Forest wastes	150–500

organisms. On the other hand, residues with C/N ratios below about 20, or N contents greater than about 2.5%, lead to an increase in mineral N levels through *net mineralization*. The N contents and C/N ratios of some organic materials are recorded in Table 5.3.

The relationship between the C/N ratio of crop residues and mineral N levels is as follows:

<20	20–30	>30
Net gain of NH_4^+ and NO_3^-	Neither gain nor loss	Net uptake of NH_4^+ and NO_3^-

Changes in NO_3 and CO_2 levels attending the decay of low-N crop residues in soil are illustrated in Fig. 5.5.[12] Under conditions suitable for microbial activity, rapid decomposition occurs with the concurrent liberation of considerable quantities of C as CO_2. To meet the N requirements of microorganisms, mineral N is consumed; that is, there is a *net immobilization* of N. However, when the C/N ratio of the decomposing material has been lowered to about 20, NO_3^- levels once again increase because of *net mineralization*.

The time required for microorganisms to lower the C/N ratio of carbonaceous plant residues to the level where mineral forms of N accumulate will depend on such factors as application rate, lignin content, degree of comminution, and level of respiration of the soil microflora. A reasonable estimate is that, under conditions favorable for microbial activity, net mineralization will commence after 4 to 8 weeks of decomposition. Accordingly, if crop residues with high C/N ratios are added to the soil immediately prior to planting, extra fertilizer N may be required to avoid N starvation of the crop.

For reasons previously outlined, results obtained for the conversion of NH_4^+ to NO_3^- by measuring increases in NO_3^- levels cannot be used to calculate a "nitrification rate," nor can decreases in NO_3^- levels by incu-

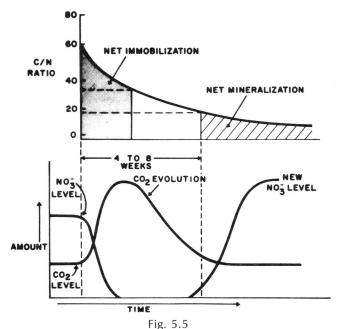

Fig. 5.5

Changes in NO_3^- levels attending the decay of crop residues in soil.

bation under anaerobic conditions be ascribed to denitrification (unless confirmed by other studies showing the absence of immobilization). A complicating factor in studying the mineralization–immobilization relationship is that NH_4^+ is the preferred form of inorganic N by soil microorganisms and that part of the NH_4^+ is subject to nitrification. Under conditions where net immobilization occurs, microorganisms consume NO_3^- (as well as NH_4^+). Following a change to net mineralization, part of the NO_3^- reappears in the NO_3^- pool now diluted with N from other sources that participated in the turnover.

Mineralization and Immobilization Rates

A prime objective of studies on N transformations in soil is to assign quantitative values for mineralization and immobilization rates. Elaborate mathematical formulas based on ^{15}N data have been derived for this purpose and have been used in some studies. As Jansson[13] pointed out, the calculations are based on some assumptions that have not proved to be valid and that give only a rough approximation of the scope of the transformation process.

Mineralization–immobilization is an important component of many N-cycle models (discussed later). Several pools for organic C are often included in these models, such as native humus, the soil biomass, and components of plant residues. The various C substrates decompose at different rates, the order for plant residue components being proteins > carbohydrates > cellulose and hemicellulose > lignin. Levels of NH_4^+ and NO_3^- in the soil increase or decrease, depending on the C/N ratio of the substrate. As noted previously, net mineralization would be expected during decomposition of substrates having low C/N ratios.

Transformations in Wet Sediments

The mineralization–immobilization reaction is modified in zones where O_2 is absent, such as wet sediments. Mineralization does not proceed beyond the NH_3 stage, although further oxidation may occur at the surface-oxidized layer where sufficient O_2 is present to meet the needs of nitrifying bacteria. In some cases, ammonification (conversion of organic N to NH_3) may also be curtailed or otherwise modified. Since anaerobic decomposition provides little energy, larger amounts of substrate must be oxidized per unit of C assimilated, which means less assimilation of mineral N into microbial tissue than for aerobic decomposition. Accordingly, competition of microorganisms with plants for available N may be less severe in aquatic environments than in aerobic soils.

Wetting and Drying

Cycles of wetting and drying have been shown to cause flushes in mineral N levels, with each successive cycle causing a slightly smaller flush.[15,16] The magnitude of the flush, or "partial sterilization" effect, is related to organic matter content and the length of time the soil has remained dry. Some of the reasons given for the net release of mineral N upon drying follow.[17]

1   Microorganisms are killed by the drying treatment (partial sterilization) and N is released as the dead cells undergo autolysis and decay.
2   Drying leads to conversion of organic N to more soluble compounds, which are attacked by microorganisms with release of mineral N.
3   Wetting and drying leads to the breakdown of water-stable aggregates, with exposure of new surfaces for microbial attack.

Freezing and thawing may affect mineralization in the same way as drying and wetting, but to a lesser degree.

MINERAL N ACCUMULATIONS

Only a small fraction of the N in soils, generally less than 0.1%, exists in available mineral compounds at any one time (as NO_3^- and exchangeable NH_4^+).[18] Thus only a few kilograms of N per hectare may be immediately available to the plant even though 6,000 kg/ha may be present in combined forms. Ammonium and NO_3^- can be considered to represent temporary accumulation products of biological transformations involving organic and gasious forms of N. Factors influencing levels of NO_3^- and exchangeable NH_4^+ have been considered in detail by Harmsen and van Schreven[19] and Harmsen and Kolenbrander.[20]

Levels of exchangeable NH_4^+ and NO_3^- vary from day to day and from one season to another and will depend on a variety of environmental factors, as discussed here.

Seasonal variation   Levels of exchangeable NH_4^+ and NO_3^- are greatly affected by temperature and rainfall. The amounts found in the surface layer of soils of the temperate humid climatic zone are lowest in winter because of leaching, rise in spring as mineralization of organic matter commences, decrease in summer through consumption by plants, and increase once again in the fall when plant growth ceases and crop residues start to decay. The level in winter seldom exceeds 10 μg/g but may increase four- to sixfold during the spring.[19]

Mineralization and immobilization   Biological turnover leads to the interchange of NH_4^+ and NO_3^- with N of the organic matter, as noted earlier. Accordingly, mineral N levels represent a delicate balance between mineralization and immobilization and are affected by the activities of soil microorganisms and the C/N ratios of plant residues.

Growing plants   As one would expect, plants exert a depressing effect on mineral N levels in soil. In addition to direct uptake, NH_4^+ and NO_3^- levels may be altered by immobilization and denitrification in the root zone, and possibly by inhibition of nitrification by root excretion products.[19,20] The influence of biochemical processes in the rhizosphere on biochemical N transformations has been reviewed by Rovira and McDougall.[21]

Leaching of NO_3^-   Nitrate is the form of N that is most mobile and that is subject to leaching and movement into water supplies. The magnitude of NO_3^- leaching is difficult to estimate and depends on a number of variables, including quantity of NO_3^-, amount and time of rainfall, infiltration and percolation rates, evapotranspiration, water-holding capacity of the soil, and presence of growing plants. Leaching is generally greatest during cool seasons when precipitation exceeds evaporation; downward movement in summer is restricted to periods of heavy rainfall.

Volatilization of NH_3   Rapid changes in NH_4^+ levels can occur as a consequence of chemical volatilization of NH_3. As noted in Chapter 4, losses are greatest on calcareous and saline soils, especially when NH_4^+-forming fertilizers are used. Only slight losses occur in soils with pH values less than 7.0, but losses increase markedly as the pH increases. For any given soil, losses increase with an increase in temperature, and they can be appreciable when neutral or alkaline soils containing NH_4^+ in the surface layer are dried

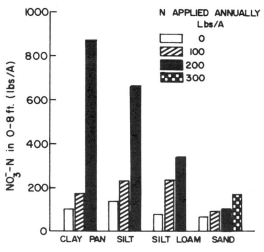

Fig. 5.6

Buildup of NO_3^--N in the upper 8 ft (2.44 m) of four soil types after 8 years of application of N fertilizers to continuous corn in Missouri. (Reproduced with permission from *Agricultural Practices and Water Quality,* edited by T. L. Willrich and G. E. Smith. ©1971 by the Iowa State University Press, Ames, Iowa 50010.)

out. The presence of adequate moisture reduces volatilization, even from alkaline soils.

Losses of NO_3^- through denitrification   Significant loss of NO_3^-–N can and does occur as a consequence of denitrification. Under anaerobic conditions, such as occur frequently in soils following a heavy rain, NO_3^- can be volatilized quantitatively in a comparatively short time, particularly when energy is available in the form of organic residues. Losses are generally negligible at moisture levels below about two-thirds of the water-holding capacity of the soil; above this value the magnitude of loss is correlated directly with moisture regime. The process may also occur in anaerobic microenvironments of well-drained soils, such as small pores containing water and in the vicinity of roots and decomposing residues. Other aspects of denitrification are discussed in Chapter 4.

Buildup of NH_4^+ and NO_3^- by fertilizer applications   In soils of humid and semihumid regions, any fertilizer N added in excess of plant or microbial needs will be lost through leaching and/or denitrification. Thus mineral forms of N seldom carry over from one season to the next. However, where leaching and denitrification are minimal, such as in soils of arid and semiarid regions, some carry-over occurs and repeated annual applications of N fertilizer can lead to a buildup of NO_3^- in the soil profile. Figure 5.6 shows the buildup in NO_3^-–N in the upper 2.44 m (8 ft) of four different soil types following application of N fertilizer for seven years to continuous corn in

Missouri.[22] The relatively lower levels in the silt loam and sandy soils can be ascribed to greater leaching of NO_3^-.

Increasing attention is now being given to NO_3^- accumulations in soil, particularly below the rooting zone, because of possible movement into water supplies. Extremely high levels of NO_3^-–N (2,000 to 4,000 kg/ha) have been observed in soil under cattle feedlots.[23]

NITROGEN AVAILABILITY INDEXES

Most temperate zone soils will contain several thousand kilograms of N per hectare but only a small fraction of this N, estimated at no more than 1 to 2%, will become available to the plant during any given growing season. Various soil tests have been proposed from time to time in attempts to predict the soils' contribution of N to the crop, but, in general, these tests have not attained the same success as those for available P and K. The subject of tests for available N and their limitations have been covered in several reviews.[17,19,24,25]

Before discussing the various soil tests, it should be stated that there is usually a sound scientific basis for many fertilizer N recommendations. Hundreds of field experiments have been conducted throughout the world for the purpose of establishing the relationship between yield and N fertilizer rate for numerous crops under a broad range of soil, climatic, and management conditions. The results of these trials, along with information on crop requirement, yield potential, level of management, and nature of the previous crop (legume or nonlegume), have been of immense value in making N fertilizer recommendations. Given the uncertainty in yields resulting from soil and weather variations, fertilizer N applications based on previous cropping history, as described previously, may prove as reliable, or more so, than chemical or biological soil tests.

A variety of tests have been proposed as indexes of soil N availability, including soil NO_3^- levels, mineralizable N by aerobic and anaerobic incubations, hot-water or hot-salt extractable total N or NH_4^+, amount of NH_3 recovered by soil distillation with alkaline $KMnO_4$, reducing sugars in soil extracts, and total N or organic matter content. Discussions of some of the more promising indexes follow.

Residual NO_3^- in the Soil Profile

In climates of low rainfall, and where very little leaching or denitrification occurs, residual NO_3^- in the root zone is approximately equivalent in availability to fertilizer N and can be taken into account when making fertilizer N recommendations.[17,24,25] The test has had wide application to soils of arid and semiarid regions, including large areas of western USA and Canada, where NO_3^- leaching below the root zone is minimal.

Incubation Methods

These methods typically involve short-term incubation (7 to 25 days) of the soil under aerobic or anaerobic (waterlogged) conditions. Most soil test correlations (with yield or plant uptake of N) have been done under greenhouse conditions. Modifications in the test include preleaching to remove residual NO_3^-, the use of vermiculite or sand to improve aeration and leachability, and additions of $CaCO_3$ or a nutrient solution. Advantages of anaerobic incubation methods are (1) only NH_4^+ needs to be measured, (2) problems with establishing optimum water content are eliminated, (3) more N is mineralized in a given period than under aerobic conditions, and (4) higher temperatures (which lead to more rapid mineralization) can be used.

Stanford and his associates[26,27] have developed an incubation approach designed to define the mineralizable (or labile) soil N pool. In this approach, measurements are made for the amounts of N mineralized from the soil over an extended time period (up to 30 weeks), with inorganic N being removed at various time intervals. Incubation is continued for as long a time as is deemed necessary to describe adequately the relationship between cumulative N mineralization (N_t) and time of incubation, as depicted in Fig. 5.7. The N mineralization potential (N_0) was assumed to follow first-order kinetics ($dN/dt = kN$) and was estimated as log ($N_0 - N$) = Log $N_0 - kt/$ 2.303. The accepted value of N_0 was that which gave the best fit for the linear relationship between log ($N_0 - N_t$) versus time (t). Both k and N_0 were believed to be definitive soil characteristics upon which N mineralization potential could be based.

Limitations of incubation methods are (1) the results are affected by conditions prevailing in the soil at the time of sampling, (2) high results are attained with soils of poor structure, and (3) reliable results, sufficiently correlated with the N requirement of field crops, can only be expected when the technique is calibrated to a given soil type in a given climatic zone and when all samples are collected at the same time during the season.[19]

Chemical Methods

A number of chemical tests have been proposed for estimating soil N availability, but they are empirical in nature and for the most part are too expensive and time-consuming for routine use. They range in severity of extraction from drastic (e.g., total organic N and hydrolyzable N) to intermediate (e.g., alkaline $KMnO_4$ distillation) to mild (e.g., hot-water and hot-salt extractions). The desirability of developing tests based on relatively mild extractants was emphasized by Stanford.[25]

The mild extraction technique of Stanford and his associates[28,29] was se-

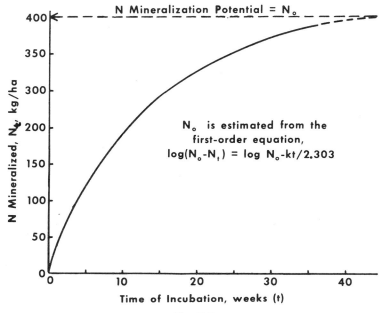

Fig. 5.7

Potentially mineralizable soil organic N based on accumulated amounts of N mineralized during consecutive incubations.[25] [Reproduced from *Nitrogen in Agricultural Soils* (1982) by permission of the American Society of Agronomy.]

lected as one of the recommended soil N availability indices for the American Society of Agronomy Monograph on Methods of Soil Analysis.[24] This method involves the determination of the NH_4^+–N that is formed when the soil is heated with 0.01 M $CaCl_2$ in an autoclave at 121° for 16 hr. The procedure is as follows:

> Place 10 g of soil in a 50-ml Pyrex test tube, add 25 ml of 0.01 M $CaCl_2$, and swirl the tube gently to mix the contents. Cap with Al foil, and place the tube in an autoclave. Bring the autoclave to 121°C, and continue autoclave treatment for 16 hr. Then allow the tube and its contents to cool to room temperature, and quantitatively transfer the soil–$CaCl_2$ mixture to a 150-ml distillation flask. Use about 25 ml of 4 M KCl from a wash bottle to facilitate the transfer. Add 0.2 to 0.3 g of MgO, and determine the amount of NH_4^+-N liberated by steam distillation for 4 min. Determine the amount of NH_4^+-N present in the soil before incubation by the procedure used for analysis of the autoclaved mixture, and calculate available N from the difference in the results of these two analyses.

The ultimate value of any given chemical method of determining soil N availability will depend on the degree to which it can be correlated with crop yield or N uptake under conditions existing in the field. Progress has been made in recent years in developing reliable and rapid methods of assessing soil N availability, but none of the proposed indexes have been adequately tested under a broad range of field conditions and none has been put to general use.[25]

STABILIZATION AND COMPOSITION OF SOIL ORGANIC N

Nitrogen entering the soil system through biological N_2 fixation, as fertilizer, or through other means is subjected to various chemical and biological transformations. Some of the N is utilized by the crop, some is lost by leaching, denitrification, or NH_3 volatilization, and some is incorporated into stable humus forms.[1] Results of studies using the stable isotope ^{15}N have shown that from 20 to 40% of the N applied as fertilizer remains behind in the soil in organic forms after the first growing season.[13,30–32] Only a small portion of this immobilized N (<15%) becomes available to plants during the subsequent growing season, and availability decreases even further in subsequent years.[31,33–37]

The Humification Process

A key to understanding the process whereby fertilizer N is converted to biologically resistant forms may be provided by consideration of the pathways whereby humic substances are formed. As noted in Chapter 1, the classical lignin-protein theory of humus formation is now believed to be obsolete. The modern view is that the so-call humic and fulvic acids are formed by multiple-stage process that includes (1) decomposition of all plant polymers, including lignin, into simple monomers; (2) metabolism of the monomers by microorganisms with an accompanying increase in the soil biomass; (3) repeated recycling of the biomass C (and N) with synthesis of new cells; and (4) concurrent polymerization of reactive monomers into high-molecular-weight polymers.[38] The consensus is that polyphenols (quinones) derived from lignin, together with those synthesized by microorganisms, polymerize in the presence or absence of amino compounds (amino acids, NH_3, etc.) to form brown-colored polymers.

The reaction between amino acids and polyphenols involves simultaneous oxidation of the polyphenol to the quinone form, such as by polyphenol oxidase enzymes. The addition product readily polymerizes to form brown nitrogenous polymers, according to the general sequence shown in Fig. 5.8.

The net effect of the humification process is conversion of the N of amino acids into the structures of humic and fulvic acids. Thus whereas much of the N in microbial tissues (biomass) occurs in known biochemical compounds (e.g., amino acids, amino sugars, purine and pyrimidine bases), most of the organic N in soil (>50%) occurs in unknown forms (see the next

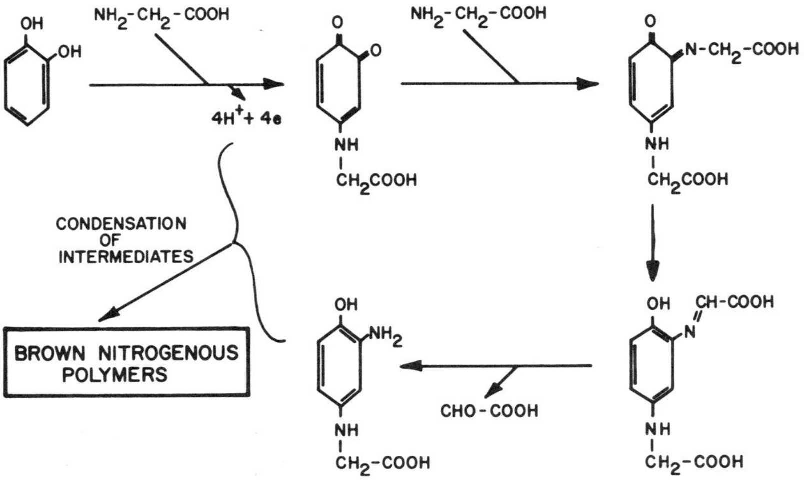

Fig. 5.8

Formation of brown nitrogenous polymers in soil as depicted by the reaction between amino acids and polyphenols. (From Stevenson,[38] by permission of John Wiley and Sons.)

section). The N associated with the structures of humic substances may exist in the following types of linkages.[39]

1   As a free amino ($-NH_2$) group
2   As an open chain ($-NH-$, $=N-$) group
3   As a part of a heterocyclic ring, such as an $-NH-$ of indole and pyrrole or the $-N=$ of pyridine
4   As a bridge constituent linking quinone groups together
5   As an amino acid attached to aromatic rings in such a manner that the intact molecule is not released by acid hydrolysis

Chemical Distribution of the Forms of Organic N in Soils

Current information regarding organic N in soil has come largely from studies utilizing acid hydrolysis to release the organic N to soluble forms.[39] In a typical procedure, the soil is heated with 3 N or 6 N HCl for from 12 to 24 hrs, after which the N is separated into the fractions outlined in Table 5.4. Methods described by Bremner[40] permit recovery of the different forms of N as NH_3, thereby facilitating the use of ^{15}N as a tracer.

The main identifiable organic N compounds in soil hydrolysates are the *amino acids* and *amino sugars*. Soils contain trace quantities of nucleic acids and other nitrogenous biochemicals, but specialized techniques are required

Table 5.4
Fractionation of Soil N by Acid Hydrolysis

Form	Definition and Method	% of Soil N (usual range)
Acid-insoluble N	Nitrogen remaining in soil residue following acid hydrolysis. Usually obtained by difference (total soil *N*-hydrolyzable N).	20–35
NH_3–N	Ammonia recovered from hydrolysate by steam distillation with MgO.	20–35
Amino acid N	Usually determined by the ninhydrin-CO_2 or ninhydrin-NH_3 methods. Recent workers have favored the latter.	30–45
Amino sugar N	Steam distillation with phosphate–borate buffer at pH 11.2 and correction for NH_3-N. Colorimetric methods are also available (see text). Also referred to as hexosamine N.	5–10
Hydrolyzable unknown N (HUN fraction)	Hydrolyzable N not accounted for as NH_3, amino acids, or amino sugars. Part of this N occurs as non α-amino-N in arginine, histidine, lysine, and proline.	10–20

for their separation and identification. Only from one-third to one-half of the organic N in most soils can be accounted for in known compounds.

The N that is not solubilized by acid hydrolysis is usually referred to as *acid-insoluble N;* that recovered by distillation with MgO is *NH_3–N.* The soluble N not accounted for as NH_3 or known compounds is the *hydrolyzable unknown (HUN) fraction.*

A rather large proportion of the soil N, usually of the order of 25 to 35%, is recovered as acid-insoluble N. Some of this N may occur as a bridge constituent linking quinone groups together, such as shown by structures VII and VIII.

VII   VIII

Another possibility is that part of the acid-insoluble N occurs as amino acids that are attached to aromatic rings in such a manner that the amino acid is not released by the acid treatment. Acid hydrolysis would be expected to remove amino acids bound by peptide bonds (IX), as well as those linked to quinone rings (X). On the other hand, amino acids bonded directly to phenolic rings (XI) may not be released.

Another unique feature of N fractionation is that a large proportion of the soil N (usually 20 to 30% for surface soils) is recovered as NH_3. Some of the NH_3 is of inorganic origin; part comes from amino sugars and the amino acid amides, asparagine and glutamine. An accounting of all potential sources of NH_3 in soil hydrolysates shows that the origin of approximately one-half of the NH_3, equivalent to 10 to 12% of the total organic N, is still obscure.

The percentage of the soil N accounted for in the HUN fraction is also appreciable, often >20% of the total N. Some of this N, estimated at from one-fourth to one-half, may occur as the non-α-amino acid N of such amino acids as arginine, histidine, lysine, and proline. The non-α-amino N in these amino acids is not included with the amino acid N values as determined by the ninhydrin-NH_3 or ninhydrin-CO_2 methods (see Table 5.4).

Typical data showing the distribution of the forms of N in soils from different climatic regions of the world are recorded in Table 5.5. Wide differences in N distribution patterns have been observed for soils of the same climatic zone. Nothwithstanding, the data indicate that more of the N in soils from the warmer climates occurs as amino acid N and as amino sugar N.[41] Lower percentages occur as hydrolyzable NH_3. No consistent differences are evident for the acid-insoluble and HUN fractions.

It is a well-known fact that the N content of most soils declines when land is subjected to intense cultivation and attains a new equilibrium level characteristic of the cropping system used (see Chapter 2). This loss of N is not spread uniformly over all N fractions, although it should be pointed out that neither long-term cropping nor the addition of organic amendments to the soil greatly affects the relative distribution of the forms of N. All forms of soil N, including the acid-insoluble fraction, appear to be biodegradable.[42,43]

Table 5.5
Nitrogen Distribution in Soils from Widely Different Climatic Zones[a]

Climatic Zone[b]	Total Soil N, %	Form of N, % of Total Soil N				
		Acid Insoluble	NH$_3$	Amino Acid	Amino Sugar	HUN
Arctic (6)	0.02–0.16	13.9 ± 6.6	32.0 ± 8.0	33.1 ± 9.3	4.5 ± 1.7	16.5
Cool temperate (82)	0.02–1.06	13.5 ± 6.4	27.5 ± 12.9	35.9 ± 11.5	5.3 ± 2.1	17.8
Subtropical (6)	0.03–0.30	15.8 ± 4.9	18.0 ± 4.0	41.7 ± 6.8	7.4 ± 2.1	17.1
Tropical (10)	0.24–1.61	11.1 ± 3.8	24.0 ± 4.5	40.7 ± 8.0	6.7 ± 1.2	17.6

[a] As recorded by Sowden et al.[41]
[b] Numbers in parentheses indicate number of soils examined.

Data showing the effect of cultivation and cropping systems on the distribution of the forms of N in soil are recorded in Table 5.6. Cultivation generally leads to a slight increase in the proportion of the N as hydrolyzable NH$_3$, but this effect is due in part to an increase in the percentage of the soil N as fixed NH$_4^+$. The proportion of the soil N as amino acid N generally decreases with cultivation, whereas the percentage as amino sugar N changes very little or increases.

Taken as a whole, results of N fractionations show that cultivation has only a negligible effect on the relative distribution of the forms of N, indicating that chemical fractionation of soil N following acid hydrolysis is of

Table 5.6
Effect of Cropping on the Distribution of the Forms of N in Soils

Location[a]	Acid Insoluble	NH$_3$	Amino Acid	Amino Sugar	HUN
Illinois (Morrow plots)[b]					
Grass border and C–O–C1 rotation (2)	20.3	16.6	42.0	10.5	10.7
Continuous corn and C–O rotation (2)	20.2	16.7	35.0	14.4	13.9
Iowa					
Virgin (10)	25.4	22.2	26.5	4.9	21.0
Cultivated (10)	24.0	24.7	23.4	5.4	22.5
Nebraska					
Virgin (4)	20.8	19.8	44.3	7.3	7.8
Cultivated (4)	19.3	24.5	35.8	7.0	13.4

[a] Values in parentheses indicate number of soils analyzed. The data from Iowa and Nebraska are from Keeney and Bremner[42] and Meints and Peterson,[43] respectively.
[b] C–O–Cl = corn, oats, clover rotation with lime and P additions; C–O = corn–oats rotation.

little practical value as a means of testing soils for available N or for predicting crop yields (see review of Stevenson[39]).

DYNAMICS OF SOIL N TRANSFORMATIONS AS REVEALED BY ^{15}N TRACER STUDIES

Because of the intricate nature of soil N transformation, many facets of the soil N cycle are poorly understood. The introduction of ^{15}N techniques has provided a means for reexamining earlier concepts and for developing sound principles from which efficient N management systems in soil can be devised. In a typical experiment, a known amount of ^{15}N-labeled fertilizer is applied to the soil, and measurements are made for the relative amounts of the soil and fertilizer N consumed by plants and for the quantities of fertilizer N retained in mineral and organic forms. Any fertilizer not accounted for in the crop or in the soil is assumed to have been lost by leaching, denitrification, or NH_3 volatilization.

Research carried out thus far using ^{15}N has shown that the reactions occurring in soil are even more complex than at first envisioned; in this respect, ^{15}N tracer research has raised more problems than have been solved. Notwithstanding, the ^{15}N approach has provided information that could not have been achieved in any other way. The use of ^{15}N in soil N studies has been discussed in several reviews.[13,30,44]

Nitrogen-15 has been used in the following types of studies:

1 Nitrogen balance of the soil, including gains and losses from the soil–plant system.
2 Stabilization of N through immobilization.
3 Uptake of soil and fertilizer N by plants.
4 Fate of residual fertilizer N in soil.
5 Fixation of NH_4^+ by clay and of NH_3 by organic matter and the availability of the fixed N to plants and microorganisms.
6 Bacterial denitrification and chemodenitrification.
7 Relative use of NH_4^+ and NO_3^- by microorganisms and higher plants.
8 Biological N_2 fixation.

Those aspects related to transformations occurring within the soil through mineralization–immobilization are emphasized herein, namely, items 1 through 5.

Nitrogen Balance Sheets

A critical problem relative to the cycling of N in agricultural soils and natural plant communities is the difficulty of preparing reliable balance sheets in which all gains and losses of N are accounted for. In those cases where fertilizer N has been applied, unexplained losses typically have occurred; complete recovery in the crop or soil has been the exception rather than the rule.

Two general approaches have been used in N balance studies. One involves a complete budget and documents the appropriate inputs and outputs, including crop removal. The second uses a specific ^{15}N-labeled input, from which a balance is calculated for the labeled input. The latter documents the fate of the labeled ^{15}N but provides little information regarding overall gains and losses for the entire system.

Only limited success has been achieved thus far in obtaining a balance for total N in the soil–plant system. Errors in N balances are accumulative. Also, the budget for any particular ecosystem is the product of numerous complex transformations that interact with each other over time. In most instances, quantitative data are lacking for one or more major items required for the calculations.

In an early review, Allison[45] summarized N balance sheets for a large number of lysimeter experiments conducted in the United States. He reported the following findings:

1 Crops commonly recovered only 50 to 75% of the N that was added or made available from the soil. Low recoveries were usually obtained where large additions of N were made, where the soils were very sandy, and where the crop was not adequate to consume the mineral N.

2 The N content of soils decreased regardless of how much N was added as fertilizer unless the soil was kept in uncultivated crops.

3 A large proportion of the N not recovered in the crop was found in the leachate, but substantial unaccounted for losses occurred. Nitrogen gains were few.

4 The magnitude of the unaccounted for N was largely independent of the form in which the N was supplied, whether as NH_4^+, NO_3^-, or organic N.

5 Unaccounted for N was commonly slightly higher in cropped soils than in fallow soils.

Many of the preceding conclusions were substantiated in the recent review of Legg and Meisinger,[32] who arrived at the following conclusions regarding N fertilizer practices.

1 Leaching and N losses are greatest when N inputs exceed crop requirements, a situation that leaves excess NO_3^-–N in the soil and permits leaching and denitrification after the growing season.

2 Nitrogen can be used effectively and efficiently by plants, provided N inputs do no exceed crop assimilation capacity and the N is applied in phase with plant uptake.

An advantage of the ^{15}N approach is the ability to trace a given N input (e.g., ^{15}N-labeled NH_4^+, NO_3^-, or urea) through the various pools, including plant uptake. In practice, precision is limited because of spatial variability of the soil and accompanying sampling errors. As Legg and Meisinger[32] pointed out, the use of ^{15}N can markedly improve accuracy in fertilizer N balance work, but only if the following conditions are met: (1) reliable estimates are obtained for the size of the various N pools, (2) a representative sample is collected for analysis, and (3) the experiment is performed with adequate replication and proper local controls. A problem also arises because of biological discrimination in the use of ^{14}N and ^{15}N by microorganisms and because not all fractions of the native N contain the same background level of ^{15}N, as noted later.

Complete balance sheets have seldom been obtained for ^{15}N-labeled fertilizer N because of the difficulty of obtaining accurate estimates for leaching and denitrification losses. In general, ^{15}N studies have shown that only a portion of the fertilizer N applied to soil, estimated at from 30 to 70%, is consumed by plants and that a significant fraction (up to 40%) is retained in the organic matter after the first growing season.[32] From 30 to 40% of the applied N is usually lost by leaching or denitrification.

Balance sheets based on ^{15}N studies have been given by Hauck[46] and Kundler.[47] The one of Kundler was based on a 10-year international study using ^{15}N-labeled fertilizers and is as follows:

Fertilizer N recovered in the crop	30 to 70%
Nitrogen retained in organic matter the first year	10 to 40%
Nitrogen lost from the system	10 to 30%

Most of the N losses were believed to occur as gaseous escape. Leaching losses under regular farming conditions were estimated to be not more than 5 to 10%.

The survey of Hauck[46] produced the following overall N balance sheet:

Applied fertilizer N recovered by the crop	55%
Immobilization (N tied up in organic matter)	Up to 45%
Denitrification	15%
Leaching and runoff	Negligible to considerable, depending on soil porosity and rainfall.

The balance sheets prepared by Hauck and Kundler represent the range of recoveries for a variety of soils and environmental conditions. They apply

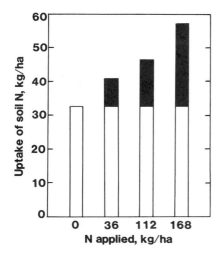

Fig. 5.9
Influence of fertilizer N application rate on the uptake of soil N by Sudan grass (*Sorghum bicolor*) in a pot experiment. The solid bars indicate the additional amount of soil N taken up by the plant in the presence of increasing amounts of fertilizer N. (Adapted from Legg and Allison.[50])

in only a general way to a specific agricultural system. Results of specific N balance studies for a variety of crops (small grain, corn, rice, grassland systems, and forests) grown under a range of soil and environmental conditions have been discussed at length by Legg and Meisinger.[32] Whereas N recoveries in the plant or soil vary considerably from one ecosystem to another, values for plant uptake and immobilized N generally follow the trends noted earlier. In this respect, it should be noted that the efficiency with which fertilizer N is used by plants, as well as the extent to which losses occur through denitrification and leaching may be related to fertilizer application rate.[32,48,49]

Soil N budgets provide a realistic basis for estimating N gains, losses, and transformations for many agricultural systems. In the final analysis, each class or type of ecosystem will have its own characteristic N balance sheet. Thus a balance sheet for fertilizer N applied to a poorly drained, heavy-textured soil where conditions are suitable for denitrification will be entirely different from the balance sheet for an irrigated sandy loam soil, where extensive leaching of NO_3^- may occur. Both leaching and denitrification losses can range from nil to over 50%.

Influence of Fertilizer N on Uptake of Native Humus N

A major objective of many ^{15}N experiments has been to determine the relative contribution of soil and fertilizer N to the N economy of plants. A unique feature of this work is that additions of fertilizer N invariably lead to increases in the amount of soil N taken up by the plant. Typical data showing this effect are given in Fig. 5.9. Explanations given for the increased consumption include (1) the increased uptake is a special feature of the mineralization–immobilization process, (2) the fertilizer N causes enhanced mineralization of native humus N through a "priming" action, and (3) plants growing on treated soil develop a more extensive root system, thereby permitting better utilization of untagged soil N by the plant.

The last explanation fails to account for results obtained in pot experiments, where the volume is limited and the soil is readily penetrated by roots. Thus most workers have attributed the increase either to turnover through mineralization–immobilization (item 1) or to a priming action (item 2). This writer agrees with the conclusion of Jansson,[13] Jansson and Persson,[1] and others that the so-called priming action has been overemphasized and that the major cause of the increased uptake is substitution of the fertilizer N for the native humus N.

Efficiency of Fertilizer N Use by Plants

The interchange between mineral and organic forms of soil N has a bearing upon the use of ^{15}N to determine the efficiency of fertilizer N use of plants and the capacity of the soil to provide available N. Because of mineralization–immobilization turnover, the conventional method of determining fertilizer use efficiency from the difference in crop uptake between the fertilized soil and the untreated plot gives higher recoveries of fertilizer N than does the tracer method. The basic reason for this effect can be seen by considering the following hypothetical case, as outlined by Clark.[51]

Assume that during the course of the growing season a total of 60 kg of N is mineralized but that over this same period 30 kg of N is utilized by microorganisms. A total of 30 kg of N will thereby be available to the plant (60 kg mineralized N − 30 kg immobilized N = 30 kg available N).

Assume now that 60 kg of ^{15}N-labeled fertilizer is applied and that microorganisms draw indiscriminately from the fertilizer and the mineralized soil N to meet their requirements. Of the 30 kg of N needed, 15 kg will come from the soil (^{14}N) and 15 kg from the applied fertilizer (^{15}N). A total of 90 kg of N thus will be available to the plant (120 kg total mineral N − 30 kg immobilized N = 90 kg of plant available N). As shown in the following diagram, 45 kg will come from the soil and 45 kg from the fertilizer.

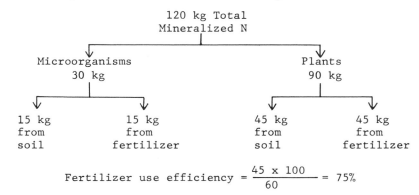

Even if plants consumed all the available N, isotopic analysis for ^{15}N in the plant would indicate 75% recovery of applied fertilizer N. The conventional method of measuring fertilizer N efficiency from the difference between the N uptake of plants growing on fertilized and unfertilized soil would have shown 100% recovery in this example.

The preceding line of reasoning is an oversimplification of the complex reactions that occur in soils, but it serves to emphasize that plant uptake of applied ^{15}N does not necessarily provide a true measure of fertilizer N efficiency. Because of turnover by mineralization–immobilization, some of the soil N not otherwise available is taken up by the plant and a corresponding amount of fertilizer N is immobilized.

The limitation outlined previously for using ^{15}N to measure fertilizer N efficiency also applies when the ^{15}N isotope is used to evaluate leaching and denitrification losses. In this case, substitution of fertilizer N for mineralized soil N during turnover would lead to a low estimate for the true effect of the fertilizer in contributing to environmental pollution (NO_3^- through leaching and N_2O by denitrification).

When the ^{15}N isotope was introduced into soil–plant studies about 25 years ago, it was anticipated that the new tracer technique would provide an expedient and accurate method for evaluating N fertilizers. This has not been the case. Instead, tracer studies have revealed that the soil is a highly dynamic system in which a change of one phase directly affects all other phases. Despite these complications, the ^{15}N tracer technique is a valuable approach for obtaining basic information about soil N transformations. Such knowledge will be indispensable in the long run for improving present-day empirical methods of managing fertilizer N, as well as biologically fixed N.

Losses of Fertilizer N Through Leaching and Denitrification

Although adequate data are not available, there is evidence to indicate that losses of fertilizer N through leaching and denitrification will not be proportional to the amount applied but will be greatest when the amount exceeds the optimum rate for maximum yield.[32] Field data obtained by Broadbent and Carlton,[48] using ^{15}N-depleted fertilizer, serve to illustrate this point (Fig. 5.10). According to Broadbent and Carlton,[48] the key to minimizing excess NO_3^-–N in the soil (and subsequent losses) is to adjust N fertilizer rates to reflect both crop N requirements and the soil's ability to provide available. N.

Losses of fertilizer N at low application rates (percentage basis) will be minimal because of net immobilization by microorganisms that cause the decay of plant residues from the previous crop. Obviously, any N added in excess of microbial requirements will not be immobilized and consequently will be highly susceptible to leaching and denitrification. The mineralization–immobilization sequence provides a natural safeguard against N losses be-

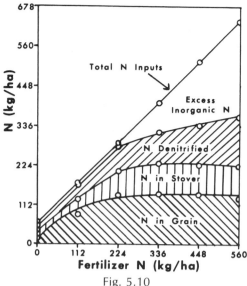

Fig. 5.10

Average annual distribution of total N inputs on an irrigated sandy loam soil. From field data of Broadbent and Carlton[48] as depicted by Legg and Meisinger.[32] [Reproduced from *Nitrogen in Agricultural Soils* (1982) by permission of the American Society of Agronomy.)

cause of net immobilization during periods when the potential for leaching and denitrification will be highest.

The magnitude and consequences of fertilizer N loss are currently subjects of considerable debate. Because of environmental concerns and the need to conserve energy, a rational approach to fertilizer N use is required. Since the point of maximum economic return from fertilizer N is generally below the point of maximum yield, it may be possible to adjust application rates for optimum efficiency while at the same time reducing losses of N to the environment.

Nitrogen losses from soil cannot be eliminated completely but they can be minimized by proper management practices. As our understanding of the N cycle in soil increases, better ways will be found to manage N for maximum efficiency and minimal pollution.

COMPOSITION, AVAILABILITY, AND FATE OF IMMOBILIZED N

Several approaches (using the stable isotope ^{15}N) have been applied in attempts to characterize the immobilized N in soil, including fractionations based on acid hydrolysis, partition into humic and fulvic acids by classical alkali extraction, and extractability with chemical reagents proposed as indexes of plant-available soil N. Other approaches have included (1) release of the organic N to mineral forms (NH_4^+ + NO_3^-) by incubation under ideal conditions in the laboratory and (2) uptake by plants grown in the greenhouse or field.

Results of studies on the uptake or extractability of applied ^{15}N are often expressed in terms of an availability or extractability ratio. For plant uptake the equation is:

$$\text{Availability ratio} = \frac{^{15}\text{N in plant/total N in plant}}{^{15}\text{N in soil/total N in soil}}$$

A similar equation is used to express results of incubation experiments for mineralizable N (NH_4^+ and NO_3^-).

$$\text{Availability ratio} = \frac{\text{residual } ^{15}\text{N mineralized/total N mineralized}}{\text{residual } ^{15}\text{N in soil/total N in soil}}$$

For chemical extraction, the ratio becomes:

$$\text{Extractability ratio} = \frac{\text{tagged } ^{15}\text{N in extract/total N in extract}}{\text{tagged } ^{15}\text{N in soil/total N in soil}}$$

A ratio of unity indicates that the residual fertilizer N and total soil N are equally available to plants, microorganisms, or chemical extraction. Values greater than or less than 1.0 represent enhanced or reduced susceptibility, respectively, of the residual fertilizer N to biological utilization or chemical extraction. Application of availabilty ratios will be made in subsequent sections.

The Soil Biomass

The decay of organic residues in soil is accompanied by conversion of C and N into microbial tissue. The biomass, along with the decomposing residues, represents the active phase of soil organic matter.

Several approaches have been used in attempts to estimate the size of the microbial biomas in soil, such as counts for microorganisms and measurements for metabolic activities. In recent years, respiratory methods have been applied using sterilized or partially sterilized soil.[52-55] In the method employed by Jenkinson[54] and Jenkinson and Powlson,[55] the soil is sterilized with chloroform and biomass C is estimated from the amount of CO_2 produced after inoculation with nonsterile soil. Attempts have been made to use this approach for estimating biomass N, but as Ladd et al.[56] pointed out, the results are affected by immobilization of N within the biomass.

Anderson and Domsch[53] estimated the quantities of N in the microbial biomass of 29 soils using respiratory and selective inhibition methods (see Chapter 1). Data for biomass C were converted to N values by measuring (1) the C/N ratio of pure cultures of 24 species of soil microorganisms, and (2) the relative contribution of bacterial and fungal cell populations to the microbial biomass. From 0.5 to 15.5% of the total N was found in the microbial biomass. The higher percentages are unusual. In most soils, no more than 3% of the N occurs in the biomass at any one time.[35,52,53,56-61]

Information on the composition and availability of biomass N has been provided by studies in which microbial tissue has been labeled by short-term incubation of soil with ^{15}N-labeled NH_4^+ and a suitable energy source. The variables have included form of applied N, type of C substrate, soil properties, and effects of moisture content and soil drying. Under optimum conditions for microbial activity, and in the presence of a readily available C source, net immobilization of added ^{15}N proceeds rapidly and reaches a maximum at incubation periods as short as 3 days with a simple substrate (e.g., glucose) to as much as 2 months or more for a complex substrate such as mature crop residue.[62,63]

Mineralization studies have shown that the turnover of N from dead microbial cells is about five times higher than that for the native soil organic N.[57,64-66] Cytoplasmic constituents are easily broken down while cell wall components are slowly mineralized. Fungal melanins resist attack by microorganisms and may persist in soil for a long time.

Drastic changes in soil moisture or temperature (i.e., drying and rewetting, freezing and thawing) result in a flush of mineral N during subsequent incubation.[64-67] This flush is normally of short duration and roughly proportional to the temperature and duration of the drying conditions. The organic material made more available to biological attack is probably derived from (1) microbial cells killed during the treatment, (2) humus materials rendered soluble because of moisture removal, and (3) previously inaccessible organic matter exposed through destruction of stable soil crumbs.

Results of studies where the soil biomass has been labeled with ^{15}N have shown that air or oven drying leads to increased net mineralization of both the newly immobilized N and the native soil N, with the effect being somewhat greater for the biomass.[64-67]

Results of a study[68] in which the biomass was labeled with ^{15}N using glucose as a C source are shown in Fig. 5.11. In agreement with results using ^{14}C-labeled substrates (see Chapter 1), incorporation of a portion of the applied substrate C into microbial tissue was accompanied by rapid immobilization of applied ^{15}N. Soil samples obtained at the point of maximum conversion of glucose C into the biomass (see Fig. 5.11) were extracted with several reagents proposed as indexes of plant-available N, the objective being to test the hypothesis that, for mild extractants, the biomass is the primary source of the extracted N. The mild extractants included hot water, hot 0.01 M $CaCl_2$, cold 0.01 M $NaHCO_3$, and hot 0.005 M $NaHCO_3$. For comparison purposes, a more drastic extractant was used (acid $KMnO_4$).

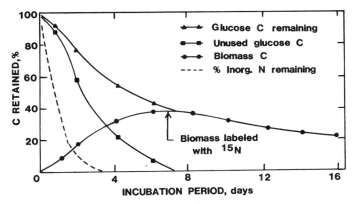

Fig. 5.11
Microbial utilization of added glucose and incorporation of substrate C and N into the biomass. Soil in which the biomass was labeled with ^{15}N (7-day incubation period) was used to determine the source of the N that is extracted by reagents proposed as indexes of plant-available N (see Table 5.7).

Data recorded in Table 5.7 show that, although the mild extractants removed two to three times more of the biomass N than the native humus N, most of the extracted N (58 to 68%) was derived from the soil. Only about 7% of the biomass N was removed by the mild extractants. The selectivity ratio of 0.90 for acid $KMnO_4$ shows that this reagent removed almost equal percentages of the biomass and soil N. Findings of the study suggest that, for the soil used, a significant portion of the biomass N was no more available

Table 5.7
Sources of N in Some Extractants Proposed as Indexes of Plant-Available Soil N[a,b]

Extractant	Source of Extracted N, %		Biomass N Recovered, %	Humus N Recovered, %	Extractability Ratio
	Biomass	Native Humus			
Hot water	33.7	66.3	6.8	2.4	2.4
Hot 0.01 M CaCl$_2$	32.1	67.9	6.9	2.8	2.2
0.01 M NaHCO$_3$	41.2	58.8	9.7	2.4	3.0
Hot 0.005 M NaHCO$_3$	32.3	67.8	10.9	3.8	2.4
Acid KMnO$_4$(NH$_3$)	13.2	86.8	8.8	11.5	0.9

[a] From Kelley and Stevenson.[68]
[b] The soil was one in which the biomass was labeled with ^{15}N.

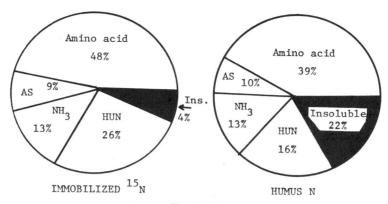

Fig. 5.12

Comparison between the distribution of the forms of organic N between newly immobilized N and the native humus N. (Adapted from data of Kai et al.[62])

to plants than the native soil N, a result that is in agreement with the mineralization data mentioned earlier. Juma and Paul[69] also found that several extractants for available N were not selective in removing N from the biomass.

Kai et al.[62] determined the distribution of the forms of immobilized N in a soil incubated for periods of up to 20 weeks with $^{15}NO_3^-$ and glucose, straw, or cellulose (C/N = 32). For all C substrates an initial net immobilization of ^{15}N was followed by a period of net mineralization. At the point of maximum incorporation of N into the biomass, there was a distinct difference in the percentage distribution of the forms of organic N between the newly immobilized ^{15}N and the native humus N. As shown in Fig. 5.12, the immobilized N was higher in amino acid N and in hydrolyzable unknown forms (HUN fraction) but lower in acid-insoluble N; percentages of the N as NH_3–N and as amino sugar N were approximately the same. The distribution of N for the labeled biomass was similar to that observed by Marumoto et al.[65,66] for hydrolysates of some microbial cells.

The results of Kai et al.[62] also demonstrated that, following maximum tie-up of N, there was a net release of immobilized ^{15}N to available mineral forms (NH_4^+, NO_3^-). However, not all N forms were mineralized at the same rate. With time the percentage of the organic N as amino acids and in the HUN fraction decreased while the percentage as NH_3 – N and acid-insoluble N increrased. The changes are shown in Table 5.8.

Newly Immobilized N

The decay of organic residues in soil is accompanied by conversion of C and N into microbial tissue. In the process, part of the C is liberated as CO_2. As the C/N ratio is lowered and as microbial tissues are attacked (with

Table 5.8

Percentage Distribution of Immobilized ^{15}N in Various Soil Organic N Fractions
After Various Incubation Periods using Glucose as Substrate[a]

Source of N and Incubation Period	Form of N, % of Total N				
	Acid Insoluble	NH$_3$	Amino Acid	Amino Sugar	HUN
Native humus N	20.8–24.1	12.4–15.5	37.5–38.6	9.1–10.4	12.9–19.3
Applied ^{15}N					
3 days	4.5	12.6	48.2	9.1	25.6
1 week	5.7	12.4	43.5	14.0	24.4
8 week	10.7	17.2	41.4	10.7	20.1
20 weeks	8.5	20.0	39.2	9.2	23.1

[a] Adapted from Kai et al.[62]

synthesis of new biomass), part of the N is converted to more stable forms (see Fig. 5.1). Accordingly, plant recovery of fertilizer N is invariably reduced when plant residues are applied to the soil. It should be noted, however, that a proportionally higher amount of the native humus N may become available because of mineralization–immobilization, especially when the "active" fraction has been maintained at a high level by keeping a rotating fund of decomposable organic matter in the soil.

Availability ratios of newly immobilized N in soil, as calculated from mineralization data, have ranged from 0.4 to 10.0, depending on source and rate of applied ^{15}N, type and amount of residue added, number of successive crops grown, and length of incubation period.[34,57,64,70,71] Low ratios were observed by Broadbent and Nakashima[70] for a clay loam soil sampled after 40 days of incubation in the presence of barley straw. More than one-half of the applied ^{15}N was present in the soil after 2 years of incubation and periodic removal of mineralized N by leaching. Chichester et al.[64] also found that applied fertilizer ^{15}N was rapidly converted to relatively stable forms during decay of organic residues in soil.

The high stabilization of immobilized ^{15}N has been confirmed by plant uptake studies. In general, relatively little of the immobilized ^{15}N is taken up by the first crop and availability decreases by consecutive cropping.[70,72,73]

Several attempts have been made to determine the availability of residual fertilizer N in soils using chemical extraction techniques. Legg et al.[72] repeatedly extracted ^{15}N-labeled soil by autoclaving with 0.01 M CaCl$_2$. Extractability ratios were compared with availability ratios based on N mineralization during incubation and by N uptake by a series of oat (*Avena sativa* L.) crops. Whereas the biologically based availability ratios indicated that residual ^{15}N was twice as available to plants as the total soil N, the extractability ratios indicated equal susceptibility for both forms. This result

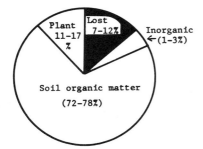

Fig. 5.13

Recovery in wheat (*Triticum aestivum* L.) of ^{15}N in decomposing legume residue. (Adapted from Ladd[75] and Ladd et al.[60])

suggests that chemical extraction is not selective in recovering plant-available N, a conclusion also reached by Stanford et al.[74]

Biologically Fixed (Legume) N

In many parts of the world, adequate N for sustained crop production depends on N_2 fixation by legumes. In the wheat belt of south Australia, for example, the conventional practice is to grow legumes, interspersed as a grazed pasture phase or as a grain legume cropping phase, in rotation with the cereal.

Ladd[75] and Ladd et al.[60] have summarized results of an elaborate field study in which doubly labeled ^{14}C and ^{15}N legume material (e.g., *Medicago littoralis*) was mixed with the top soil at three field sites in south Austrialia and allowed to decompose about eight months before the soils were sown to wheat (*Triticum aestivum* L.). As shown in Fig. 5.13, only 11 to 17% of the ^{15}N was taken up by the wheat, with most of the remainder (72 to 78%) being recovered in the organic phase of the soil. Even after 4 years, nearly one-half of the added ^{15}N was still present in the soil in stable organic forms.[60,75] The conclusion was reached that the value of legumes as a source of N was due not so much to their capacity to provide relatively large amounts of immediately available N as to long-term effects whereby soil organic N levels are maintained or increased, thereby insuring an adequate supply of N by slow decompositon of the stable organic N.

Obviously, more data are required regarding the fate of biologically fixed N in soils.

Residual Fertilizer N

As noted earlier, from 10 to 40% of the fertilizer ^{15}N applied to the soil remains behind in organic forms after the first growing season. Not more than about 15% of this residual N is available to plants during the second growing season, and availability decreases even further for succeeding crops.[31,33-37] In the study conducted by Westerman and Kurtz,[37] the percentage of the residual fertilizer ^{15}N recovered in the plant tops (three harvests) of a sorghum-sudan hybrid (*Sorghum sudanense*) was of the order of

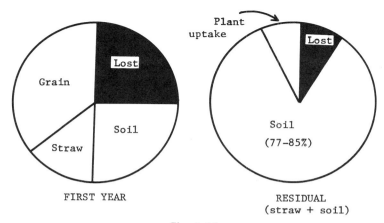

Fig. 5.14
Recovery of applied $^{15}NH_4^+$ by rice (*Oryza sativa* L.) in a flooded soil. The left side shows uptake during the first cropping season while the right side shows recovery of residual ^{15}N (straw and soil) during the second cropping season. (Adapted from Reddy and Patrick.[76])

13 to 18%, equal to from 4 to 6% of the ^{15}N originally applied. The residual ^{15}N remaining after the second growing season, representing approximately 24% of the initial fertilizer, was even less available (of the order of 1.5%) and was assumed to be in equilibrium with the native soil N.[77]

The low availability of residual fertilizer N has been demonstrated for a number of cropping systems, as noted in Fig. 5.14 for rice (*Oryza sativa* L.) for uptake of labeled $^{15}NH_4^+$ from a flooded soil.[76] In this case, approximately 50% of the applied ^{15}N was recovered in the crop (grain plus stover) during the first growing season (left side of Fig. 5.14); approximately 26% was accounted for in the soil (roots + soil). Following harvest, the rice straw was returned to the soil and a second rice crop was grown, with some plots receiving supplemental unlabeled fertilizer N. Less than 10% of the residual labeled ^{15}N was recovered in the second rice crop (right side of Fig. 5.14). The addition of unlabeled N had only a slight influence on the uptake of residual fertilizer ^{15}N.

Chemical Characterization

Clues as to the high stability of residual fertilizer N in soils have come from fractionations based on acid hydrolysis. In the case of the field study of Allen et al.,[77] an average of one-third of the fertilizer N initially applied was accounted for in the surface soil after the end of the first growing season, the remainder having been consumed by plants or lost through leaching and denitrification. Isotope-ratio analyses revealed that the residual N had been incorporated into the organic matter. Comparision of the distribution pattern

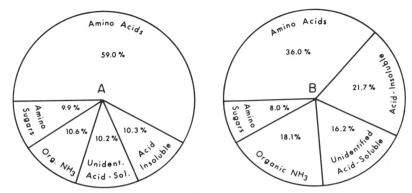

Fig. 5.15

Comparison between the percentage distribution of fertilizer-derived organic N (A) and the native humus N (B).[77] [Reproduced from *J. Environ. Qual.* **2,** 120–124 (1973) by permission of the American Society of Agronomy.]

for the fertilizer-derived N with that of the native humus N (Fig. 5.15) shows that a considerably higher proportion of the fertilizer N occurred in the form of amino acids (59.0 versus 36.0%) and amino sugars (9.9 versus 8.0%); lower proportions occurred as hydrolyzable NH_3 (10.6 versus 18.1%), acid-insoluble N (10.3 versus 21.7%), and the HUN fraction (10.2 versus 16.2%). When the plots were resampled four years later, the fertilizer N remaining in the soil, representing one-sixth of that initially applied, had a composition very similar to that of the native humus N.

Results similar to those noted earlier were obtained by Smith et al.,[78] who also found that fertilizer N was initially incorporated into such compounds as amino acids and subsequently into more stable forms. As compared to the native soil N, more of the residual fertilizer N occurred in an amino acid-containing fraction, with lower amounts being accounted for as hydrolyzable NH_3 and insoluble N. Equilibrium with the native soil N had not been achieved in 3 years.

Residual fertilizer N has been partitioned into the classical humus fractions (e.g., humic acid, fulvic acid, and humin) by alkali extraction.[36,79] The relative amounts of N recovered in the humic and fulvic acids were found by Wojcik–Wojtkowiak[79] to depend on a variety of factors, including form of applied N. McGill and Paul[80] and McGill et al.[81] isolated humic and fulvic acids from ^{15}N-labeled soils and observed a higher degree of labeling in the fulvic acid fraction.

Long-Time Effects

The postulated long-time fate of residual fertilizer N in Mollisol soils of the corn-belt region of the USA is shown in Fig. 5.16. The mean residence time (MRT), or average life, for the N retained after the first season is about 5

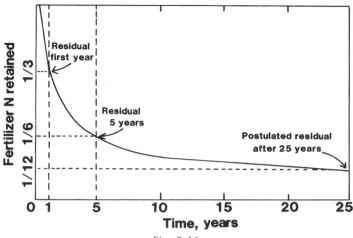

Fig. 5.16

Postulated long-time fate of residual fertilizer N in Mollisol soils of the corn-belt section of the United States. The amount retained is compensated for by mineralization of an equivalent amount of humus N, thereby maintaining steady-state levels of organic N in the soil.

years. Because of increased humification, any N retained after this period will have a MRT of about 25 years. Thereafter the MRT will be the same as that for the native humus, estimated at from 200 to 800 years. From this it can be seen that a small amount of the fertilizer N will remain in the soil for a long time, perhaps centuries.

Humus as a Source of N

Essentially all of the N in the plow layer of the soil, of the order of 93 to 97%, occurs in organic combinations. Most of the remainder can be accounted for as clay-bound (fixed) NH_4^+. Only trace quantities of N are found in available mineral forms (NO_3^- and exchangeable NH_4^+) at any one time. The assumption is often made that 1 to 3% of the soil organic N is mineralized during the course of the growing season, and presumably available to plants. However, as explained in the following discussion, this statement must be accepted with considerable reservation, for the reason that the humus content of most soils is in a state of quasi-equilibrium (i.e., level does not change significantly from one year to another).

The absolute amount of organic N in soils varies greatly and is influenced by those factors affecting the organic matter content of the soil, namely, the soil-forming factors of climate, topography, vegetation, parent material, and age. As noted in Chapter 2, organic matter levels usually decline when soils are first placed under cultivation, with establishment of new steady-state

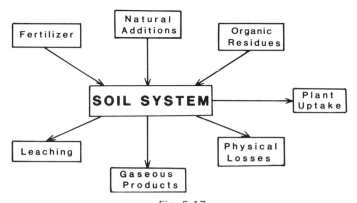

Fig. 5.17
"Black-box" approach for modeling the soil N cycle.

levels that are substantially lower than the virgin soil, depending on the cropping system. The effectiveness of soil management practices in maintaining soil N reserves depends on such factors as rate of application of organic residues, kind of residue and its N content (C/N ratio), manner of incorporation into the soil, soil characteristics, and seasonal variations in temperature and moisture.

At steady-state, the amount of N released during the year and either recovered in the harvested portion of the crop or lost through leaching and denitrification is balanced by incorporation of N from other sources into newly formed humus. A *net annual release* of N occurs only when organic matter levels are declining, a condition that is to be avoided as exploitation generally leads to a reduction in the productive capacity of the soil.

When the organic N level of the soil is at equilibrium and the cropping sequence does not include a legume, transformations occurring within the soil can be ignored when modeling the N cycle (e.g., only incomes and outgoes of N need to be considered). A simplified "black-box" approach for modeling the soil N cycle is shown in Fig. 5.17. The approach has been used by Tanji et al.[49] to predict N losses from soil.

For most soils the quantity of N returned in crop residues, in rain water, or through biological N_2 fixation is sufficient to meet the needs of microorganisms in synthesizing new humus. Thus a reasonably good estimate of the quantity of fertilizer *N required by the crop can be taken as equal to the amount removed in the harvested portion* divided by the efficiency with which the fertilizer N is used:

$$N_{\text{fertilizer}} = \frac{N_H}{E_f} \qquad (1)$$

where N_H is the amount of N expected to be removed in the harvested portion and E_f is the efficiency factor. The assumptions are made that the amount of N contained in the stover will be compensated for from that returned in residues from the previous crop, and that any excess mineral N will have been lost through leaching, denitrification, or NH_3 volatilization. When the stover is not returned to the soil or is burned (e.g., sugar cane production), additional fertilizer N must be added to compensate for the extra N removed. The efficiency factor will vary with the crop and environmental conditions but will generally range from 0.50 to 0.75. Better ways are needed to increase the efficiency of fertilizer N use by plants.

The use of equation 1 to estimate fertilizer N requirements can be illustrated by considering N removal in the grain for a good yield of corn (*Zea mays* L.). From 120 to 150 kg N per hectare will be removed. For an assumed efficiency factor of 0.66, from 180 to 230 kg of fertilizer N per hectare will be required to attain the desired yield when corn follows corn in a rotation. In the corn-belt region of the USA, corn often follows soybeans (*Glycine max* L.) in a rotation, in which case the recommended fertilizer rate is reduced by 25 to 45 kg N/ha. For those soils where leaching and denitrification is not a problem (e.g., many unirrigated arid and semiarid zone soils), an adjustment in fertilizer N rate can be made for residual mineral N in the rooting zone.[25,83]

A more elaborate equation based on soil test results for mineralizable N (laboratory incubation) has been given by Parr[84] for predicting the amount of fertilizer N to be applied to corn:

$$N_{\text{fertilizer}} = \frac{N_{\text{crop}} - (N_{om} + N_R)}{E_f} \tag{2}$$

where N_{crop} is the total amount of N removed (grain + stover), N_{om} is the N mineralized from the soil organic matter in a laboratory incubation soil test, and N_R is the residual mineral N (NO_3^- + exchangeable NH_4^+). The assumption is made that the mineralized N will be utilized by the crop at the same efficiency as the fertilizer N. Limitations of incubation tests for assessing soil N availability have been discussed by Stanford.[25]

When crop residues are returned to the soil, only that portion of the N contained in the harvested portion represents a net loss from the soil–plant system. In general, more N is contained in the harvested portion than in the stover, vines, straw, or roots, as noted earlier in Fig. 4.4. Grasses or forages harvested for hay will remove large quantities of N. However, not all of the N removed by leguminous crops will come from the soil; instead it will be derived from the air through biological N_2 fixation. In sugar cane production, very little N is removed in the portion of the plant from which sugar is extracted, but huge losses occur when the tops and leaf trash on the cane field are burned to facilitate harvest. Over 100 kg N/ha is removed in the harvested portion of many nonleguminous crops, thereby accounting for the large amounts of fertilizer needed for sustained production.

Crop yields on continuously cropped arable land are often related to the N supply. Since surplus inorganic N is subject to losses through leaching and denitrification, efficiency in the utilization of soil and fertilizer N is dependent to some extent on the effective management of crop residues. The ideal situation is to have adequate mineral N in the soil during periods of active uptake by plants and to avoid excesses that will be subject to losses through leaching and denitrification.

NONBIOLOGICAL REACTIONS AFFECTING THE INTERNAL N CYCLE

Not all transformations of N in soil are mediated by microorganisms but are chemical in nature. These nonbiological reactions play a prominent role in the internal cycle of N in soils. Chemical reactions of inorganic forms of N are of three main types.

1 Fixation of NH_4^+ on interlamelar surfaces of clay minerals, notably vermiculite and the hydrated micas.
2 Fixation of NH_3 by the soil organic fraction, which supplements biological immobilization in that the fixed N is not readily available to plants or microorganisms.
3 Reactions of NO_2^- with organic constituents, including humic and fulvic acids. Part of the NO_2^-–N is converted to organic forms and part is lost from the soil system as N gases.

The fate of inorganic forms of N in soils, including uptake by plants, is influenced by the magnitude of these processes, which vary from one soil to another and with cultural practice. Nitrite reactions (item 3) are discussed in Chapter 4 in connection with N losses from soil and will not be covered here.

Ammonium Fixation by Clay Minerals

It has long been known that many soils are capable of retaining considerable amounts of NH_4^+ in nonexchangeable forms. Fixation can be regarded as resulting from the substitution of NH_4^+ for interlayer cations (Ca^{2+}, Mg^{2+}, Na^+) within the expandable lattice of clay minerals. According to one popular theory, the reason NH_4^+ is fixed is that the ion fits snugly into hexagonal holes or voids formed by oxygen atoms on exposed surfaces between the sheets of 2:1 lattice-type clay minerals. These voids have a diameter of about 2.8 nm, which is of the order of the diameter of NH_4^+ (as well as K^+). When occupied by a fixable cation, the lattice layers contract and are bound together, thereby preventing hydration and expansion. Cations which have hydrated diameters greater than 2.8 nm, such as Na^+, Ca^{2+}, and Mg^{2+}, cannot enter the voids and are thereby able to move more freely in and out of the clay sheets. The subject of NH_4^+ fixation has been covered in considerable detail in reviews by Nommik[85] and Nommik and Vahtras.[86]

Biological Availability of Fixed NH_4^+

The availability of NH_4^+ to microorganisms and higher plants can be reduced by fixation. However, the review of Nommik[85] suggests that fixation is usually not a serious problem under normal fertilizer practices. As will be noted, K^+, being a fixable cation, is effective in blocking the release of fixed NH_4^+. Thus the application of large amounts of K^+ simultaneously or immediately following NH_4^+ additions may reduce the availability of the fixed NH_4^+ to higher plants. Ammonium fixation cannot be considered entirely undesirable because N losses through leaching may be reduced. In addition, fixation may ensure a more continuous supply of available N throughout the growing season.

Ammonium-Fixing Capacity

Individual soils vary in their capacity for fixing NH_4^+ from only a few kilograms to several hundred kilograms per hectare, depending on the factors outlined in the following paragraphs.[85-87]

TYPE OF CLAY MINERAL.   Vermiculite, illite, and montmorillonite are the main minerals that retain NH_4^+ in a nonexchangeable form. These clays have a crystal lattice that is unique in that the individual sheets can expand and contract. Soils that contain significant amounts of vermiculite or weathered illite will fix 6 meq or more of NH_4^+ per 100 g (1,880 kg of NH_4^+–N per hectare to plow depth); those soils where the clay fraction is dominated by montmorillonite will fix only 0.2 to 0.3 meq per 100 g (62 to 108 kg of NH_4^+–N per hectare). Practically no fixation will occur in those soils where kaolinite is the predominant clay mineral.

The differential behavior of $2:1$ type clay minerals in fixing NH_4^+ is due in part to the source of the negative charge, that is, whether the main part of the charge originates from the octahedral Al layer or the tetrahedral Si layer. The attraction is greatest when the charge arises from isomorphous substitution of Al for Si in the tetrahedral layers, which gives a shorter distance between the interlayer cation and the negative site of the lattice. Substitution in the tetrahedral Si sheet accounts for 80 to 90% of the substitution in vermiculites, about 65% or more in illites, and less than 20% in montmorillonites.

POTASSIUM STATUS OF THE SOIL.   Ammonium-fixing capacity of the soil depends to some extent on K^+ content. Fixation by micaceous minerals (such as illite) occurs only when the lattice has been impoverished of K^+ by weathering. When the interlattice charge is balanced completely by a fixable cation (such as K^+), the sheets are fully contracted and NH_4^+ cannot enter. Soils containing large amounts of illite may or may not fix NH_4^+, depending on

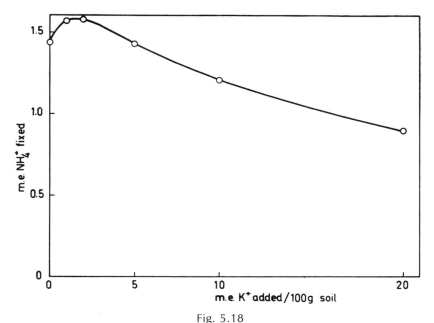

Fig. 5.18

Fixation of NH_4^+ by a vermiculite-containing clay soil as influenced by the simultaneous addition of K^+. (Adapted from Nommik.[87])

degree of weathering and saturation of the lattice by K^+. The capacity of many soils to fix NH_4^+ may be increasing because of removal of K^+ by cropping.

In general, surface soils have a much lower capacity for fixing NH_4^+ than subsoils, which may be due to a combination of the blocking effect of native K^+ and interference from organic matter.

Potassium fertilization can have a depressing effect on NH_4^+ fixation.[85] Additions of K^+ prior to or concurrently with fertilizer NH_4^+ reduces fixation of the NH_4^+ in proportion to the amount of K^+ applied, as shown in Fig. 5.18. Continuous fertilization with K^+ may reduce the ability of soil to fix NH_4^+. The blocking effect of K^+ can be avoided, if desired, by applying the K^+ well in advance of the NH_4^+.

CONCENTRATION OF NH_4^+. The amount of NH_4^+ fixed by soil increases with rate of application. However, the percentage of the added NH_4^+ that is fixed decreases with an increase in the amount applied. Although fixation can occur over a prolonged period, most fixation occurs within a few hours.

MOISTURE CONDITIONS. Drying out of the soil after addition of NH_4^+ increases both the rate and magnitude of fixation. Vermiculite and illite can fix NH_4^+ under most conditions, but drying appears to be essential for fixation by montmorillonite. Alternate drying and wetting may be particularly effective in enhancing fixation. Freezing and thawing of the soil may influence NH_4^+ fixation in the same way as drying.

SOIL PH.   The relationship between soil acidity and NH_4^+ fixing capacity is not pronounced. However, fixation tends to increase slightly with increasing pH, such as through liming. Highly acidic soils (<pH 5.5) generally fix little NH_4^+.

Naturally Occurring Fixed NH_4^+

Virtually all soils that contain K^+-bearing silicate minerals (e.g., illite) will also contain naturally occurring fixed NH_4^+.[88]

Data given in Table 5.9 show that a wide range of values has been recorded for fixed NH_4^+ in soils. The surface layer often contains 200 µg/g or more of N as fixed NH_4^+, equivalent to 448 kg/ha, or 400 lb/acre. Clay and clay loam soils contain larger amounts of fixed NH_4^+ than silt loams, which in turn contain larger amounts than sandy soils. Spodosols contain rather low amounts of fixed NH_4^+, which can be attributed to their very low clay contents. Organic soils contain very little fixed NH_4^+.

The highest value thus far reported for fixed NH_4^+ in the surface layer of the soil appears to be that of Dalal,[89] who recorded a value of 1,300 µg/g in a Trinidad soil formed from micaceous schist and phyllite. Martin et al.[90] also recorded high values for fixed NH_4^+ in some subtropical soils derived from phyllite in Australia; in one soil, the content of fixed NH_4^+-N ranged from 415 µg/g in the surface layer to over 1,000 µg/g at a depth of 120 cm (4 ft).

The amount of N in the soil as fixed NH_4^+ on an acre-profile basis can be appreciable, as illustrated by Fig. 5.19. The soil volume occupied by plant roots may contain over 1,700 kg N/ha (1,520 lb/acre) of fixed NH_4^+.

Such factors as drainage, type of vegetative cover, and extent of leaching of the profile by percolating water have little effect on the fixed NH_4^+ content of the soil; the amounts are more closely related to the kinds and amounts of clay minerals that are present.[91] Regarding clay mineral type, fixed NH_4^+ content follows the order vermiculite > illite > montmorillonite > kaolinite.

Results obtained for the vertical distribution of fixed NH_4^+ in representative forest and prairie grassland soils of the United States are given in Table 5.10. The rather high levels in the lower horizons of some of the profiles can be explained by the fact that the parent material from which the soils were formed contain large amounts of illite.[91]

In contrast to the sharp decline in total N with depth, the soil's content of fixed NH_4^+ changes very little or increases. Thus the proportion of the soil N as fixed NH_4^+ increases with depth (see Table 5.10). In general, less than 10% of the N in the plow layer of the soil occurs as fixed NH_4^+, although

Table 5.9
Some Typical Values for Fixed NH_4^+ in Soils[a]

Location	Range, µg/g	Comments
Australia	41–1,076	From 221 to 1,076 µg/g (5–90% of N) in profiles developed on Permian phyllite and from 41 to 315 µg/g (5–82% of N) in soils formed on other parent materials
Canada		
Saskatchewan	158–330	From 7.7 to 13.3% of N in surface soil and up to 58.6% in subsoil; cultivation did not affect fixed NH_4^+ content
Alberta	110–370	From 7 to 14% of N in a wide variety of surface soils; percentage increased with depth
England	52–252	From 4 to 8% of N in surface soils and from 19 to 45% in subsoils
Nigeria	32–220	From 2 to 6% of N in surface layers and from 45 to 63% in the subsoil
Russia	14–490	From 2 to 7% of the N in surface soil but the percentage increases with depth
Sweden	10–17	Values are for a Spodosol profile low in clay
Taiwan	140–170	From 10.6 to 32.6% of the N in surface layer of nine soils
United States		
North Central	7–270	A wide range has been recorded, the lowest being in Spodosols and the highest in silt loams and soils rich in illite; from 4 to 8% of the N in the surface layer with the proportion increasing with depth
Pacific Northwest	17–138	From 1.1 to 6.2% of N in surface soils with the proportion increasing with depth in some soils but not others
Hawaii	0–585	Volcanic ash soils contained less (4–178 µg/g) than soils from basalt (up to 585 µg/g)

[a] For references see Young and Aldag.[88]

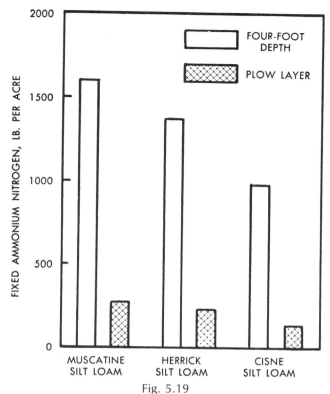

Fig. 5.19

Content of fixed NH_4^+-N in the plow layer and to a depth of 1.22 m (4 ft) in three agricultural soils of Illinois. (From Stevenson,[38] reproduced by permission of John Wiley and Sons.)

Table 5.10
Distribution of Fixed NH_4^+ in Some Soil Profiles

Horizon	Mollisols (5)[a]		Alfisols (5)		Others[b]	
	μg/g	% of N	μg/g	% of N[c]	μg/g	% of N
A_1	96.4–128.8	4.3– 5.6	112.6–154.6	5.6	57.4–140.6	3.5– 7.9
A_2	105.6–131.2	8.2– 9.3	109.2–168.0	9.4	50.4–154.0	7.8–10.8
B_2	99.4–155.4	11.5–18.4	142.8–210.0	20.1	54.6–182.0	12.7–17.6
B_3	103.6–173.6	21.4–27.7	184.8–224.0	26.5	84.0–210.0	13.1–17.0
C	98.0–187.6	28.4–44.7	102.2–224.0	34.9	89.6–210.0	27.2–36.1

[a] Numerals refer to number of profiles examined.
[b] Includes an Ultisol and two Mollisols with argillic horizons.
[c] For one profile only.

higher percentages are not uncommon. In some subsurface soils, over two-thirds of the N may occur as fixed NH_4^+. Narrowing of the C/N ratio with increasing depth in soil is due in part to fixed NH_4^+.

Changes have been observed in the fixed NH_4^+ content of the soil through long-time cropping, but they have been slight and variable. Notwithstanding, the proportion of the total N as fixed NH_4^+ tends to increase with cropping, the reason being that organic N is usually lost at a rapid rate when soils are placed under cultivation while fixed NH_4^+ levels change very little. Walsh and Murdock[92] concluded that even under the most advantageous cropping conditions very little native fixed NH_4^+ was available to crops, which was attributed to the blocking effect of K^+.

Fixation of NH_3 by Organic Matter

The chemical reaction of NH_3 with soil organic matter is frequently referred to as "NH_3 fixation," and this convention will be adopted herein. The term should not be confused with retention or "fixation" of the NH_4^+ ion by clay minerals (see previous section).

The ability of lignin and soil organic matter to react chemically with NH_3 has been known for more than 50 years, and numerous patents have been issued over this period for the conversion of peat, sawdust, lignaceous residues (e.g., corn cobs), and coal products into nitrogenous fertilizers by treatment with NH_3. Fixation is associated with oxidation (uptake of O_2) and is favored by an alkaline reaction. Thus the application of alkaline fertilizers to soil, such as aqueous or anhydrous NH_3, may result in considerable fixation. Injection of anhydrous NH_3, for example, results in a pronounced increase in soil pH, with the highest pH being along the injection line with a gradient extending outward from that line. Similarly, the highest concentration of NH_3 will be found in the injection zone. These conditions are highly favorable for NH_3 fixation by organic matter.

The relative importance of clay and organic colloids in retaining NH_3 will depend upon the nature of the soil, but at pH values above 7.0, the organic fraction appears to be more reactive in relation to the amount present than is the clay.

While greatest attention has been given to fixation of fertilizer NH_3, the possibility should not be overlooked that the reaction occurs under natural soil conditions. Grassland soils have higher N contents and lower C/N ratios than forest soils, and the suggestion has been made that conditions in the former (e.g., higher pH's) are more conducive to NH_3 fixation and thereby enrichment of the organic matter with N.[93]

For complete coverage of NH_3 fixation by organic matter, the reader is directed to several reviews on the subject.[85,94,95]

Mechanisms of Fixation

Very little is known concerning the NH_3 fixation reaction, although several plausible mechanisms have been proposed. These are based on the obser-

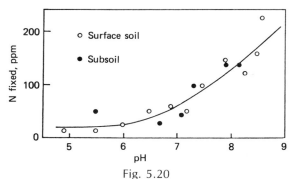

Fig. 5.20
Effect of pH on NH₃ fixation by a muck soil. (From Broadbent et al.[95])

vation that fixation proceeds most favorably at high pH values (Fig. 5.20) and that fixation is accompanied by the uptake of O_2 (Fig. 5.21). The reaction occurs rapidly under optimum conditions, but fixation can continue for a prolonged period, though at a diminishing rate. Burge and Broadbent[96] estimated the NH_3-fixing capacity of a series of organic soils and concluded that, under aerobic conditions, one molecule of NH_3 was fixed for every 29 C atoms. As one might expect, NH_3 fixation bears a close relationship to the organic matter content of the soil (Fig. 5.22).

Flaig[97] suggested a mechanism for NH_3 fixation by phenolic compounds

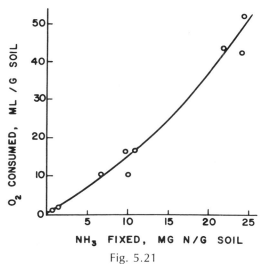

Fig. 5.21
Relationship between NH_3 fixation and O_2 consumption.[94] [Reproduced from *Agricultural Anhydrous Ammonia* (1966) by permission of the American Society of Agronomy.]

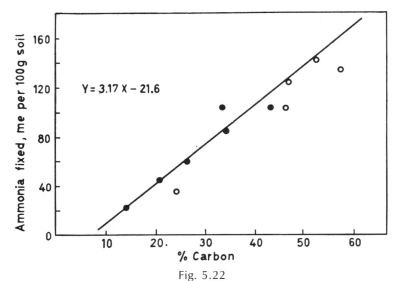

Fig. 5.22

Relationship between NH_3 fixation and the C content of several soils high in organic matter content. (Adapted from Burge and Broadbent.[96])

in which polymers are formed containing N in heterocyclic linkages. The initial step was believed to involve O_2 consumption by the phenol to form a quinone, which subsequently reacted with NH_3 to form a complex polymer. Catechol (XII), for example, was believed to be converted by the action of O_2 under alkaline conditions to the o-quinone (XIII), which was then hydrated to form benzentriol (XIV). Further oxidation was postulated to produce a mixture of o-hydroquinone (XV) and p-hydroxy-o-quinone (XVI), both capable of reacting with NH_3.

The incorporation of NH_3 into p-hydroxy-o-quinone (XVI) was believed to produce structures of the types represented by XVII and XVIII.

Other possible types of structures are of the following type.

XIX

An interesting feature of structure XIX is that a relatively small amount of substrate will fix a considerable amount of NH_3. As yet there is no sound basis for selecting any one structure as the correct one or for assuming that only one mechanism is involved. Fixation may involve the incorporation of NH_3 into a wide variety of compounds, including some of a refractory nature (such as lignin). Also, NH_3 may participate in polymerization or polycondensation reactions of small molecules to form polymers with properties similar to natural humic substances.

Availability of Chemically Fixed NH₃ to Plants

An important question with respect to NH_3 fixed by organic matter is whether the N is available to plants. Most research indicates that availability is relatively low.[85,86,94] In a greenhouse experiment using ^{15}N-labeled NH_3, Burge and Broadbent[96] observed that over 95% of the fixed NH_3 was no more available to plants than the indigenous soil N. It should be pointed out that the application of anhydrous (or aqueous) NH_3 to soil may result in the solubilization of organic matter because of alkaline conditions that are created, thereby enhancing the availability of native humus N to plants and microorganisms.

NATURAL VARIATIONS IN ^{15}N ABUNDANCE

Slight variations occur in the N isotope composition of soil. These variations result from isotopic effects during biochemical and chemical transformations, such as nitrification[99] and denitrification.[100] The overall effect of these isotope effects in soil is a slight increase in the average ^{15}N content of soil N and its fractions, as compared to atmospheric N_2.

Natural variations in isotopic abundance are usually expressed in terms of the per mil excess ^{15}N, or delta ^{15}N (δ^{15}N). The equation is

$$\delta^{15}N = \frac{atom\ \%\ ^{15}N\ in\ sample\ -\ atom\ \%\ ^{15}N\ in\ standard}{atom\ \%\ ^{15}N\ in\ standard} \times 1,000$$

Thus a δ^{15}N value of $+10$ indicates that the experimental sample is enriched by 1% compared with the atom % ^{15}N of the standard. A negative value indicates that the sample is depleted in ^{15}N relative to the standard. The δ^{15}N value of the soil generally falls within the range of $+5$ to $+12$, although higher and lower values are by no means rare.[101,102] For any given soil, variations exist in the $\delta^{15}N$ of the various N fractions.

Natural variations in ^{15}N content have been used to determine the fate of fertilizer N in soils, including consumption by plants and movement within the soil and into surface waters.[103-108] The approach has also been used to obtain evidence for N_2 fixation by plants and transfer of legume-fixed N_2 to nonlegumes.

Kohl et al.[108] used natural variations in ^{15}N content to estimate the relative contribution of soil and fertilizer N to the NO_3^- in surface waters, the basis being that the ^{15}N enrichment of soil-derived NO_3^- will be higher than that of NO_3^- originating from the fertilizer. Since fertilizer N is obtained by chemical fixation of atmospheric N_2, it is usually depleted in ^{15}N as compared to the soil N. The conclusions of Kohl et al.[108] regarding the contribution of fertilizer N to surface waters generated considerable interest and were challenged by some soil scientists;[109] for a response to these criticisms, the reader is referred to Kohl et al.[110] One problem in evaluating the contribution from the soil is that the ^{15}N content of mineralized N derived from humus is not constant and can be lower than that for the total soil N. Data obtained by Feigin et al.[111] show that the ^{15}N content of soil-derived NO_3^- increases with incubation time; long-term incubation often gave $\delta^{15}N$ values twice as large as those measured after a short incubation. In evaluating the relative contribution of soil and fertilizer N to the NO_3^- in drainage waters, the question naturally arises as to the proper $\delta^{15}N$ value to select for soil-derived NO_3^-.

Measurements of natural ^{15}N abundance require high precision and are subject to variable and significant errors due to variations in the ^{15}N content of different organic and inorganic N components of the soil. Bremner and Hauck[44] concluded that the many sources of error in sample collection and processing, together with uncertainties in data interpretation, militate against use of variations in ^{15}N abundance as a means of obtaining reliable information about N cycle processes in complex biological systems.

MODELING OF THE SOIL N CYCLE

Modeling represents an attempt to describe in mathematical terms the dynamic aspects of the soil N cycle.[112-119] Many models are *simulation* models in that they attempt to forecast how a system will behave or perform without actually using the physical system or its prototype. *Mathematical* models, on the other hand, utilize empirical or observational data to provide quantitative values for gains, losses, and transfers of N, as well as for the amounts of N contained in one or more pools as a function of time. Models can be local, regional, or global in scope.

Mathematical models are of three main types, as follows:[115]

Stochastic   This type is based on the assumption that the processes to be modeled obey the laws of probability.

Empirical Based on observational data. Input and output processes are expressed in terms of regression equations.
Mechanistic Based on well-established physical, chemical, and biological laws that describe the various processes.

Mechanistic models are more versatile than other types in that historical data are not required for their development. However a complete understanding of the system is required. Davidson et al.[115] have tabulated rate coefficients for mineralization, immobilization, nitrification, and denitrification in soil, all of which are necessary for mechanistic models.

One major objective of modeling is to provide quantitative data for the fate of fertilizer N in soil. Appraisals for plant uptake and NO_3^- losses through leaching or denitrification involve consideration of the many sources and sinks of N, as well as flow pathways of both NO_3^- and water. Other objectives of modeling are (1) to obtain a better understanding and increased insight into complex problems, (2) to evaluate available information and ascertain the adequacy of published data for problem evaluation, (3) to test existing as well as new concepts and hypotheses, (4) to estimate by difference an unknown output or input, (5) to obtain a better evaluation or prediction of an observed phenomena, (6) to identify research needs, and (7) to help develop guidelines for best management practices.[112]

Models are usually illustrated by a flow diagram showing one or more pools of N with inputs and/or outputs for each. Values are assigned to pool sizes and rates for inputs and outputs. The various models differ greatly in complexity, ranging from those designed to similate a single process, such as leaching or denitrification, to those that include transformations occurring within the soil. Finally, there are models that involve interactions with other components of the ecosystem.

The degree of sophistication of any particular model is determined by the understanding of the system to be modeled, the availability of reliable data to serve as inputs, and the intended application.[115] Those models involving transformations within the soil usually include mineralization, immobilization, nitrification, denitrification, and leaching, although one or more of these are sometimes ignored. The more complex models take into account differences in the decomposition rate of various constituents of plant residues (proteins, sugars, cellulose, and lignin), as illustrated in Fig. 5.23. Molina et al.[119] have developed and calibrated a model for the short-term dynamics of organic N, NH_4^+, and NO_3^- in which the soil organic phase is divided into two pools that decompose independently of each other. Each pool contained a labile and resistant component.

The absence of reliable data for modeling is a major obstacle in most modeling studies. Most processes are transient; some occur simultaneously. Some pools and fluxes can be evaluated with a reasonable degree of accuracy whereas others cannot. A major difficulty with models that include mineralization of the soil organic N is that this pool is very large compared to the

Table 5.11

Some Modeling Approaches for Soil N Transformations

Reference	Modeling Approaches[a]
Davidson et al.[115]	Nitrification, mineralization, and immobilization by first-order kinetics and modified by soil water pressure head; denitrification by first-order kinetics considering pressure head; denitrification by first order kinetics considering pressure head, water content, and organic matter content; NH_4^+ sorption by linear partition model; NH_4^+ and NO_3^- uptake by Michaelis–Menten kinetics.
Dutt et al.[116]	Nitrification, mineralization–immobilization, and urea hydrolysis modeled by regression equation; NH_4^+ sorption by equilibrium cation exchange equation; N plant uptake proportional to water uptake.
Frissel and van Veen[113]	Growth of *Nitrosomonas* and *Nitrobacter* by Michaelis–Menten kinetics including considerations of O_2 level; denitrification by a physical–biological model involving O_2 diffusion; mineralization–immobilization considers organic matter grouped into freshly applied organic matter such as animal manures, straw and waste water, and resistant biomass residues; NH_3 volatilization by physiochemical model including dissociation of NH_4OH; NH_4^+–clay fixation by reversible first-order kinetics.
Mehran and Tanji[117]	Nitrification, denitrification, mineralization, immobilization, and N plant uptake by irreversible first-order kinetics; NH_4^+ exchange by reversible first-order kinetics.
Molina et al.[119]	Computes short-term dynamics of organic N, NH_4^+, and NO_3^- that result from the process of residue decomposition, mineralization, immobilization, nitrification, and denitrification. Organic pools include the residues and two pools of the soil organic fraction, each containing a labile and resistant component.

[a] Derived in part from Tanji.[112]

other pools and a slight mistake in estimating the mineralization rate introduces a major error in estimates for the amount of mineral N available to the plant. When processes occurring within the soil cannot be evaluated, a "black box" approach to modeling is sometimes used (see Fig. 5.17).

Some selected modeling approaches for soil N transformations are listed in Table 5.11. Data used for the various transformations (e.g., nitrification,

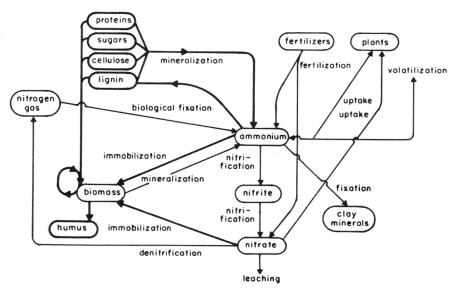

Fig. 5.23

Soil N transformations and components of crop residues and soil organic matter. (From Frissel and van Veen,[113] reproduced by permission of Academic Press.)

mineralization, immobilization, denitrification, plant assimilation, etc.) are modeled in different ways: multiple regression, zero and first-order chemical kinetics, and Michaelis–Menten kinetics. A detailed discussion of the various approaches is beyond the scope of the present text and the reader is referred to several reviews for more detailed information.[112,114,115]

SUMMARY

Nitrogen undergoes a wide variety of transformations in soil, most of which involve the organic fraction. Although considered individually, each process is affected by others occurring sequentially; in some cases, opposing processes operate simultaneously.

An internal "N cycle" exists in soil apart from the overall cycle of N in nature. Even if N gains and losses are equal, such as may occur in a mature ecosystem, the "N cycle" is not static. Continuous turnover of N occurs through mineralization–immobilization, with incorporation of N into microbial tissue, the so-called active fraction of organic matter. Whereas much of the newly immobilized N is recycled through mineralization, some is converted to stable humus forms. The manifestation of mineralization–immobilization turnover has not been fully understood and has caused considerable confusion regarding the interpretation of tracer data on soil N transformations.

Fixation of NH_4^+ by clay minerals is of some importance in select soils and significant amounts of the soil N often occurs in the form of naturally occurring fixed NH_4^+. Ammonia fixation by organic matter will be particularly important when urea or anhydrous NH_3 is applied to soils rich in organic matter, an effect that has not been fully investigated under field conditions.

Data obtained using the stable isotope ^{15}N have shown that only a portion of the fertilizer N applied to soil, estimated at from 30 to 70%, is consumed by plants and that a significant fraction (to 40%) is retained in the organic matter after the first growing season. This residual N is relatively unavailable to plants during the second growing season, and availability decreases even further in subsequent years because of conversion to stable humus forms. A similar effect has been noted for the N of crop residues. The value of keeping a rotating fund of decomposable organic matter in the soil through the frequent and periodic return of crop residues (including legumes) is due, to a large extent, to their effect in maintaining or increasing organic matter levels, thereby insuring a continuous supply of N through mineralization–immobilization turnover.

Some goals for research on N transformations in soil are as follows:

1 Develop methods for distinguishing between N in the various soil N pools.
   a. Biomass.
   b. Active fraction of humus.
   c. Stable fraction of humus.
2 Define and quantify reaction mechanisms whereby N becomes stabilized.
3 Establish the relationship between mineralization–immobilization turnover and gaseous loss of soil and fertilizer N.
4 Quantify the relationship between soil organic and inorganic N pools as influenced by soil type, cropping system, climate and residue management practice.
5 Determine the long-time fate of immobilized N under different crop management practices and climates.
6 Devise management strategies.
   a. For effective use of biologically and chemically fixed N.
   b. For efficient use of N from decomposing plant and animal residues under conditions existing in the field.
   c. To maintain and efficiently utilize N resources of the soil for efficient crop production.

REFERENCES

1 S. L. Jansson and J. Persson, "Mineralization and Immobilization of Soil Nitrogen," in F. J. Stevenson, Ed., *Nitrogen in Agricultural Soils,* American Society of Agronomy, Madison, Wis., 1982, pp. 229–252.

2 J. N. Ladd and R. B. Jackson, "Biochemistry of Ammonification," in F. J. Stevenson, Ed., *Nitrogen in Agricultural Soils,* American Society of Agronomy, Madison, Wis., 1982, pp. 173–228.

3 S. Winogradsky, *Ann. Inst. Pasteur* **4,** 213, 257, 760 (1890).

4 L. W. Belser, *Ann. Rev. Microbiol.* **33,** 309 (1979).

5 D. D. Focht and W. Verstraete, *Adv. Microbiol. Ecol.* **1,** 135 (1977).

6 H. A. Painter, *Prog. Water Tech.* **8,** 3 (1977).

7 E. L. Schmidt, "Nitrification in Soil," in F. J. Stevenson, Ed., *Nitrogen in Agricultural Soils,* American Society of Agronomy, Madison, Wis., 1982, pp. 253–288.

8 S. W. Watson, "Gram-negative Chemolithrotrophic Bacteria. Family 1," in R. E. Buchanan and N. E. Gibbons, Eds., *Bergey's Manual of Determinative Bacteriology,* 8th ed., Williams and Wilkins, Baltimore, 1974, pp. 450–456.

9 J. M. Bremner and A. M. Blackmer, *Science* **199,** 295 (1978).

10 A. M. Blackmer, J. M. Bremner, and E. L. Schmidt, *Appl. Environ. Microbiol.* **40,** 1060 (1980).

11 C. A. I. Goring and D. A. Laskowski, "The Effects of Pesticides on Nitrogen Transformations in Soils," in F. J. Stevenson, Ed., *Nitrogen in Agricultural Soils,* American Society of Agronomy, Madison, Wis., 1982, pp. 689–720.

12 F. J. Stevenson, "Soil Nitrogen," in V. Sauchelli, Ed., *Fertilizer Nitrogen,* Amer. Chem. Soc. Monograph Series No. 161, Reinhold, New York, 1964, pp. 18–39.

13 S. L. Jansson, "Use of ^{15}N in Studies on Soil Nitrogen," in A. D. McLaren and J. Skujins, Eds., *Soil Biochemistry,* Vol. 2, Dekker, New York, pp. 129–166.

14 M. J. Frissel and J. A. van Veen, "A Critique of Computer Simulation Modeling for Nitrogen in Irrigated Croplands," in D. R. Nielsen and J. G. MacDonald, Eds., *Nitrogen in the Environment,* Vol. I, Academic Press, New York, 1978, pp. 145–162.

15 H. F. Birch, *Plant Soil* **12,** 81 (1960).

16 A. S. Agarwal, B. R. Singh, and Y. Kanehiro, *Soil Sci. Soc. Amer. Proc.* **35,** 96, 455 (1971).

17 C. A. Campbell, "Soil Organic Carbon, Nitrogen, and Fertility," in M. Schnitzer and S. U. Khan, Eds., *Soil Organic Matter,* Elsevier, New York, 1978, pp. 173–271.

18 F. J. Stevenson, Ed., *Nitrogen in Agricultural Soils,* American Society of Agronomy, Madison, Wis., 1982.

19 G. W. Harmsen and D. A. van Schreven, *Adv. Agron.* **7,** 299 (1955).

20 G. W. Harmsen and G. J. Kolenbrander, "Soil Inorganic Nitrogen," in W. V. Bartholomew and F. E. Clark, Eds., *Soil Nitrogen,* American Society of Agronomy, Madison, Wisc., 1965, pp. 43–92.

21 A. D. Rovira and B. M. McDougall, "Microbiological and Biochemical Aspects of the Rhizosphere," in A. D. McLaren and G. H. Petersen, Eds., *Soil Biochemistry,* Dekker, New York, 1967, pp. 417–463.

22 F. J. Stevenson and G. H. Wagner, "Chemistry of Nitrogen in Soils," in T. L. Willrich and G. E. Smith, Eds., *Agricultural Practices and Water Quality,* Iowa State University Press, Ames, 1970, pp. 125–141.

23 B. A. Stewart, F. G. Viets Jr., G. L. Hutchinson, and W. D. Kemper, *Environ. Sci. Tech.* **1,** 736 (1967).

24 D.R. Keeney, "Nitrogen-Availability Indices," in A. L. Page et al., Eds., *Methods of Soil Analysis,* Part 2, 2nd ed., American Society of Agronomy, Madison, Wis., 1982, pp. 711–733.

25 G. Stanford, "Assessment of Soil Nitrogen Availability," in F. J. Stevenson, Ed., *Nitrogen in Agricultural Soils,* American Society of Agronomy, Madison, Wis., 1982, pp. 651–688.

26 G. Stanford and S. J. Smith, *Soil Sci. Soc. Amer. Proc.* **36,** 465 (1972).

27 G. Stanford, J. N. Carter, and S. J. Smith, *Soil Sci. Soc. Amer. Proc.* **38,** 99 (1974).

28 S. J. Smith and G. Stanford, *Soil Sci.* **111,** 228 (1971).

29 G. Stanford and S. J. Smith, *Soil Sci.* **122,** 71 (1976).

30 R. D. Hauck and J. M. Bremner, *Adv. Agron.* **28,** 219 (1976).

31 S. L. Jansson, *Soil Sci.* **95,** 31 (1963).

32 J. O. Legg and J. J. Meisinger, "Soil Nitrogen Budgets," in F. J. Stevenson, Ed., *Nitrogen in Agricultural Soils,* American Society of Agronomy, Madison, Wis., 1982, pp. 503–566.

33 F. E. Broadbent, *Agron. J.* **72,** 325 (1980).

34 J. O. Legg and F. E. Allison, *Soil Sci. Soc. Amer. Proc.* **31,** 403 (1967).

35 E. A. Paul and N. G. Juma, "Mineralization and Immobilization of Soil Nitrogen by Microorganisms," in F. E. Clark and T. Rosswall, Eds. *Terrestrial Nitrogen Cycles: Processes, Ecosystem Strategies and Management Impacts,* Ecol. Bull. 33, Stockholm, 1981, pp. 179–195.

36 Ye. V. Rudelov, *Soviet Soil Sci.* **14,** 40 (1982).

37 R. L. Westerman and L. T. Kurtz, *Soil Sci. Soc. Amer. Proc.* **36,** 91 (1972).

38 F. J. Stevenson, *Humus Chemistry: Genesis, Composition, Reactions,* Wiley-Interscience, New York, 1982.

39 F. J. Stevenson, "Organic Forms of Soil Nitrogen," in F. J. Stevenson, Ed., *Nitrogen in Agricultural Soils,* American Society of Agronomy, Madison, Wis., 1982, pp. 67–122.

40 J. M. Bremner, "Organic Forms of Nitrogen," in C. A. Black, Ed., *Methods of Soil Analysis,* Part 2, American Society of Agronomy, Madison, Wis., 1965, pp. 1238–1255.

41. F. J. Sowden, Y. Chen, and M. Schnitzer, *Geochim. Cosmochim. Acta* **41,** 1524 (1977).

42 D. R. Keeney and J. M. Bremner, *Soil Sci. Soc. Amer. Proc.* **28,** 653 (1964).

43 V. W. Meints and G. A. Peterson, *Soil Sci.* **124,** 334 (1977).

44 J. M. Bremner and R. D. Hauck, "Advances in Methodology for Research in Nitrogen Transformations in Soils," in F. J. Stevenson, Ed., *Nitrogen in Agricultural Soils,* American Society of Agronomy, Madison, Wis., 1982, pp. 467–502.

45 F. E. Allison, "Evaluation of Incoming and Outgoing Processes that Affect Soil Nitrogen," in W. V. Bartholomew and F. E. Clark, Eds., *Soil Nitrogen,* American Society of Agronomy, Madison, Wis., 1965, pp. 573–606.

46 R. D. Hauck, "Quantitative Estimates of Nitrogen-Cycle Processes: Concepts and Review," in *Nitrogen-15 in Soil Plant Studies.* Proc. Research Coordination Meeting, Sofia, Bulgaria, International Atomic Energy Agency, Vienna, 1971, pp. 65–80.

47 P. Kundler, *Albrecht-Thaer-Arch.* **14,** 190 (1970).

48 F. E. Broadbent and A. B. Carlton, "Field Trials with Isotopically Labeled Nitrogen Fertilizer," in D. R. Nielsen and J. G. MacDonald, Eds., *Nitrogen in the Environment,* Vol. 1, Academic Press, New York, 1978, pp. 1–41.

49 K. K. Tanji, M. Fried, and R. M. Van De Pol, *J. Environ. Qual.* **6,** 155 (1979).

50 J. O. Legg and F. E. Allison, *Trans. 7th Intern. Congr. Soil Sci. (Madison)* **II,** 545 (1960).

51 F. E. Clark, "A Re-evaluation of Microbial Concepts Concerning Nitrogen Transformations in the Soil," in *Soil and Fertilizer Nitrogen Research: A Projection into the Future,* Tennessee Valley Authority, Muscle Shoals, Alabama, 1964, pp. 18–23.

52 J. P. E. Anderson and K. H. Domsch, *Soil Biol. Biochem.* **10,** 215 (1978).

53 J. P. E. Anderson and K. H. Domsch, *Soil Sci.* **130,** 211 (1980).

54 D. S. Jenkinson, *Soil Biol. Biochem.* **8,** 203 (1976).

55 D. S. Jenkinson and D. S. Powlson, *Soil Biol. Biochem.* **8,** 209 (1976).

56 J. N. Ladd, M. Amato and J. W. Parsons, "Studies of Nitrogen Immobilization and Mineralization in Calcareous Soils," in *Soil Organic Matter Studies,* International Atomic Energy Agency, Vienna, Austria, 1977, pp. 301–311.

57 M. Amato and J. N. Ladd, *Soil Biol. Biochem.* **12,** 405 (1980).

58 A. Faegri, V. L. Torsvik, and J. Goksöyr, *Soil Biol. Biochem.* **9,** 105 (1977).

59 D. S. Jenkinson and J. N. Ladd, "Microbial Biomass in Soil—Measurement and Turnover," in E. A. Paul and J. N. Ladd, Eds. *Soil Biochemistry,* Vol. 5, Dekker, New York, 1981, pp. 415–417.

60 J. N. Ladd, J. M. Oades and M. Amato, *Soil Biol. Biochem.* **13,** 119 (1981).

61 J. M. Oades and D. S. Jenkinson, *Soil Biol. Biochem.* **11,** 201 (1979).

62 H. Kai, Z. Ahmad, and T. Harada, *Soil Sci. Plant Nutr.* **19,** 275 (1973).

63 E. A. Paul and J. A. van Veen, *Trans. 11th Intern. Congr. Soil Sci. (Edmonton)* **III,** 61 (1978).

64 F. W. Chichester, J. O. Legg, and G. Stanford, *Soil Sci.* **120,** 455 (1975).

65 T. Marumoto, H. Kai, T. Yoshida, and T. Harada, *Soil Sci. Plant Nutr.* **23,** 9, 23 (1977).

66 T. Marumoto, J. P. E. Anderson, and K. H. Domsch, *Soil Biol. Biochem.* **14,** 461, 469 (1982).

67 R. Hayashi and T. Harada, *Soil Sci. Plant Nutr.* **15,** 226 (1969).

68 K. R. Kelley and F. J. Stevenson, *Soil Biol. Biochem.* **17,** 517 (1985).

69 N. G. Juma and E. A. Paul, *Soil Sci. Soc. Amer. J.* **48,** 76 (1984).

70 F. E. Broadbent and T. Nakashima, *Soil Sci. Soc. Amer. Proc.* **31,** 648 (1967).

71 F. E. Broadbent and T. Nakashima, *Soil Sci. Soc. Amer. Proc.* **38,** 313 (1974).

72 J. O. Legg, F. W. Chichester, G. Stanford, and W. H. DeMar, *Soil Sci. Soc. Amer. Proc.* **35,** 273 (1971).

73 F. E. Broadbent and T. Nakashima, *Soil Sci. Soc. Amer. Proc.* **29,** 55 (1965).

74 G. Stanford, J. O. Legg, and F. W. Chichester, *Plant and Soil* **33,** 425 (1970).

75 J. N. Ladd, *Plant and Soil* **58,** 401 (1981).

76 K. R. Reddy and W. H. Patrick, *Soil Sci. Soc. Amer. J.* **42,** 316 (1978).

77 A. L. Allen, F. J. Stevenson, and L. T. Kurtz, *J. Environ. Qual.* **2,** 120 (1973).

78 S. J. Smith, F. W. Chichester, and D. E. Kissel, *Soil Sci.* **125,** 165 (1978).

79 D. Wojcik-Wojtkowiak, *Plant and Soil* **49,** 49 (1978).

80 W. B. McGill and E. A. Paul, *Can. J. Soil Sci.* **56,** 203 (1976).

81 W. B. McGill, J. A. Shields, and E. A. Paul, *Soil Biol. Biochem.* **7,** 57 (1975).

82 F. J. Stevenson, *Trans. 12th Intern. Congr. Soil Sci., New Delhi, Symposia Papers* **1,** 137 (1982).

83 R. A. Olson and L. T. Kurtz, "Crop Nitrogen Requirements, Utilization, and Fertilization," in F. J. Stevenson, Ed., *Nitrogen in Agricultural Soils,* American Society of Agronomy, Madison, Wis., 1982, pp. 567–604.

84 J. F. Parr, *J. Environ. Qual.* **2,** 75 (1973).

85 H. Nommik, "Ammonium Fixation and Other Reactions Involving a Nonenzymatic Immobilization of Mineral Nitrogen in Soil," in W. V. Bartholomew and F. E. Clark, Eds., *Soil Nitrogen,* American Society of Agronomy, Madison, Wis., 1965, pp. 198–258.

86 H. Nommik and K. Vahtras, "Retention and Fixation of Ammonium in Soils," in F. J.

Stevenson, Ed., *Nitrogen in Agricultural Soils*, American Society of Agronomy, Madison, Wis., 1982, pp. 123–171.

87   H. Nommik, *Acta Agric. Scand.* **7**, 395 (1957).

88   J. L. Young and R. W. Aldag, "Inorganic Forms of Nitrogen in Soil," in F. J.Stevenson, Ed., *Nitrogen in Agricultural Soils,* American Society of Agronomy, Madison, Wis., 1982, pp. 43–66.

89   R. C. Dalal, *Soil Sci.* **124**, 323 (1977).

90   A. E. Martin, R. J. Gilkes and J. O. Skjemstad, *Aust. J. Soil Res.* **8**, 71 (1970).

91   F. J. Stevenson and A. P. S. Dhariwal, *Soil Sci. Soc. Amer. Proc.* **23**, 121 (1959).

92   L. M. Walsh and J. T. Murdock, *Soil Sci.* **89**, 183 (1960).

93   S. Mattson and E. Koutler-Andersson, *Lantbruks, Hogskol. Ann.* **11**, 107 (1943).

94   F. E. Broadbent and F. J. Stevenson, "Organic Matter Reactions," in M. H. McVickar et al., Eds., *Agriculture Anhydrous Ammonia,* American Society of Agronomy, Madison, Wis., 1966, pp. 169–187.

95   F. E. Broadbent, W. D. Burge, and T. Nakashima, *Trans. 7th Intern. Congr. Soil Sci. (Madison)* **2**, 509 (1960).

96   W. D. Burge and F. E. Broadbent, *Soil Sci. Soc. Amer. Proc.* **25**, 199 (1961).

97   W. Flaig, *Z. Pflanzenähr. Dung. Bodenk.* **51**, 193 (1950).

98   M. R. Lindbeck and J. L. Young, *Anal Chim Acta* **32**, 73 (1965).

99   C. C. Delwiche and P. L. Steyn, *Environ. Sci. Tech.* **4**, 929 (1970).

100   A. M. Backmer and J. M. Bremner, *Soil Biol. Biochem.* **9**, 73 (1977).

101   D. A. Rennie, E. A. Paul, and L. E. Johns, *Can. J. Soil Sci.* **56**, 43 (1976).

102   G. Shearer, D. H. Kohl, and S-H. Chien, *Soil Sci. Soc. Amer. Proc.* **42**, 899 (1978).

103   R. E. Karamanos and D. A. Rennie, *Soil Sci. Soc. Amer. J.* **44**, 57 (1980).

104   D. H. Kohl, G. B. Shearer, and B. Commoner, *Soil Sci. Soc. Amer. Proc.* **37**, 888 (1973).

105   V. W. Meints, L. V. Boone, and L. T. Kurtz, *J. Environ. Qual.* **4**, 486 (1975).

106   R. E. Karamanos and D. A. Rennie, *Can. J. Soil Sci.* **61**, 553 (1981).

107   G. Shearer and J. O. Legg, *Soil Sci. Soc. Amer. Proc.* **39**, 896 (1975).

108   D. H. Kohl, G. B. Shearer, and B. Commoner, *Science* **174**, 1331 (1971).

109   R. D. Hauck et al., *Science* **177**, 453 (1972).

110   D. H. Kohl, G. B. Shearer, and B. Commoner, *Science* **177**, 454 (1972).

111   A. Feigin, D. H. Kohl, G. Shearer, and B. Commoner, *Soil Sci. Soc. Amer. Proc.* **38**, 90 (1974).

112   K. K. Tanji, "Modeling of the Soil Nitrogen Cycle," in F. J. Stevenson, Ed., *Nitrogen in Agricultural Soils,* American Society of Agronomy, Madison, Wis., 1982, pp. 721–772.

113   M. J. Frissel and J. A. van Veen, "A Critique of Computer Simulation Modeling for Nitrogen in Irrigated Croplands," in D. R. Nielsen and J. G. MacDonald, Eds., *Nitrogen in the Environment,* Vol. 1, Academic Press, New York, 1978, pp. 145–162.

114   M. J. Frissel, Ed., *Cycling of Mineral Nutrients in Agricultural Ecosystems,* Elsevier, New York, 1978.

115   J. B. Davidson, D. A. Graetz, S. C. Rao, and H. M. Selim, *Simulation of Nitrogen Movement, Transformation, and Uptake in Plant Root Zone,* U.S. Environmental Protection Agency, Athens, Georgia, 1978.

116   R. G. Dutt, M. J. Shaffer, and W. J. Moore, *Computer Simulation Model of Dynamic Bio-physicochemical Processes in Soils,* Arizona Agric. Exp. Sta. Tech. Bull. 196, 1972.

117   M. Mehran and K. K. Tanji, *J. Environ. Qual.* **3**, 391 (1974).

118 K. K. Tanji and S. K. Gupta, "Computer Simulation Modeling for Nitrogen in Irrigated Cropland," in D. R. Nielsen and J. G. MacDonald, Eds., *Nitrogen in the Environment,* Vol. 1, Academic Press, New York, 1978, pp. 79–130.

119 J. A. E. Molina, C. E. Clapp, M. J. Shaffer, F. W. Chichester, and W. E. Larson, *Soil Sci. Soc. Amer. J.* **47,** 85 (1983).

SIX

IMPACT OF NITROGEN ON
HEALTH AND THE ENVIRONMENT

Concern regarding the integrity of the soil N cycle has a multiple basis that includes ecological and health hazards associated with the presence of NO_3^- in natural waters, excess NO_3^- and NO_2^- in human food and animal feed, and the possible adverse effect of soil and fertilizer N in increasing the concentration of nitrous oxide (N_2O) in the atmosphere, thereby contributing to atmospheric ozone (O_3) depletion. The possible effect of N on promoting the unwanted growth of algae in lakes and streams (eutrophication) has been debated for decades while concern over depletion of the O_3 layer is a rather recent development. Potential adverse health and environmental effects of N are listed in Table 6.1.

The well-publicized need for using greater amounts of fertilizer N for worldwide production of food and fiber is, to some extent, in conflict with a need for controlling levels of NO_3^- in ground waters, and possibly N_2O in the atmosphere.

NITRATES IN WATER AND FOOD

The problem of NO_3^- in food and water arises from the fact that, when consumed in large amounts by animals and humans, NO_3^- has the potential for causing a health problem called methemoglobinemia, a condition that results in impairment of O_2 transport in the blood caused by the presence of NO_2^-, or indirectly from NO_3^-. Nitrate itself is relatively nontoxic. The reaction of NO_2^- with hemoglobin to produce methemoglobin is of little consequence in adults but, although rare, can be fatal in infants. Nitrate poisoning can also affect cattle, particularly ruminants, where microorganisms in the rumen reduce NO_3^- to NO_2^-. The NO_3^- problem as it relates to animal and human health has been covered in several reviews.[2-7]

With regard to the problem of NO_3^- poisoning, U.S. Public Health Service Standards specify that the NO_3^-–N content of drinking water should not exceed 10 mg/liter (ppm), or 45 mg/liter on a NO_3^- basis. Some streams and reservoirs that are being used as sources of city water seem to be approaching

Table 6.1
Potential Adverse Health and Environmental Impacts of N[a]

Impact	Causative Agents
Human health	
Methemoglobinemia in infants	Excess NO_3^- and NO_2^- in water and food.
Cancer	Nitrosamines from NO_2^- and secondary amines
Respiratory illness	Peroxyacyl nitrates, alkyl nitrates, NO_3^- aerosols, NO_2^-, and HNO_3 vapor in urban atmospheres
Animal health	
Loss of livestock	Excess NO_3^- in feed and water
Plant growth	
Stunted growth	High levels of NO_2^- in soil
Excessive growth	Excess available N
Environmental quality	
Eutrophication	Inorganic and organic N in surface waters
Stratospheric ozone depletion	Nitrous oxide from nitrification, denitrification, and stack emissions
Materials and ecosystem damage	HNO_3 aerosols in rainfall

[a] Adapted from Keeney.[1]

this value. Farmers using well water need to be particularly aware of the potential danger of high NO_3^- levels, from the standpoint of poisoning of both infants and cattle. It has long been known that well water in many localities of the United States contains NO_3^- in excess of 100 mg/liter. The Public Health Service has also established NO_3^- limits for certain foods, as well as meat and fish products.

Agricultural soils, particularly overfertilized ones, are known sources of NO_3^- in natural waters. However, soils are not the sole culprit. In many instances the soils' contribution is of minor importance as compared to other sources, which include municipal and rural sewage, feedlots or barnyards, food processing wastes, septic tank effuents, natural NO_3^- accumulations (caliche of semiarid regions), sanitary facilities of recreational areas, landfills, miscellaneous industrial wastes, and biological N_2 fixation.

The relative contribution of each source will depend on conditions existing at the particular location or ecosystem under consideration. The contribution from municipal sewage can be appreciable when raw or digested sewage is discharged directly into lakes or streams, as has frequently been done in the past. To use one example, domestic sewage accounts for about one-half of the N loading of San Francisco Bay and its tributaries; fertilizer N from irrigated agricultural land accounts for less than 2%.[5]

Methemoglobinemia

Nitrate becomes toxic (i.e., causes methemoglobinemia) only under conditions in which the NO_3^- is reduced to NO_2^- by microorganisms. The NO_2^- is absorbed into the bloodstream where it oxidizes oxyhemoglobin to methemoglobin, which is incapable of transporting O_2. Both acute (death) and subacute toxicities are recognized. Situations under which NO_3^- reduction occurs are as follows:

1   In the stomach of infants under about 3 months of age, where high stomach acidity permit the growth of microorganisms capable of reducing NO_3^- to NO_2^-. From time to time, deaths from this cause have been reported, although very few in recent years.[2,5]
2   In the rumen of cattle, as well as the secum and colon of the horse.
3   In vegetables or prepared foods that contain high amounts of NO_3^- and that are stored under conditions that permit the proliferation of microorganisms.[4] Most cases have dealt with spinach.
4   In damp forage materials of high NO_3^- content. Ingestion by livestock has been shown to be toxic.

A second hazard of excess NO_3^- arises when forage crops are ensiled. During ensiling, NO_3^- is converted to NO_2^- by denitrifying bacteria, from which poisonous yellow and brown gases (NO, NO_2, and N_2O_4) are formed. Under certain circumstances these N oxides accumulate in silo chutes and attached buildings and are lethal when inhaled by animals and humans.

Nitrates in Surface Waters

Historically, man has increased the NO_3^- content of surface waters by removing native vegetation, by tilling the soil, by increasing livestock numbers, and, more recently, by heavy fertilization with N.

The NO_3^- concentration of surface waters within any given watershed depends on the supply of available NO_3^-, rainfall, extent to which NO_3^- is removed by growing plants, and gaseous losses through denitrification. In the central United States, NO_3^- levels tend to be highest in spring, lowest in mid- to late summer, and intermediate in fall and winter (Fig. 6.1). Aldrich[2] summarized data from 16 sampling sites in 13 major U.S. rivers for the period from 1960 to 1975. Nitrate-N levels were generally less than 3 mg/liter, the main exception being the Santa Ana River in California, where NO_3^- –N levels of the order of 5 to 77 mg/liter were observed. No consistent trend has been observed between the NO_3^- content of the 13 major U.S. river waters and increased fertilizer N use over the past 20 years.

Aldrich[2] arrived at the following conclusions regarding changes in NO_3^- levels for the rivers of Illinois.

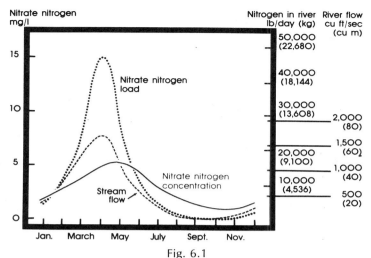

Fig. 6.1

Seasonal flow, $NO_3^- -N$ concentration, and $NO_3^- -N$ load for a typical midwestern river of the United States. (From Aldrich.[2])

1   Annual variations in NO_3^- content are relatively large, owing mainly to climatic factors that affect the amount of NO_3^- present in soils in the fall and extent of leaching during winter and early spring.

2   Annual average NO_3^- levels show an upward trend from the earliest to more recent years of sampling, with the greatest increase occurring before the 1965 to 1968 period. Trends since that time have leveled out or decreased.

A typical result showing an initial increase in the NO_3^--N content of an Illinois river with an increase in fertilizer N use is shown in Fig. 6.2. Note that $NO_3^- -N$ levels have leveled off or decreased in recent years despite greater fertilizer N use. The unexpected reduction in NO_3^- levels in recent years for the Illinois rivers has been attributed to:[2]

1   An increase in the amount of N removed in harvested plants due to yield increases over the same period.

2   A leveling off in the acreage of row crops, with the result that less NO_3^- is released from the soil organic matter.

3   A temporary interruption in fertilizer N use during the 1966 to 1974 period.

Trends for other U.S. rivers follow closely those noted above for the rivers of Illinois, some showing increases with time, some showing decreases, and some showing no change. Rivers showing recent increases

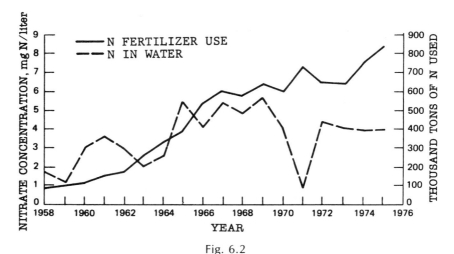

Fig. 6.2

Trends in N fertilizer use in Illinois and in the average $NO_3^- - N$ concentrations in the Kaskaskia River at Shelbyville, Illinois. (From Keeney[1] as modified from NRC.[5])

should be carefully monitored so as to establish future trends that might indicate an adverse effect on the environment.

As one might expect from the foregoing, considerable controversy exists as to the contribution of fertilizer N to the NO_3^- in lakes and streams. In soils of the humid and semihumid zones, where conditions are favorable for leaching, any fertilizer N added in excess of crop requirements will not remain long in the soil (unless immobilized). The main problem in developing a predictive model for the amount of NO_3^- that will reach surface and ground water is that there are no reliable estimates for the amounts converted to N gases (N_2 and N_2O) through denitrification. One fact that is frequently ignored is that substantial amounts of NO_3^- in drainage waters can be lost during movement and transport to lakes and reservoirs. In addition to losses by denitrification, NO_3^- can be stripped from water by algae and higher plants, and it can be immobilized during decay of dead plant remains in drainage channels and streams.

In considering the contribution of fertilizer to the NO_3^- in surface waters, and hence lakes and streams, two relationships are important, namely, plant uptake, or yield, and leaching losses as influenced by application rate. The quantity of fertilizer N lost through leaching (or denitrification) cannot be regarded as being directly proportional to the amount applied for the reason that the first increment of applied N will be consumed by microorganisms and incorporated into organic forms, particularly when residues with high C/N ratios have been applied to the soil.

A hypothetical relationship between crop yields, fertilizer application rate, and the $NO_3^- - N$ content of drainage waters is illustrated in Fig. 6.3

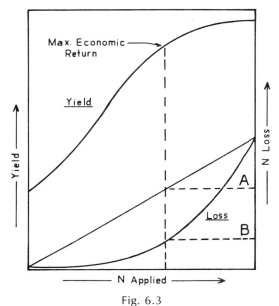

Fig. 6.3
Hypothetical relationship between crop yields, fertilzer application rate, and the
NO_3^- –N content of drainage waters.

for a crop with a high N requirement (corn). There is reason to believe that the NO_3^- content of drainage waters will follow line B rather than line A. Thus leaching losses would be minimal when fertilizer N is applied in amounts below that required by plants. Should restrictions in fertilizer N use become necessary to keep the NO_3^- concentration in lakes and streams below some desired upper limit, the consequences will be substantially less severe if the type B relationship is followed, as compared to type A. It should be noted that type A describes a situation in which leaching losses would be directly related to the quantity of fertilizer N applied.

Nitrates in Wells

Attention has frequently been given to the NO_3^- content of shallow wells and numerous examples can be cited where NO_3^- –N levels have exceeded by 10-fold or more the upper limit of 10 mg/liter regarded as safe for human consumption. Data for NO_3^- –N levels in the well waters of several states are recorded in Tables 6.2 and 6.3. As can be seen from Table 6.3, NO_3^- – N levels are particularly high in shallow wells. Whereas the water in lakes and streams may or may not be used for human consumption, most wells are drilled specifically as sources of private or public drinking water.

As noted earlier, the NO_3^- in well water can be derived from numerous

Table 6.2
Nitrate Content of Rural Wells in Several States (mg/liter)[a]

State	Type of Well	No. of Samples	Samples within Range, %			
			0–10	11–20	21–50	50+
California	Domestic	168	69	13	12	6
	Domestic and irrigation	67	55	13	16	16
	Irrigation	306	75	10	11	4
Illinois	(See Table 6.3)					
Iowa	Dug	454	66.5	10.6	14.3	8.6
	Drilled	647	95.5	2.5	1.5	0.4
	Bored	278	79.5	7.9	7.2	5.4
Kansas	–	–	—— 67+ ——		—— 33+ ——	
Michigan	–	2,847	95	—— 7 ——		
Nebraska	Farm	275	78.0	—— 22.0 ——		
	Nonmunicipal	2,687	79.8	—— 20.2 ——		
Ohio	Rural	10,500	90.5	5.0	3.5	1.0

[a] From Aldrich.[2]

sources. In general, the soils' contribution must be regarded as minor, for the reason that the NO_3^- content of percolating waters is normally too low to account for the observed high NO_3^- contents of contaminated wells. Where high NO_3^- levels have been observed, a source other than the soil is suspect, such as a barnyard, livestock feeding area, septic tank, or other source of animal or human contamination. High $NO_3^- -N$ levels in shallow wells of certain counties in Illinois and Missouri have been attributed to proximity of the wells to livestock feeding areas or to septic tank tile fields. For private wells and wells used for municipal water in Long Island, New York, infiltrated sewage and domestic waste disposal systems are regarded

Table 6.3
Relationship of Well Depth to $NO_3^- -N$ Concentration in 8844 Illinois Wells[a]

Depth, cm	No. of Wells Sampled	No. Exceeding 10 mg N/liter	% Exceeding 10 mg N/liter
0–20	480	134	28
23–38	926	185	20
41–76	1,568	78	5
79–152	2,042	61	3
1,930	3,828	23	0.6

[a] Adapted from NRC.[5]

as major sources. When possible, shallow wells should not be located within or adjacent to areas where contamination can occur.

Concern for NO_3^- in shallow wells can be traced to early studies indicating a high incidence of infant deaths as a result of drinking high-NO_3^- water. The first report of infant death due to methemoglobinemia was recorded in Iowa in 1945 by Comly[8] and other reports soon followed (for documentation, see Aldrich[2] and NRC[5]). Most early reports of methemoglobinemia involved farm wells rather than wells used for municipal water supplies.

For reasons that are not clear, the early experience has not been substantiated in recent years. Only a few cases of methemoglobinemia have been reported since 1950.[2,5] A NO_3^- problem, when it does exist, appears to be limited largely to infants under 3 months of age (or slightly older in case of diarrhea or digestive upset). One preventive measure is to use an alternate source of water for infants where the usual water supply is known to have a high NO_3^- content. As was the case for infants, early reports of widespread NO_3^- toxicity to farm animals through drinking water from farm wells have not been confirmed. For information on this subject, the review of Deeb and Sloan[3] is recommended.

Nitrates in Animal Feed

Nitrates often accumulate in such high amounts in plants that they constitute a hazard when fed to cattle and sheep.[7] Annual weeds, grasses, and the cereal crops are more likely to contain high amounts of NO_3^- than perennial grasses and legumes.

Factors affecting NO_3^- levels in plants have been discussed in detail elsewhere.[4,7] They include a high content of available N in the soil and adverse conditions for plant growth, such as drought, shade, and cloudy weather. Nitrogen fertilization at high rates does not constitute a major problem so long as conditions are favorable for plant growth. Areas where toxicities have been reported include the Northern Plains and the corn-belt section of the United States.

The biochemical basis for NO_3^- accumulation in plants is a loss of NO_3^- reductase activity under growth-limiting conditions. Nitrate reductase is one of the enzymes responsible for converting NO_3^- to NH_3 in plants (see Chapter 5) and is a necessary step for protein synthesis.

Nitrates in Food Crops

Nitrates are often present in very high concentrations in certain vegetables, notably lettuce, celery, beets, spinach, radishes, and turnip greens.[3,4,9] Nitrate contents as high as 3,000 $\mu g/g$ have been reported. Nitrates are added to many meat products, but, notwithstanding, a person is likely to consume as much or more NO_3^- from the vegetables he eats.[6] Another potential hazard arising from NO_3^- additions to meat products is ingestion of preformed nitrosamines.[5]

The NO_3^- content of vegetables varies greatly, depending on sampling date, variety, and environmental conditions. Factors that favor high accumulations include (1) high levels of NO_3^- in the soil, such as may be brought about by fertilization, (2) reduced light intensity during maturation, (3) soil moisture deficiency, and (4) nutrient deficiencies in the plant.

ADVERSE ENVIRONMENTAL ASPECTS

In recent years there has been increased concern over the possible connection between soil (and fertilizer) N and enhanced eutrophication of lakes, depletion of the stratospheric ozone (O_3) layer, toxicity of NO_2^- to plants, and ecosystem damage due to acid rain. These subjects are discussed in this section.

Possible Connection Between Fertilizer N and Depletion of Ozone in the Stratosphere

Ozone (O_3) exists high up in the atmosphere (i.e., the stratosphere), where it provides a protective shield against excessive ultraviolet radiation reaching the earths' surface. The O_3 layer exists as a 16.1 km (10 mile) band between 16.1 and 33 km (10 to 20 miles) above the earth. Over 99% of the incoming ultraviolet radiation in the most hazardous wavelength (<300 nm) is adsorbed by the O_3 layer.

Ozone is continuously being formed and destroyed in the stratosphere and a decrease in the average amount present may cause a variety of problems for plant and animal life, including skin cancer in man. It should be noted that although the O_3 layer is quite thick (about 16.1 km), the layer would only be a millimeter or so thick at sea-level pressure and temperature. According to an NAS[5] report, each 1% reduction in the O_3 level of the stratosphere might lead to a 2% increase in skin cancer as a result of additional ultraviolet rays reaching the earth surface. The conclusion was also reached that reduction in O_3 content might cause changes in global temperature and rainfall distribution patterns.

The basic reactions leading to the formation of O_3 in the stratosphere are as follows:

$$O_2 \xrightarrow[<242\ nm]{\mu v\ rays} O + O \tag{1}$$

$$O + O_2 \longrightarrow O_3 \tag{2}$$

This process is in dynamic equilibrium with processes that destroy O_3

$$O_3 \xrightarrow[<310\ nm]{uv\ rays} O + O_2 \atop XO \quad X + O_2 \tag{3}$$

where X is NO, a halogen, hydroxyl, or hydrogen.

A natural source of NO for reaction 3 is N_2O, which diffuses into the stratosphere over long time periods from the earth's surface. Thus an increase in the amount of N_2O entering the atmosphere has the potential for decreasing O_3 levels.

The O_3 layer is vulnerable to a variety of products introduced into the atmosphere through the activities of man, including spray can propellants (chlorofluorocarbons), products from the burning of fossil fuels, and exhaust gases from supersonic airplanes. Over decades these constituents may reach the upper troposphere (layer immediately below the stratosphere) and enhance the destruction of O_3. Since exhaust fumes of supersonic aircraft are injected at the lower boundary of the stratosphere, they can affect the O_3 layer rather quickly.

Among the many gases that tend to deplete the O_3 layer is N_2O, which is produced in soil by bacterial denitrification and nitrification. The fear has been expressed that an increase in fertilizer N use will be accompanied by a concomitant increase in the amount of N_2O released into the atmosphere, ultimately causing a reduction of O_3 in the stratosphere and the ability to absorb ultraviolet rays.[5,10-13]

The main cycle involving N_2O and other N oxides in the stratosphere is illustrated in Fig. 6.4. Nitrous oxide, transported to the stratosphere through turbulent flow, undergoes photolysis by absorbing ultraviolet radiation, yielding a singlet oxygen (O) and N_2. The singlet oxygen (also produced by

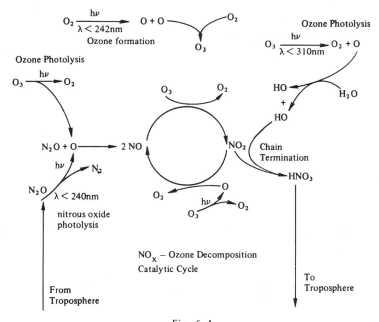

Fig. 6.4
Main cycle involving N_2O and other N oxides in the stratosphere. (From NRC.[5])

O_3 photolysis) reacts with more N_2O to produce NO, which along with NO_2 serves as a sink for O_3. The final reaction product, HNO_3, is ultimately returned to the troposphere.

Prediction of the potential impact of soil-derived N_2O on O_3 in the stratosphere is fraught with uncertainties. They include a lack of information as to the fraction of the denitrification product that occurs as N_2O (versus N_2) and the residence time of N_2O in the troposphere. The magnitude of a fertilizer N effect may well prove to be negligible when compared to other causes of O_3 depletion.

A summary of the probable effects of stratojets, chlorofluorocarbons (spray can propellants), and fertilizer N on O_3 in the stratosphere has been prepared by NRS.[5] The summary is reproduced in Table 6.4. The estimate for fertilizer N is a 1.5 to 3.5% reduction in O_3 by the year 2100. Aldrich[2] concluded that this estimate is high, as the soils' contribution has been declining over the years because of a reduction in mineralization losses. He pointed out that organic matter levels decline rather rapidly when soils are first placed under cultivation and that a new equilibrium level is reached characteristic of each agricultural system (see also Chapter 2). Since much of the arable land of the earth has now reached equilibrium, soil organic matter has been a declining source of N oxides during the time that fertilizers have become an increasing source. According to Aldrich,[2] fertilizer has not made a net contribution because a significant portion has been needed to fill the gap created by declining mineralization of soil humus. The fallacy of this line of reasoning is that N losses during cultivation were extremely rapid over the first few years, during which time NO_3^- would have been produced in excess of crop needs and was itself an environmental pollutant. Furthermore, fertilizer N is often applied to soils in amounts that greatly exceed crop requirements.

Table 6.4
Postulated Decrease in Atmospheric O_3 by Three Different Mechanisms[a]

Mechanism	Ozone Response Time	Steady-State Global O_3 Reduction	
		Long-Term Use at Current Level, %	Long-Term Use at Possible Future Level, %
Stratospheric aviation	Years	<0.1	6.5
Chlorofluorocarbons	Decades	14	Large
Fertilizer N[b]	Decades to centuries	1.5	3.5

[a] From NRC.[5]
[b] Current level is 79×10^9 kg/yr. Projected level is 200×10^9 kg/yr.

The long response time for fertilizer N to reach the stratosphere (decades to centuries, as shown in Table 6.4) can be attributed to the long residence time of the fertilizer in soil or water, followed by many additional years in the troposphere. In contrast, chlorofluorocarbons are added directly to the atmosphere and can reach the stratosphere within a decade or so.

With regard to fertilizer N use, the conclusion reached by NRC[5] was that "the current value to society of those activities that contribute to global N fixation far exceeds the potential cost of any moderate . . . postponement of action to reduce the threat of future ozone depletion of N_2O." Research currently underway is expected to provide information on which valid conclusions can be reached.

Eutrophication

Eutrophication can be defined as the process of enrichment of waters of lakes, ponds, and streams with nutrients, thereby giving rise to an increase in the growth of aquatic plants, including plankton and algae. Eutrophication detracts from the recreational value of lakes, and it can lead to depletion of O_2 so that fish and other water-dwelling animals cannot survive. Odor and taste of water are affected, thus increasing the cost of providing high-quality water for urban use.

The process of eutrophication is not new but has taken place since early geological time. For example, the vast peat bogs in cooler regions of the earth are a direct manifestation of eutrophication. There is little doubt, however, that the process has in many instances been enhanced through the activities of man.

The suggestion has been made that NO_3^- in drainage waters from agricultural lands leads to man-induced eutrophication of lakes and streams, a claim that has yet to be substantiated. Many investigators believe that other factors outweigh any possible effect of NO_3^- in drainage waters.

Phosphorus, rather than N, is usually the limiting nutrient for growth of aquatic plants. Also, N_2 fixation by blue-green algae, along with N in atmospheric precipitation, may provide sufficient N to meet the needs of the common algae and aquatic macrophytes. However, the productivity of estuaries is sometimes limited by N.[1]

Ecosystem Damage to Acid Rain

Acid rain (i.e., defined as rainfall of pH <5.7) occurs over broad areas of northern Europe, eastern Canada, and the northeastern United States. The trend is toward a decrease in pH and an expansion of the affected area with time. Acidity is due mainly to the mineral acids H_2SO_4 and HNO_3, with minor contribution from HCl.

The most noticeable effect from acid rain has been a lowering of pH in numerous lakes in eastern North America and in Scandinavia. Accompanying the decrease in pH has been an increase in dissolved Al, which is toxic to plants. As a consequence, some lakes are virtually lifeless; others are approaching this state.

Ecological effects of acid rain are difficult to evaluate but may include disruption of biogeochemical cycling, nutrient turnover, organic matter decomposition, damage to growing plants, a decline in soil fertility, and acidification of lakes, with an accompanying effect on algae, macrophytes, invertebrates, and fish.[14-16]

A combination of many human activities is responsible for the H_2SO_4 and HNO_3 in acid rain, including stack gases from the burning of coal, gases from oil-burning facilities, and emissions from motor vehicles. The soil is not considered to be a significant contributor to the acid-forming constituents in rainwater.

FORMATION OF NITROSAMINES

The potential hazard of nitrosamines as toxicants in certain foodstuffs, or formed following ingestion, has been subject of considerable interest. Nitrosamines, which are carcinogenic, mutagenic, and acutely toxic at very low concentrations, are formed through the reaction of NO_2^- with amines.

Primary aliphatic amines react with NO_2^- to form diazonium salts, which are unstable and break down to yield N_2 and a complicated mixture of organic compounds.

$$R-NH_2 + NaNO_2 + HX \longrightarrow [RN_2^+] \xrightarrow{H_2O} N_2 + \text{alcohols and alkenes}$$

Primary   Nitrite   Acid
aliphatic
amine

In contrast, secondary aliphatic and aromatic amines react with NO_2^- to form nitrosamines, according to the following reaction:

$$\begin{matrix} R \\ R \end{matrix} \!\!\! \diagdown N-H + NaNO_2 + HX \longrightarrow \begin{matrix} R \\ R \end{matrix} \!\!\! \diagdown N-N=O + NaX + H_2O$$

Secondary   Nitrite   Acid   Nitrosamine
amine

As was the case for methemoglobinemia (discussed earlier), NO_3^- only constitutes a problem when the NO_3^- is reduced to NO_2^-, such as can occur in vegetables or prepared foods that are stored under conditions that permit microbial growth. The subject of nitrosamines in foods has been discussed elsewhere.[5,6] Wolff and Wasserman[6] pointed out that evidence for the presence or formation of nitrosamines in foods is limited and that some earlier reports indicating the occurrence of nitrosamines in foods may have resulted from inadequacies in analytical procedures for determining nitrosamines.

A potential hazard to the health of man and animals would exist if nitrosamines were formed in soil from pesticide degradation products or from precursors present in manures and sewage sludge. A health hazard would become a reality, however, only if the nitrosamines thus formed were leached into water supplies or taken up by plants used as food by livestock or humans. Trace quantities of nitrosamines have been detected in soils amended with known amines (dimethylamine, trimethylamine) and NO_2^- or NO_3^-, but, for the most part, this work has been done under ideal conditions for nitrosamine formation in the laboratory, such as high additions of reactants.[17-20] Evidence is lacking that the synthesis of nitrosamines in field soils represents a threat to the environment. Mosier and Torbit[20] were unable to detect N-dimethylnitrosamine and N-diethylnitrosamine in manures even through the necessary precursors were known to be present.

SUMMARY

Several real and potential health and environmental impacts of N are known to exist. The soil N cycle affects primarily those that involve excess NO_3^- in drinking water, eutrophication, and possibly O_3 depletion of the stratosphere. Where high NO_3^- levels are observed in shallow wells, a source other than the soil is suspect, such as a barnyard or domestic waste disposal system. Nitrate levels in most U.S. rivers have remained rather constant over time, and, contrary to some reports, do not appear to be increasing with an increase in N fertilizer use by farmers.

The worldwide demand for food and fiber will require continued use of fixed N in agriculture, much of which will need to come from fertilizer. Better management practices are desired to maximize production while at the same time minimizing adverse health effects and pollution of the environment.

REFERENCES

1 D. R. Keeney, "Nitrogen Management for Maximum Efficiency and Minimum Pollution," in F. J. Stevenson, Ed., *Nitrogen in Agricultural Soils,* American Society of Agronomy, Madison, Wis., 1982, pp. 605–649.

2 S. R. Aldrich, *Nitrogen in Relation to Food, Environment, and Energy.* Illinois Agric. Exp. Sta. Special Publ. 61, Urbana, 1980, pp. 1–452.

3 B. S. Deeb and K. W. Sloan, "Nitrates, Nitrites, and Health," Illinois Agric. Exp. Sta. Bull. 750, 1975.

4 O. A. Lorenz, "Potential Nitate Levels in Edible Plant Parts," in D. R. Nielsen and J. G.

MacDonald, *Nitrogen in the Environment: Soil–Plant–Nitrogen Relationships,* Academic Press, New York, 1978, pp. 201–219.

5   National Research Council (NRC), "Nitrates: An Environmental Assessment," *National Academy of Sciences,* Washington D.C., 1978.

6   I. A. Wolff and A. E. Wasserman, *Science* **177,** 15 (1972).

7   M. J. Wright and K. L. Davidson, *Adv. Agron.* **16,** 197 (1964).

8   H. J. Comly, *J. Amer. Med. Assoc.* **129,** 112 (1945).

9   W. J. Corre and T. Breimer, *Nitrate and Nitrite in Vegetables,* Unipub, New York, 1979.

10   Council for Agricultural Science and Technology (CAST), *Effect of Increased Nitrogen Fixation on Stratospheric Ozone.* Report No. 53, Iowa State University, Ames. 1976.

11   P. J. Crutzen and D. H. Emhalt, *Ambio* **6,** 112 (1977).

12   H. S. Johnson, *Proc. Nat. Acad. Sci.* **69,** 2369 (1972).

13   M. B. McElroy and J. C. McConnell, *J. Atmos. Sci.* **28,** 1095 (1971).

14   J. N. Galloway, G. E. Lickens, and M. E. Hawley, *Science* **226,** 829 (1984).

15   G. E. Likens, *Chem. Eng. News.* **54,** 29 (1976).

16   R. Patrick, V. P. Binetti, and S. G. Halterman, *Science* **211,** 446 (1981).

17   A. L. Mills and M. Alexander, *J. Environ. Qual.* **5,** 437 (1976).

18   S. K. Pancholy, *Soil Biol. Biochem.* **10,** 27 (1978).

19   R. L. Tate and M. Alexander, *Soil Sci.* **118,** 317 (1974).

20   A. R. Mosier and S. Torbit, *J. Environ. Qual.* **5,** 465 (1976).

SEVEN

THE PHOSPHORUS CYCLE

The P cycle in soil is a dynamic system involving soils, plants, and microorganisms. Major processes include uptake of soil P by plants, recycling through return of plant and animal residues, biological turnover through mineralization–immobilization, fixation reactions at clay and oxide surfaces, and solubilization of mineral phosphates through the activities of microorganisms. In the natural state, essentially all the P consumed by plants is returned to the soil in plant and animal residues; under cultivation, some P is removed in the harvest and only part is returned. Losses of soil P occur through leaching and erosion.

The general cycle of P in soils has been diagrammed in a number of ways, depending on the particular aspect to be emphasized (e.g., biological transformations, soil–plant interrelationships, phosphate reactions). In the diagram shown in Fig. 7.1, emphasis is given to plant–soil–microbiological relationships in which the P is partitioned into "pools" based on availability of various organic and inorganic forms to plants. Soil solution P is shown to be in equilibrium with a given quantity of labile inorganic P such that as P is taken up by plants, or immobilized by microorganisms, additional inorganic P is solubilized. Stewart[1] has depicted microbial activity as a "wheel" that rotates in the soil, simultaneously consuming and releasing P to the soil solution. Studies using the radioactive isotope ^{32}P to follow changes in the forms of P in soils have shown that transformations between the pools are relatively rapid.

In many respects the P cycle in soil is analogous to the N cycle. However, the former is less spectacular in that no valency changes occur during assimilation of inorganic phosphate by living organisms or during breakdown of organic P compounds by microorganisms. Next to N, P is the most abundant nutrient contained in microbial tissue, making up as much as 2% of the dry weight. Partly for this reason, P is the second most abundant nutrient in soil organic matter.

Chemical and biochemical aspects of the P cycle have been reviewed from several standpoints, including fluxes of P on a global scale,[2–5] interactions with the C, N, and S cycles,[6–8] pedogenesis,[9–10] plant nutrition,[11–13] inorganic forms and fixation reactions,[12,14,15] soil organic P and associated

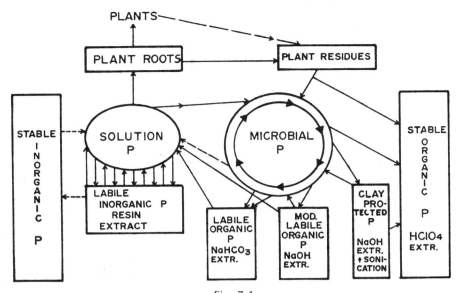

Fig. 7.1

The P cycle in soils, showing the partition of organic and inorganic forms of P into pools based on availability to plants. (From Stewart.[1] Reprinted by permission from Winter 1980–81 issue of *Better Crops with Plant Food*. Copyright 1981 by Potash & Phosphate Institute, Atlanta, Georgia 30329.)

transformations,[16-25] and environmental pollution.[26,27] A soil and plant computer model has been developed for simulating P transformations in soil.[28]

GLOBAL ASPECTS OF THE P CYCLE

Major reserves of P in the earth (Table 7.1) are marine sediments (840,000 $\times$ 10^{12} kg), terrestrial soils (96 to 160 $\times$ 10^{12} kg), dissolved inorganic PO_4^{3-} in the ocean (80 $\times$ 10^{12} kg), crustal rocks as apatite (19 $\times$ 10^{12} kg), and the biota or biomass (2.7 $\times$ 10^{12} kg). The amount of P in the terrestrial biota (2.6 $\times$ 10^{12} kg) far exceeds that of the marine biota (0.05 to 0.12 $\times$ 10^{12} kg).

An overview of fluxes of P on a global scale is illustrated in Fig. 7.2. The annual uptake of P by terrestrial plants amounts to 200 $\times$ 10^9 kg, a portion of which is removed in the harvested portion of the crop (4 to 7 $\times$ 10^9 kg). Losses of terrestrial P to the oceans through erosion is about 17 $\times$ 10^9 kg/year; a smaller amount (4.3 $\times$ 10^9 kg/yr) is carried by wind into the atmosphere. Unlike the C, N, and S cycles, the P cycle does not have a gaseous component; accordingly, movement of P to and from the atmosphere is of minor importance because of the small amounts that are circulated as at-

Table 7.1

Major Reservoirs of P in the Earth[a]

Reservoir	Total P $\times$ 10^{12} kg
Land	
Soil	96–160
Mineable rock	19
Biota	2.6
Fresh water (dissolved)	0.090
Ocean	
Sediments	840,000
Dissolved (inorganic)	80
Detritus (particulates)	0.65
Biota	0.050–0.12

[a] From Richey.[4]

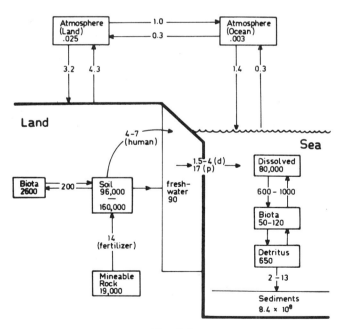

Fig. 7.2

Pools and annual fluxes of P for the global P cycle. Reservoirs are expressed in 10^9 kg and fluxes are in 10^9 kg/year. (From J. E. Richey, The Phosphorus Cycle, in SCOPE 21, *The Major Biogeochemical Cycles and Their Interactions*, edited by B. Bolin and R. B. Cook, John Wiley and Sons, Ltd., 1983.)

mospheric particulates. Inspection of Table 7.1 shows that P eventually finds its way to the sea, where it is deposited as ocean sediment and thereby removed from the cycle. Man has intervened in the P cycle through the mining of phosphates for fertilizer use, as well as by release of P into the environment in domestic and industrial effluents.

In natural ecosystems the P cycle is virtually closed, and most plant P is recycled by microbial breakdown of litter and organic debris. A typical example of this is the Brazilian rain forest, where essentially all the P exists as living and dead organic matter and where the underlying soil contains such a low amount of P that good yields of crops cannot be obtained following clearing of the forest.[22] As will be noted later, much of the P in grassland soils resides in the biomass.

Chemical Properties of Soil P

Phosphorus is an element with a mass of 30.98. Six isotopes are known, one of which (^{32}P) has a sufficiently long half-life to be of value in soil investigations. Although P exhibits coordination numbers of 1, 3, 4, 5, and 6, the vast number of P compounds have coordination numbers of 3 and 4. In the pedosphere, P is found largely in its oxidized state as orthophosphate, mostly as complexes with Ca, Fe, Al, and silicate minerals. Some secondary minerals of phosphate are wavellite $[Al_3(PO_4)_2(OH)_3 \cdot 5H_2O]$, vivianite $[Fe_3(PO_4)_2 \cdot 8H_2O]$, dufrenite $[FePO_4 \cdot Fe(OH)_3]$, strengite $[Fe(PO_4) \cdot H_2O]$, and variscite $[Al(PO_4) \cdot 2H_2O]$. Small amounts of phosphine gas, PH_3, often occur in lakes or marshes under highly reducing conditions.

The form of the phosphate ion in solution varies according to pH. In dilute solution, phosphoric acid dissociates as follows:

$$H_3PO_4 \underset{+H^+}{\overset{-H^+}{\rightleftharpoons}} H_2PO_4^- \underset{+H^+}{\overset{-H^+}{\rightleftharpoons}} HPO_4^{2-} \underset{+H^+}{\overset{-H^+}{\rightleftharpoons}} PO_4^{3-}$$

In the pH range of most soils (5 to 8) the amounts of the undissociated H_3PO_4 and trivalent PO_4^{3-} are negligible. Thus essentially all the phosphate consumed by plants is in the $H_2PO_4^-$ and HPO_4^{2-} forms. At pH 6, about 94% of the phosphate occurs as $H_2PO_4^-$ but the percentage drops to 60% at pH 7.

Origin of P in Soils

The native P in soils was derived from the apatite of soil-forming rocks. From a mineralogical point of view, apatite is a complex compound of tricalcium phosphate having the empirical formula $3[Ca_3(PO_4)_2] \cdot CaX_2$, where X can be either Cl^-, F^-, OH^-, or CO_3^{2-}. The most common minerals are the chloro-, fluor-, hydroxy-, and carbonate-apatites.

During weathering and soil development, the P of apatite is liberated and subsequently: (1) adsorbed by plants and recycled, (2) incorporated into the organic matter of soils and sediments, and (3) redeposited as insoluble or slowly soluble mineral forms, such as Ca, Fe, and Al phosphates and the occluded P of hydrous oxides.

Reserves of Mineable Phosphate Rock

Deposits rich in apatite (e.g., rock phosphate) are found in sediments deposited at the bottom of ancient seas, such as those in extensive areas of North America (4.7×10^{12} kg P). Other phosphate-rich deposits occur in tropical Africa (5.9×10^{12} kg P) and the Kola peninsula in the USSR (0.98×10^{12} kg P). Known reserves of rock phosphate are of the order of 5×10^{13} kg, representing a total of about 5×10^{12} kg of P based on an average P content of 10% due to inclusion of low-phosphate-content rocks.[29,30] Results of recent surveys dealing with known world reserves, distribution of resources by geological deposit type, and trends in worldwide phosphate production rates can be found in the monograph edited by Khasawneh et al.[12] Rock phosphates from various sources vary widely in chemical composition but those rich in carbonate-apatite are the most commonly used fertilizer materials.

Treatment of rock phosphate with inorganic acids (e.g., sulfuric and phosphoric) is used to produce the more soluble phosphate fertilizers. Superphosphate is a product resulting from the mixing of approximately equal quantities of 60 to 70% H_2SO_4 and phosphate rock. The overall reaction is:

$$3[Ca_3(PO_4)_2] \cdot CaF_2 + 7H_2SO_4 + 3H_2O \rightarrow 3Ca(H_2PO_4)_2 \cdot H_2O$$
$$\text{Ca-dihydrogen phosphate}$$
$$+ 7CaSO_4 + 2H_2O + 2HF$$
$$\text{Gypsum}$$

The fertilizer obtained in this manner is referred to as *ordinary* or *normal superphosphate* and contains about 10% P. Reaction of rock phosphate with excess H_2SO_4 and removal of much of the gypsum gives *concentrated superphosphate*, a product containing about 20% P. The principal reaction leading to the production of H_3PO_4 by the so-called wet process is as follows:

$$3[Ca_3(PO_4)_2] \cdot CaF_2 + 10H_2SO_4 + xH_2O \rightarrow 10CaSO_4 \cdot xH_2O$$
$$+ 6H_3PO_4 + 2HF$$

Three environmental problems are encountered with the mining of phosphate ores and the production of P fertilizers, namely, emission of fluorine, disposal of gypsum, and accumulation of Cd and other heavy metals in soils and possibly crops. The latter topic will be discussed in more detail in Chapter 9.

Although total reserves of phosphate rock would appear to be appreciable, phosphate is a scarce commodity when considered in terms of worldwide requirements. Cosgrove[19] pointed out that a "phosphate problem" is likely to develop in the not too distant future caused by shortages of mineable phosphates, at which time crop yields will be limited by the rate at which phosphate is released from insoluble forms in the soil. Forecasts as to when shortages will develop vary considerably, depending on estimates that are made for unknown reserves and the P content of the rock at which mining will be profitable. At the present mining rate (about 7×10^{10} kg of rock phosphate annually), reserves will be depleted in about 700 years. However, should usage of phosphate fertilizers continue to grow at recent rates, known reserves will be used up in a much shorter time (about 100 years). Ultimately, ways may be found to exploit the soil's reserve of "fixed" phosphate, one possibility being the development of phosphate-efficient plants through genetic engineering.

Phosphorus Content of Soils

The P content of soils in their natural state varies considerably, depending on the nature of the parent material, degree of weathering, and extent to which P has been lost through leaching. The P contents of the common soil-forming rocks vary from as little as 0.01% (100 µg/g) in sandstones to over 0.2% (2,000 µg/g) in high-phosphate limestones. As an average, the percentage of P in surface soils is about one-half that of N and one-twentieth that of K.[14]

The usual range of P in soils is of the order of 500 to 800 µg/g (dry-weight basis). Total P is usually highest in the upper A horizon and lowest in the lower A and upper B horizons because of recycling by plants. A soil containing 500 µg/g P will contain 1,120 kg P/ha (1,000 lb/acre) to the plow depth.

Soils formed from acid igneous rocks are generally low in P; those derived from basic rocks contain moderate to high amounts. Unweathered calcareous soils of dry regions often have high P contents due to the general lack of leaching and the presence of appreciable amounts of P in the form of apatite. Dark-colored soils, such as the Mollisols, tend to be high in P (and other nutrients as well). It should be noted that the amount of P in plant-available forms does not necessarily bear a direct relationship to total P.

The content of native P in surface soils of the United States is shown in Fig. 7.3. The average content is of the order of 600 µg/g; most values fall within the range of 200 to 900 µg/g. The highest amounts are found in the soils of a large area in the Northwest, which typically contain from 1,000 to 1,300 µg/g P. Sandy soils of the Atlantic and Gulf coastal plains are extremely low in P (<100 µg/g). As will be noted later, the P content of many soils has been altered by cropping, additions of animal manure, and fertilization.

The amounts of P in subsurface horizons are generally of the same order as those found in the surface layer, as indicated by Table 7.2. The profiles represent markedly different soil types and the data also serve to illustrate

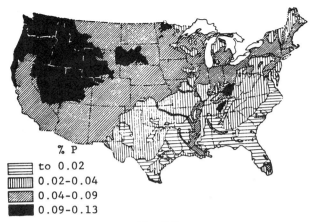

Fig. 7.3

Phosphorus content in the surface layer of U.S. soils. (Adapted from Sauchelli.[5])

the effects of parent material and climate on P content. The Mollisols are representative of the dark-colored, prairie soils of the upper midwestern United States, where weathering has been slight to moderate. These soils tend to be high in P. The Aridisol, represented by the desert soils, also contains high amounts of P. This soil is typical of those found in dry regions where leaching has been minimal. In contrast to the Mollisols and Aridisol, the Ultisols have very low P contents. The parent material from which the sandy loam Ultisol was formed contains high amounts of P (1,700 μg/g) and the relatively low amounts in the profile (400 μg/g or less) suggest that losses through leaching were extensive during weathering and soil formation.

Most of the P in soils occurs in inorganic forms, the main exception being peat (Histosols), where essentially all the P occurs in organic forms. Prairie grassland soils, forest soils, and certain tropical soils also contain relatively high amounts of P in organic forms, as will be shown later.

Losses of Soil P

Because of the strong tendency of phosphate to be adsorbed on colloidal surfaces and form insoluble complexes with di- and trivalent cations, losses of P from agricultural soils in subsurface and groundwater runoff are generally small, amounting to from 0.1 to 1.2 kg P/ha·yr. However, accumulated losses of soil P over a period of years, or over a broad area, can be appreciable. Leaching represents a major mechanism of P loss from forest lands.

Whereas short-term losses of P through leaching are usually of little significance from an agricultural standpoint, they may be of some importance when considered in terms of the P enrichment of natural waters, whereby the P contributes to eutrophication (see Chapter 6). Environmental aspects of the leaching of P into water supplies has been discussed by Ryden et al.[33]

Table 7.2
Distribution of P in the Horizons of Some Soil Profiles

Type	Description	Horizon and Parental Material	Depth, cm	P, μg/g
Aridisol[a]	California desert soil	A	0–5	1,047
		B21	5–12	655
		B22	12–27	655
		B3	27–34	1,353
		C1	34–84	916
Argiudoll[a]	Mahaska silty clay loam, Iowa	A1p	0–20	691
		A12	20–33	631
		A3	33–43	547
		B1	52–54	504
		B22	75–88	464
		B31	88–105	598
		C1	128–138	766
Mollisol[b]	Tama silt loam, Iowa	A1	15–30	790
		A3	36–61	610
		B	64–102	440
		C	127–178	390
		Loess	229 +	440
Mollisol[b]	Crete silt loam, Nebraska	A11	0–4	600
		A12	4–10	570
		A3	10–51	520
		B2	51–97	900
		C_{Ca}	97–152	920
		Loess	152–213	440
Ultisol[b]	Sandy loam, Georgia	A1	0–2.5	400
		A2	2.5–23	200
		B1	23–36	200
		B2	36–71	400
		C	71–152	100
		Fresh granite		1,700
Ultisol[b]	Silt loam soil, Virginia	A1	0–5	240
		A2	5–28	70
		B1	28–41	140
		B2	41–91	100
		C	91–142	140
		Fresh limestone		140

[a] From Finkl and Simonson.[31]
[b] From Simonson.[32]

The concentration of dissolved inorganic P in drainage waters is generally low (Table 7.3) and depends on the concentration of inorganic phosphates in the soil solution, percolation rate, surface area exposed to percolating waters, and kind and amount of P-adsorbing material in the profile. Losses of P in surface runoff will be affected by slope, P fertilizer history, cropping practice, type of soil, and amount and intensity of rainfall.

The main mechanism of P loss from most agricultural soils is through erosion. Excluding arid soils of the intermountain West, most surface soils of the United States contain from 200 to 900 μg/g P, or about 450 to 1,800 kg/ha to the depth of plowing. Thus each metric ton of soil lost by erosion will carry with it from 0.2 to 0.8 kg of P.

In contrast to the low amounts of dissolved P in drainage waters from agricultural lands, usually of the order of 0.04 to 0.06 μg P/ml, rather high loads are found in surface waters from domestic wastes (4 to 9 μg/ml). The contribution of domestic wastes to the P of surface waters of the United States (9 to 227 $\times$ 10^6 kg/yr) is of the same order of magnitude as for agricultural lands (5.5 to 545 $\times$ 10^6 kg P/year). Unusually high amounts of P (>100 μg/ml) have been found in surface runoff from cattle feedlots.[33,34]

Effect of Cropping and Fertilization on Soil P

Long-time cropping of the soil without fertilizer or manure additions invariably leads to a reduction in P content. For a soil containing 500 μg/g P, equivalent to 1,120 kg P/ha to plow depth, continuous cropping for a period of 10 years under a management system where 10 kg of P is removed annually in the harvested portion of the crop will result in a reduction of total P by nearly 10%. Organic matter is lost at a rather rapid rate when soils are first placed under cultivation (see Chapter 2) and much of the initial loss of P can be attributed to losses of organic P.

Data obtained by Haas et al.[35] for the soils of 15 dryland experiment stations in the U.S. Great Plains show that total P was reduced by an average of 8% by cropping with wheat over a period of 30 to 48 years, with most of the loss being due to organic P. Somewhat higher losses (from 12 to 29% after 60 to 90 years of cultivation) were obtained by Tiessen et al.[36] for the soils of three grassland soil associations of the Canadian prairies. Their data, given in Table 7.4, show that, for two of the associations, essentially all the P loss was accounted for by the organic fraction. It is of interest that losses of organic C exceeded those of organic P.

Several studies have shown that losses of organic P by cultivation are less than for N, a result that has been attributed to greater absorption of N by plants and greater losses through leaching and denitrification.[35] Another factor to consider is that organic P is probably stabilized by a different mechanism than for N, a point that has been emphasized in recent reviews.[8,21]

Most of the P applied to soil as fertilizer, often as high as 90% or more,

Table 7.3

Concentrations of Dissolved Inorganic P in Subsurface Runoff[a]

Location	Soil Texture	Drainage System	Crop	Dissolved Inorganic P, (μg/ml)	
				Range	Mean
Ontario Canada	Clay	Tile drains	Corn, oats	0.20–0.17	0.18
			Alfalfa, bluegrass	0.10–0.27	0.21
Snake Valley, Idaho	Calcareous silt loam	Irrigation return flow	Alfalfa, corn, root crops, pasture	0.01–0.23	0.01
Woburn, England	Sandy	Tile drains	Arable land and grassland	0–0.30	0.08
South central Michigan	Clay to sandy loams	Tile drains and ditches	Root crops	0.01–0.30	–
San Joaquin Valley, California	Heavy silty clay	Irrigation return flow drains	Cotton, rice, alfalfa	0.05–0.23	0.08
Yakima Valley, Washington	Sandy loam	Surface return flow drains	–	0.07–0.30	0.16
		Subsurface return flow drains		0.03–0.46	0.18

[a] From Ryden et al.,[33] where specific references can be obtained.

Table 7.4

Changes in P Content by Long-Time Cultivation of Soils from Three
Grassland Soil Associations of the Canadian Prairies[a]

Soil Association	Native Prairie	60–70 Years Cultivation	C or P loss, %
Blain Lake			
Organic C, mg/g	47.9 ± 10.2	32.8 ± 5.2	32
Total P, μg/g	823 ± 92	724 ± 53	12
Organic P, μg/g	645 ± 125	528 ± 54	18
Inorganic P, μg/g	178 ± 47	196 ± 8	NS
Southerland			
Organic C, mg/g	37.7 ± 6.5	23.7 ± 1.8	37
Total P, μg/g	756 ± 28	661 ± 31	12
Organic P, μg/g	492 ± 5.2	407 ± 30	17
Inorganic P, μg/g	256 ± 44	254 ± 19	NS
Bradwell			
Organic C, mg/g	32.2 ± 8	17.4 ± 1.6	46
Total P, μg/g	746 ± 101	527 ± 15	29
Organic P, μg/g	446 ± 46	315 ± 21	29
Inorganic P, μg/g	300 ± 84	212 ± 46	29

[a] From Tiessen et al.[36] Most soils were Typic Cryoborolls.

is not taken up by the crop but is retained in insoluble or fixed forms. Whereas a portion of the residual P can be used by subsequent crops, further additions of fertilizer are often required in order to maintain high crop yields. Repeated applications of P in amounts exceeding crop uptake will inevitably result in an accumulation of P in the soil. As one might expect, the extent of P accumulation will depend on both fertilizer application rate and years of application.

Results obtained by Barber[37] for the accumulation of P in soils of a rotation–fertility experiment where P was applied at several rates over a 25-year period are given in Table 7.5. In this case, accumulations of P occurred when P application rates exceeded 22 kg/ha·yr. Net losses of P at the lower rates can be accounted for by crop removal.

Cosgrove[19] pointed out that the P content of many cultivated soils in industrial countries has been increasing over the years because of extensive use of P fertilizers. To this extent the soil has served as a "sink" for P.

Increases in soil P also occur by long-time applications of animal manures.[35,36,38] In the study of Haas et al.,[35] total P was increased by an average of 14% above the virgin soils where manure had been applied at an annual rate of 5,600 to 11,200 kg/ha (2.5 to 5 tons/acre), equivalent to about 6.2 to 12.3 kg/ha (5.5 to 11 lb P/acre), over a 30- to 48-year period.

Extensive fractionation procedures involving partition of P into the various inorganic and organic pools (see Fig. 7.1) have been carried out in

Table 7.5
Influence of Continuous P Fertilization over a
25-year Period on Total P in Soil[a]

P Applied, kg/ha·year	Change in Soil P, μg/g	Total Soil P, μg/g
0	−10	400 ± 20
11	−4	455 ± 44
22	+5	492 ± 50
44	+23	589 ± 57
54	+32	632 ± 48

[a] From Barber.[37]

attempts to determine changes in soil P induced by cultivation, with and without applications of chemical fertilizers and manures. Coverage of this work is beyond the scope of this chapter and the reader is referred to several reviews[12,15] and reports[39–41] for detailed information.

PHOSPHORUS AS A PLANT NUTRIENT

Phosphorus is an essential constituent of all living organisms. Plants deficient in P are stunted in growth and maturity is delayed. The lower leaves are typically yellow and tend to wither and drop off. Leaf pigmentation in young leaf tissue is usually abnormally dark green, often shading to red and purple hues because of excess anthocyanin accumulation. Phosphorus is needed or favorable for seed formation, root development, strength of straw in cereal crops, and crop maturity.

Phosphorus exists in plants as the inorganic phosphate ion ($H_2PO_4^-$ or HPO_4^{2-}) and in combination with organic compounds, some of which will be discussed in the section on "Organic P Compounds in Soil." The organic P of plants exists as compounds in which phosphate is esterified with OH groups of sugars or alcohols or bound by a pyrophosphate bond to another phosphate group. Major P-containing compounds are the nucleic acids (see Chapter 5), phospholipids, and phytin (Ca–Mg salt of inositol hexaphosphate); other compounds include adenosine triphosphate (ATP), which functions in energy storage and transfer, and the phosphopyridine nucleotides (NAD, NADP), which serve as carriers of hydrogen.

Phosphorylated sugars are mainly intermediate compounds of carbohydrate metabolism. In cellular metabolism, sugars (e.g., glucose) are converted to phosphate esters, which subsequently undergo a series of stepwise transformations involving phosphorylated intermediates. The sequence is:

glucose $\rightleftharpoons$ glucose 6-phosphate $\rightleftharpoons$ fructose 6-phosphate $\rightleftharpoons$ fructose 1,6-diphosphate $\rightleftharpoons$ glyceraldehyde 3-phosphate (+ dihydroxyacetone phosphate) $\rightleftharpoons$ 1,3-diphosphoglyceric acid $\rightleftharpoons$ 3-phosphoglyceric acid $\rightleftharpoons$ 2-phosphoglyceric acid $\rightleftharpoons$ phosphopyruvic acid $\rightleftharpoons$ pyruvic acid. Typical examples of sugar phosphates are as follows:

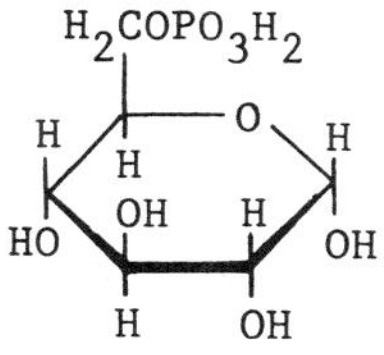

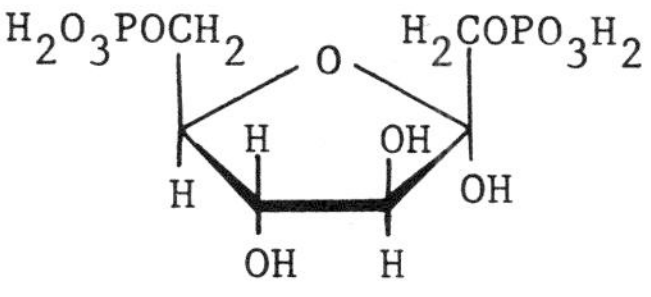

Glucose 6-Phosphate   Fructose 1,6-diphosphate

One of the more important compounds containing the pyrophosphate linkage is adenosine triphosphate (ATP), whose structure is shown in Fig. 7.4.

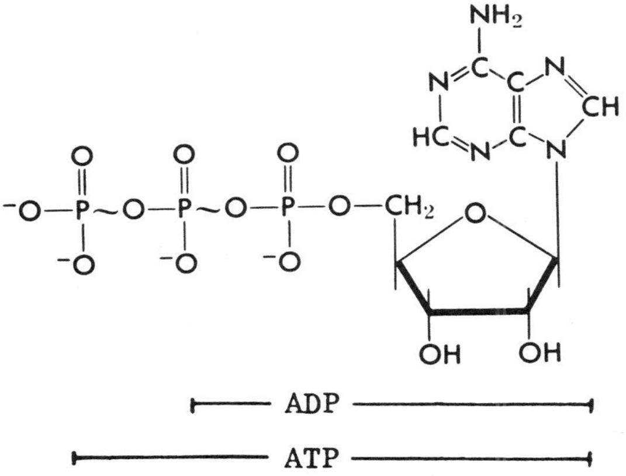

Fig. 7.4
Chemical structure of adenosine triphosphate (ATP).

The pyrophosphate linkage of ATP is an energy-rich bond indicated by the symbol ($\sim$). Energy absorbed during photosynthesis is used for the synthesis of the pyrophosphate bond of ATP. The bound energy, in turn, is used for endergonic processes, such as ion uptake and synthesis of polysaccharides (starch, cellulose, etc.), protein, fats, and lignin, among other products.

Another important class of organic P compounds are the coenzymes nicotinamide adenine dinucleotide (NAD) and nicotinamide dinucleotide phosphate (NADP), which are involved in biological oxidations. The nucleotides are sometimes abbreviated as DPN and TPN, respectively. The structure of NAD is given in Fig. 7.5.

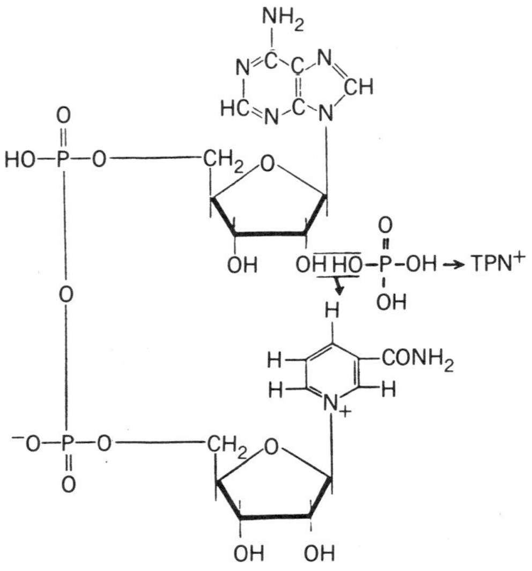

Fig. 7.5

Chemical structure of nicotinamide adenine dinucleotide (NAD or DPN).

Considrable variation exists in the P requirements of plants, as suggested by the data given in Table 7.6. For the crops shown, total P removed by the crop amounts to from 19 to 54 kg/ha. At the higher value, a soil containing 500 μg P/g will contain sufficient P to the plow depth for only a few decades of cropping without further P additions. However, much of the P in soil is not in plant-available forms (discussed later). It should also be noted that substantial amounts of the P consumed by plants will be returned to the soil in crop residues. Unlike N, which can be returned to the soil by biological N_2 fixation (Chapter 4), P cannot be replenished except from external sources once it is removed in the harvest or by erosion.

Although most plants take up P throughout the entire growing season, about 50% of the seasonal total of P is consumed by the time plants have accumulated 25% of their seasonal total of dry matter. Plant species differ in their requirements for P, ability to extract P from the soil, and response to insoluble forms of P fertilizers (e.g., rock phosphate). Crops that are relatively efficient in using relatively insoluble forms of P include alfalfa, buckwheat, millet, lupins, and sweet clover; inefficient crops include barley, cotton, corn, oats, potatoes, and wheat. For detailed information on the P nutrition of plants, the reader is referred to Khasawneh et al.[12] and Mengel and Kirkby.[13]

CHEMISTRY OF SOIL P

The chemistry of P in soils is exceedingly complex and full details are avail-

Table 7.6
Approximate Amounts of P Removed from the Soil in Specific Crops[a]

Crop	Yield		P, kg/ha
	kg/ha	units/acre	
Grains			
Corn	12,544	200 bu	52
Grain sorghum	8,064	8,000 lb	54
Wheat	5,376	80 bu	34
Barley	5,376	100 bu	28
Oats	3,584	100 bu	22
Rice	7,280	145 bu	25
Forage crops			
Alfalfa	12,544	6 tons	35
Clovers	8,064	4 tons	20
Grasses (general)	8,064	4 tons	20
Oil crops			
Soybeans (beans only)	3,360	50 bu	25
Peanuts	3,360	3,000 lb	22
Fiber crops			
Sugarcane	67,200	30 tons	19
Sugar beets (roots and tops)	67,200	30 tons	28

[a] Adapted from Khasawneh et al.[12] and Pierre and Norman.[15]

able elsewhere.[12,14,15] Only the overall aspects are discussed here.

The P compounds in soil can be placed into the following classes:

1 Soluble inorganic and organic compounds in the soil solution.
2 Weakly adsorbed (labile) inorganic phosphate.
3 Insoluble phosphates.
   a. Of Ca in calcareous and alkaline soils of arid and semiarid regions.
   b. Of Fe and Al in acidic soils.
4 Phosphates strongly adsorbed and/or occluded by hydrous oxides of Fe and Al.
5 Fixed phosphate of silicate minerals.
6 Insoluble organic forms.
   a. Of the soil biomass.
   b. In undecomposed plant and animal residues.
   c. As part of the soil organic matter (humus).

Essentially all the inorganic P in soil exists in the form of orthophosphates or derivatives of phosphoric acid (H_3PO_4). The main compounds are phosphates of Ca, Al, and Fe with trace amounts of other cations. Various forms of apatite, such as fluoroapatite, constitute the principal P minerals in calcareous and alkaline soils of arid and semiarid regions. Trace quantities of P may be present in the lattices of silicate minerals and as inclusions in minerals.

From the standpoint of plant nutrition, the P in soils can be considered in terms of "pools," as noted earlier in Fig. 7.1. Only a small fraction of the P occurs in water-soluble forms at any one time. A portion of the insoluble P appears to be somewhat more available to plants than the bulk of the soil reserves. This fraction, designated as the "labile" pool, is believed to consist of easily mineralized organic P and phosphates weakly adsorbed to clay colloids. From 1 to 2% of the soil P occurs in microbial tissue (biomass). The bulk of the soil P (>90%) occurs in insoluble or fixed forms, namely, as primary phosphate minerals, humus P, insoluble phosphates of Ca, Fe, and Al, and phosphate fixed by colloidal oxides and silicate minerals.

The ability of the soil to provide P to higher plants is determined by a variety of factors that include: (1) the quantities of $H_2PO_4^-$ and HPO_4^{2-} in the soil solution, (2) the solubilities of Fe- and Al-phosphates and phosphate complexes with hydrous oxides and clay minerals in acid soils, (3) the solubilities of Ca phosphate and P minerals in calcareous soils, (4) amount and stage of decomposition of organic residues, and (5) activities of microorganisms. As will be shown later, maximum P availability tends to occur at intermediate pH values of 6 to 7.

Chemical Fractionation Schemes

Knowledge of the forms of P in soils has come largely from chemical fractionations based on the ability of selective chemical reagents to solubilize discrete types of inorganic P compounds. Since its inception, the fractionation procedure of Chang and Jackson[42] has been widely used for investigations of the forms of native inorganic P and transformations of P applied as fertilizer. Their fractionation scheme is as follows:

1 Al phosphate   Extraction with 0.1 N NH_4F
2 Fe phosphate   Extraction with 0.1 N NaOH
3 Ca phosphate   Extraction with 0.5 N H_2SO_4
4 Occluded Fe phosphate   Extraction with dithionate–citrate
5 Occluded Al phosphate   Extraction with 0.5 N NH_4F after
                             dithionate–citrate extraction

Subsequent studies indicated that the various extractants were not as specific as at first envisioned. Also, difficulties were encountered with calcareous soils and sediments because of retention of P by CaF_2 formed from $CaCO_3$ during NH_4F extraction.[43] The readsorbed P was then recovered in other fractions. The Chang and Jackson procedure was modified by Williams et al.,[43] who included a step to correct for readsorption of phosphate. Further changes were made by Syers et al.[44] in attempts to render the procedure more suitable for use with calcareous soils. Their modification is outlined below, with fraction (5) being added to account for the unextracted P.

1	Fe and Al phosphates	Extraction with 0.1 N NaOH
2	P sorbed by carbonate during step 1	Extraction with 0.1 M NaCl and citrate–bicarbonate solution
3	P occluded within Fe oxides and hydrous oxides	Extraction with citrate–dithionite–bicarbonate solution
4	Ca phosphates	Extraction with 1 N HCl
5	Mineral phosphate	Inorganic P in final soil residue

Limitations of chemical fractionation schemes for characterizing inorganic P in soils include the following: (1) part of the released P may be readsorbed from solution by various soil components, (2) intermediate reaction products of applied fertilizer P may persist for years and the behavior of the products during successive extraction is unknown, and (3) some of the inorganic P may be derived from organic P. As indicated above, a limitation of fractionation schemes that include an NH_4F treatment to determine Al phosphates is that in calcareous soils, the formation of CaF_2 leads to errors in estimates for other fractions, such as occluded P and Ca-bound P.

Calcium phosphates constitute a series of compounds ranging from the relatively soluble mono- and dicalcium phosphates (normally present in rather small amounts) to the relatively insoluble hydroxyapatite and fluorapatite. Iron and Al phosphates also vary greatly in solubility, depending on the amount of P present. In some soils, part of the Fe and Al phosphates may exist as the minerals dufrenite and wavellite, which are very insoluble except under neutral or alkaline conditions.

Results of extensive fractionations of inorganic P in native soils, and of soils subject to long-time cropping, with and without manure and fertilizer additions, have been given in a number of reviews.[11–15] The reader is referred to these papers for detailed information.

As one might expect, considerable variation exists in the kinds and amounts of any given form of P in soils. Thus Fe and Al phosphates tend to accumulate in acid soils whereas Ca phosphates are predominant in neutral or alkaline soils. As will be noted later, a soil pH in the range 6 to 7 is best from the standpoint of P availability because this range is above the

Table 7.7

Sequential Fractionation Scheme for the Separation of Inorganic and Organic P into Fractions that Vary in Availability to Plants[a]

Fraction	Treatment	Comments
1	Extraction with an anion exchange resin, bicarbonate form	Removes the most biologically available inorganic P
2a	Extraction with 0.5 M NaHCO$_3$	Removes labile inorganic and organic P sorbed on soil minerals plus a small amount of microbial P
2b	Treatment with chloroform (CHCl$_3$) followed by extraction with 0.5 M NaHCO$_3$	The difference between 2b and 2a represents inorganic and organic P originating from lysed microbial cells
3	Extraction with 0.1 M NaOH	Removes inorganic and organic compounds held more strongly by chemisorption to Fe and Al components of soil surfaces
4	Extraction with 0.1 M NaOH following ultrasonification	Enables extraction of inorganic and organic P held at internal surfaces of soil aggregates
5	Extraction with 1 M HCl	Removes mainly apatite-type minerals but also occluded P in more weathered soils
6	Oxidation (H$_2$O$_2$) and acid digestion (H$_2$SO$_4$) of final soil residue	Stable organic P forms and relatively insoluble inorganic P compounds

[a] From Hedley et al.[47,48]

pH of maximum insolubility of Fe and Al phosphates but below the pH of maximum insolubility of Ca phosphates.

An elaborate fractionation procedure[45-48] has recently been used in attempts to separate inorganic and organic forms of P into fractions (pools) that vary in their availability to plants (Table 7.7). Biologically available inorganic P is removed first with an anion exchange resin, following which a mild extractant (0.5 M NaHCO$_3$) is used to remove "labile" inorganic and organic forms. At this step, inclusion of a chloroform treatment permits estimates to be made for organic P originating from lysed microbial cells. Stable P forms (Fe and Al phosphates, apatite-type minerals, and highly insoluble organic and inorganic P compounds) are removed by stronger extracting reagents. Application of the fractionation procedure to soils from a long-term rotation experiment (65 years of cropping in a wheat–wheat–fallow rotation) showed that the cultivated soils contained less plant-available

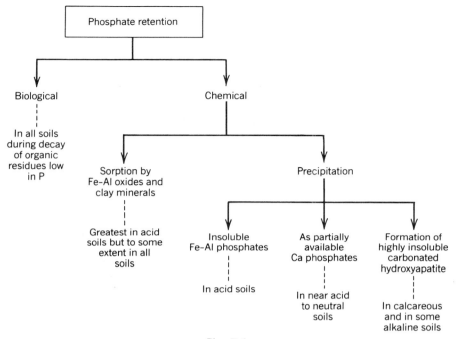

Fig. 7.6
Phosphate fixation reactions in soil. (Adapted from Sauchelli.[5])

forms of inorganic P (resin- and 0.5 M NaHCO$_3$ extractable inorganic P) than an adjacent pasture soil. Also, the cultivated soils contained less extractable organic P and residual P.[48] The method also appears suitable for measuring short-term changes in soil P, such as are brought about by organic matter additions.[45,46]

The use of 0.5 M NaHCO$_3$ to determine "labile" organic P is based on the finding that this mild extractant, originally used to extract plant-available inorganic P in calcareous soils (but later extended to noncalcareous soils), removes a small portion of organic P that is more readily mineralized than the bulk of the soil organic P.[49] As will be noted later, a portion of the biomass P can also be regarded as being readily available to plants.

Phosphate Fixation

Numerous studies have shown that much of the P applied to soils in water-soluble forms [e.g., monocalcium phosphate, Ca(H$_2$PO$_4$)$_2$] does not remain as such for long but is converted to one of many insoluble or complex forms, as illustrated by Fig. 7.6. Both biological and chemical processes are involved in fixation, with the latter being of greatest importance from the standpoint of retention of fertilizer P.

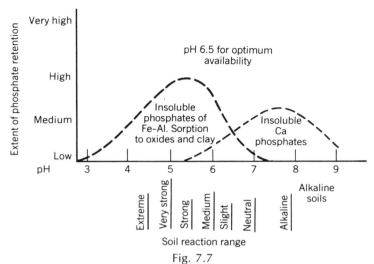

Fig. 7.7

Effect of pH on phosphate fixation reactions. (Adapted from Sauchelli.[5])

Figure 7.7 illustrates the importance of pH is governing fixation reactions and thereby the availability of phosphates to plants. For most soils, maximum availability would be expected in the slightly acid to neutral pH range. Phosphate fixation is known to be influenced by clay mineralogy and decreases in the following order: amorphous hydrous oxides > goethite—gibbsite > kaolinite > montmorillonite.

In highly acid soils, phosphate is readily precipitated as the highly insoluble Fe- or Al-phosphates or adsorbed to oxide surfaces. Both forms are poor sources of P for higher plants. This is particularly true for soils rich in Fe, such as lateritic soils (Oxisols) of the tropics and subtropics. On such soils, plants often show severe P deficiency symptoms and relatively large amounts of fertilizer P must be applied to meet plant requirements.

In calcareous soils the less soluble di- and tricalcium phosphates [$Ca_2(HPO_4)_2$ and $Ca_3(PO_4)_2$] are formed, with the latter being gradually converted to carbonate apatite, $3[Ca_3(PO_4)_2] \cdot CaCO_3$. This highly insoluble compound is not a good source of P for plants but availability may be enhanced in the presence of growing plants through the action of organic acids secreted from plant roots or synthesized by microorganisms in the rhizosphere, as will be noted later.

Adsorption of phosphate at hydrous oxide surfaces through a ligand-binding mechanism (replacement of H_2O and/or OH) is one of the important processes affecting phosphate availability to plants.[50-56] The overall reaction is as follows.

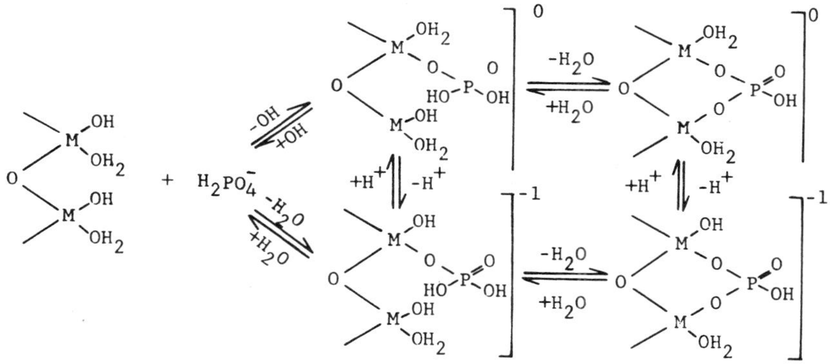

Labile P   Difficultly available P

Variations in the availability of "fixed" phosphate to plants has been attributed to the formation of two types of chemical bonds, a single coordinate linkage (labile P) and a chelate ring (difficultly available P). The latter is shown by the structures at the right side of the preceding diagram. The concept of a "labile" pool of inorganic P was noted earlier.

Three distinct mechanisms have been proposed by Ryden et al.[56] for the sorption of phosphate by soils and hydrous ferric oxide gels, namely, (1) chemisorption at protonated surface sites ($-OH_2^+$), (2) chemisorption by replacement of surface OH groups, and (3) a more physical sorption of P as a potential-determining ion. The latter was believed to operate only at high P concentrations. Chemisorption, (1) and (2), was believed to occur according to the reactions:

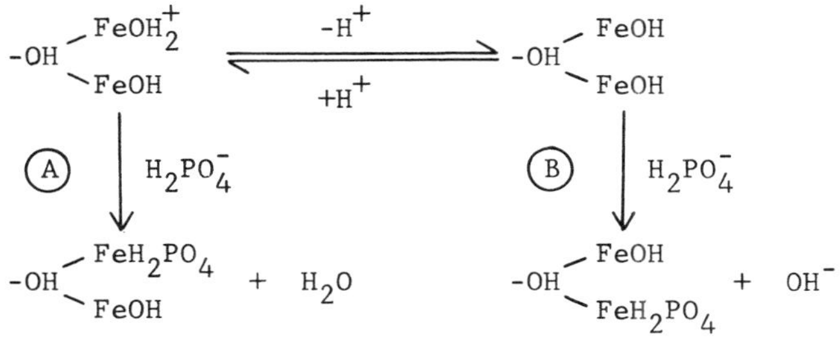

Only when neutralization of surface OH_2 sites was complete did the pH rise owing to release of OH^-, indicating that sorption followed the order A > B. Ryden et al.[56] concluded that their proposed mechanisms served to explain some of the anomalies of P fixation by soils and soil components, such as effects due to pH, ionic strength, and phosphate concentration. The postulated mechanisms do not preclude the formation of binuclear complexes from the products of reactions A and B (see reaction sequence shown earlier).

Because Fe- and Al-oxides can be positively charged at pH values below the point of zero charge, phosphate fixation is sometimes thought to result from anion exchange. However, this does not appear to be the case, as suggested by the reactions shown above.

Phosphate fixation at oxide surfaces is particularly important in intensively weathered soils of the tropics and subtropics. The clay fractions of youthful soils formed on volcanic ash also fix appreciable amounts of phosphate, which can be accounted for by their high content of allophane (chemical formula of $xSiO_2,Al_2O_3 \cdot yH_2O$). In a study of phosphate retention by Chilean allophanic soils, Borie and Zunino[57] observed an initial fast adsorption onto allophanic surfaces, following which the added phosphate was subjected to reactions leading to the formation of organic matter–P associations or complexes through Al bridges. The humus P complexes were found to comprise the major fraction of the total P in these soils and was viewed as a P sink in the overall P cycling process.

Attempts to increase the availability of P in highly weathered acid soils through liming have generally been unsuccessful, a result that has been attributed to the formation of insoluble Ca-phosphates, thereby counteracting the effect of pH on the solubilization of fixed phosphate.[58]

Another mechanism that has been advanced to explain P fixation in soil is the combination of phosphate with clay minerals. Two reactions are possible, one being replacement of a hydroxyl group from a structural Al atom and the second a linkage with Ca on the exchange complex to form a clay–Ca–PO$_4$ linkage. The phenomenon is greater for 1:1-type clay minerals than for the 2:1 type, which can be explained by their higher content of exposed Al—OH groups. Kafkafi et al.[59] identified three successive regions for adsorption of phosphate on kaolinite. The first two were associated with edge-face OH groups and a third with occlusion of phosphate into the amorphous region of the clay surface. The initial step in fixation was believed to be hydration, followed by replacement of Al—OH groups with phosphate, according to the following reactions.

$$\begin{array}{c}
\diagdown \text{Al-OH} \quad + \quad \text{H}_2\text{O} \rightleftharpoons \diagdown \text{Al} \overset{\text{H}_2\text{O}}{\underset{\text{OH}}{\diagup}} \\[2em]
\text{H}_2\text{PO}_4^- \searrow \downarrow \rightarrow \text{OH}^- \\[2em]
\diagup \overset{\delta+}{\text{Al}} \overset{\text{H}_2\text{O}}{\cdots\cdots} \overset{\delta-}{\text{H}_2\text{PO}_4} \xleftarrow{\text{wash}} \diagdown \text{Al} \overset{\text{H}_2\text{O}}{\underset{\text{H}_2\text{PO}_4}{\diagup}}
\end{array}$$

(exchangeable
phosphate)

The coordination of water by the edge Al (first step) is essential before exchange can take place (i.e., formation of Al—OH linkage). The final step is intended to show that the bond between phosphate and the surface after washing is at least partly covalent and that the $H_2PO_4^-$ may have a slight negative charge. The phosphate has the potential for linking with two Al atoms to form a stable six-membered ring, as shown below.

Water-Soluble P

Phosphorus is absorbed by plants largely as the negatively charged primary and secondary orthophosphate ions ($H_2PO_4^-$ and HPO_4^{2-}), which are present in the soil solution. Thus the water-soluble pool is of particular interest because this P has a direct effect on plant growth, and hence yields. Plants have the ability to absorb certain soluble organic phosphates but it is not known whether they serve as direct sources of P to plants.

Numerous attempts have been made to determine the minimum soil solution P concentration that will coincide with maximum crop yield. A range of from 0.01 to 0.3 μg P/ml in the soil solution has been recorded for a variety of crops.[60] Results obtained for corn grown on two dissimilar soils (Oxisol and Hydrandept) are shown in Fig. 7.8. For both soils a value of 0.05 μg P/ml was required for 95% of maximum yield. Critical concentrations for other soils will vary from this value, depending on plant uptake conditions (climate, rooting volume, etc.)

Factors affecting levels of soluble P in soils include crop removal, chemical fixation reactions (conversion to insoluble forms), immobilization by microorganisms, and leaching.

The replenishment of P in the soil solution is of considerable importance from the standpoint of plant nutrition, for the reason that the quantity of inorganic P in the solution phase at any one time, usually of the order of 0.3 to 1 μg/ml, is insufficient to meet crop requirements over the whole of the growing season. At the higher concentration (1 μg/ml), the total amount of P in the soil solution (25% water on a dry-weight basis) would be about 0.13 kg/ha (0.12 lb/acre) to a depth of 0.6 m (2 ft). Crop removal is usually between 4.5 and 22.4 kg/ha (4 and 20 lb/acre); thus total soluble P would be less than 5% of the crop requirement. Continuous renewal of P in the soil solution is thereby required to meet the needs of the growing plant. However, it should be pointed out that plants do not absorb P uniformly throughout

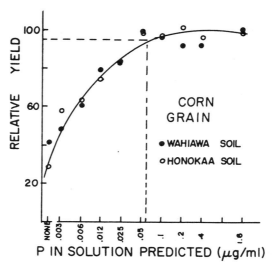

Fig. 7.8

Corn grain yields in relation to predicted soluble P in two soils with different mineralogical composition.[60] The Wahiawa soil is a Oxisol whereas the Honokaa soil is a Hydrandept. (Reproduced from *Chemistry of the Soil Environment* (1981) by permission of the American Society of America.)

the entire soil mass, and that the rate of replenishment of soil solution P would be greatest in small areas near the root tips, where the most active uptake of P occurs.

Replenishment of inorganic P in the soil solution is affected by mineralization (m) and/or immobilization (i) of organic P by microorganisms and by chemical equilibria between soluble and insoluble mineral forms of P, the postulated reactions being as follows:

$$\text{Organic P} \underset{i}{\overset{m}{\rightleftharpoons}} \text{Inorganic P in the Soil Solution} \underset{fast}{\rightleftharpoons} \text{Labile Inorganic P}$$

$$\updownarrow$$

Relatively Insoluble Ca-
Fe-, and Al-Phosphates and
Fixed Forms of P

A vital factor affecting plant uptake of soil P is the rate at which a suitable concentration of phosphate is maintained in the soil solution near the root surface. Movement of phosphate to plant roots by mass flow is believed to account for only a small portion of the phosphate consumed by plants. Accordingly, soil immediately adjacent to the root is a major source of P for

Table 7.8
Inorganic and Organic Phosphate Contents of the Displaced Solutions of Several Soils[a]

Type	No. of Soils	Moisture Content, %	Inorganic PO_4^{3-}, μg/ml	Organic PO_4^{3-}	
				μg/ml	% of P
Sand	2	7.0– 9.2	trace–0.02	0.28–0.38	93
Sandy loam	6	8.1–14.6	0.02–0.32	0.23–0.90	65–94
Silt loam	9	13.1–21.4	0.03–0.35	0.15–0.89	59–94

[a] From Pierre and Parker.[61] Data also provided for water extracts (1:5) of the same soils.

the growing plant. This is the region where microorganisms are particulary active (the rhizosphere).

Data obtained by Pierre and Parker,[61] summarized in Table 7.8, suggested that a substantial portion of the total soluble P in soils occurs in organic forms, a result confirmed by other studies (see Dalal[20]). Very little is known of the nature of the organic P or its availability to plants. Some of the organic P may exist as phosphate esters released into the soil solution during rupture of microbial cells; part may be colloidal and associated with microbial cells and cellular debris.[26] Conversion of soluble organic P to inorganic phosphate is likely because of the occurrence of phosphatase enzymes (discussed later). As Dalal[20] pointed out, identification of the organic P compounds in the soil solution is necessary in order to improve our understanding of the significance of organic P to the P nutrition of plants.

In many soils the rate of release of organic and inorganic P to soluble forms is too slow to satisfy crop requirements, in which case fertilizer P must be applied to insure optimum yields.

Soil Tests for Available P

Numerous soil tests have been developed over the years for measurement of plant available P in soils. Some procedures vary only in minor detail; others represent extremes in extraction conditions. No single test will apply for all soils and all methods have limitations that must be taken into account when making fertilizer P recommendations. A limitation common to most methods is that little, if any, information is provided on the rate at which the P in insoluble complexes is converted to plant available (soluble) forms during the course of the growing season.

Dalal and Hallsworth[62] pointed out that a description of available soil P must include an intensity factor I, a quantity factor Q, and a capacity factor $\Delta Q/\Delta I$, as well as rate and diffusion factors. As was noted earlier, the immediate source of available P for plants is that contained in the soil solution–

the intensity factor I. The labile portion of the solid phase is the quantity factor Q, which is approximated by most soil test extraction techniques. The capacity of the soil system to maintain P concentration in the solution phase as P is removed by plants ($\Delta Q/\Delta I$) is not directly determined by soil tests, nor is the rate of replenishment of soil solution P from solid phase forms.

The seven test procedures outlined below are those described in the 1982 edition of the American Society of Agronomy monograph *Methods of Soil Analysis*,[63] in which specific references can be found. Additional information can be found in the reviews of Khasawneh et al.[12] and Pierre and Norman.[15]

Phosphorus Soluble in Water

The main objective of this method is to determine the P concentration in the soil solution that is required for optimum plant growth, the intensity factor I. A soil extract more closely approximates the P concentration in the soil solution but such an extract is not easily obtained. Most tests involve extraction with water or a dilute salt solution (e.g., 0.01 M $CaCl_2$). Under some soil and crop conditions, the method provides a satisfactory prediction of P fertilizer requirement. The approach would appear to be most useful for sandy soils, where conversion of fertilizer P to insoluble forms is minimal.

Phosphorus Extractable with Dilute Acid–Fluoride (HCl·NH₄F)

This widely used extractant (0.025 N HCl : 03 N NH_4F) is designed to remove easily acid-soluble P forms, largely Ca-phosphates and a portion of the Al- and Fe-phosphates. The NH_4F dissolves the latter through formation of fluoride complexes with Al and Fe. In general, the method has been found to be most successful for acids soils (<pH 6.5). Low estimates are obtained with calcareous soils because of neutralization of the acid by $CaCO_3$.

Phosphorus Soluble in NaHCO₃

Extraction of soil with 0.5 M $NaHCO_3$ (pH near 8.5) has been highly successful for predicting P availability in calcareous, alkaline, or neutral soils. This extractant decreases the concentration of Ca in solution by forming the insoluble $CaCO_3$, with the result that the concentration of phosphate in the solution is increased.

Extraction of soil with 0.5 M $NaHCO_3$ also leads to solubilization of a portion of the soil organic P, and, as mentioned earlier, the amount thus solubilized has been regarded as a quantitative measure of the potential contribution of soil organic P to plant uptake (i.e., labile organic P).[45–48]

Phosphorus Soluble in NH₄HCO₃–DTPA

This relatively new test was developed for simultaneous extraction of P, K, and micronutrient cations. The extractant solution is 1 M NH_4HCO_3 : 0.005 M diethylenetriaminepentaacetic acid (DTPA) at pH 7.6. The chelating agent (DTPA) is used for chelation and solubilization of micronutrients (i.e., Cu,

Fe, Mn, and Zn). The reagent dissolves about half as much P as 0.5 M $NaHCO_3$.

Anion Resin–Extractable P

This mild test procedure removes P from soil without chemical alteration or changes in pH. In addition to serving as a test for available P, the approach has been used to assess the availability of residual P and to measure the rate of P released from insoluble forms.

Phosphorus Soluble in Dilute $HCl \cdot H_2SO_4$

Soil tests involving mixed acids have been found useful for predicting P availability in soils which fix appreciable amounts of P, such as those of the Piedmont region of North Carolina. The recommended extractant is 0.05 N $HCl:0.025\ N\ H_2SO_4$.

Isotopic Dilution of ^{32}P

Phosphate in the soil solution exists in equilibrium with a portion of the solid-phase P. By measuring dilution of applied ^{32}P-labeled phosphate in the aqueous phase, the combined amount of native ^{31}P in the two soil phases can be determined. The ^{31}P of the solid phase has sometimes been referred to as "labile" P.

When ^{32}P-labeled phosphate is added to a soil–water system, the following equilibrium reaction is established.

$$^{31}P_{solid} + {}^{32}P_{solution} \rightleftharpoons {}^{32}P_{solid} + {}^{31}P_{solution}$$

At equilibrium, the ratio of the two isotopes in a portion of the solid phase is equal to their ratio in solution. Therefore, $^{31}P_{solid}$ can be calculated from the relationship:

$$^{31}P_{solid} = \frac{^{32}P_{solid}}{^{32}P_{solution}} \times {}^{31}P_{solution}$$

Organic Forms

From 15 to 80% of the P in soils occurs in organic forms, the exact amount being dependent on the nature of the soil and its composition.[20,21,23] The higher percentages are typical of peats and uncultivated forest soils, although much of the P in tropical soils and certain prairie grassland soils may occur in organic forms. In fertilized temperate zone soils, however, the contribution of organic forms is likely to be rather small relative to inorganic P.

Determination of Organic P

Methods used for the determination of organic P can be divided into two main types, as follows:

Extraction methods: In this approach, organic and mineral forms of P are recovered from the soil by extraction with acid and base. Organic P is converted to orthophosphate and the content of organic P is determined from the increase in inorganic phosphate as compared to a dilute acid extract of the original soil.

Organic P = total P in alkaline extract − inorganic P in acid extract

Ignition method: Organic P is converted to inorganic P by ignition of the soil at elevated temperatures and is calculated as the difference between inorganic P in acid extracts of ignited and nonignited soil.

Organic P = inorganic P of ignited soil − inorganic P of untreated soil

Advantages and disadvantages can be pointed out for both approaches. In the extraction method, pretreatment of the soil with mineral acid (usually conc. HCl) is essential for removing polyvalent cations which render the P compounds insoluble but this treatment may cause some hydrolysis of organic P compounds. Other extractants (8-hydroxyquinoline and EDTA) have been used in attempts to eliminate this problem. The alkali used to solubilize organic matter (dilute NaOH) also causes some hydrolysis of organic P.

Of the two approaches, the ignition method is the easiest and is often the one used for survey types of investigations. Limitations of the method include alteration in the solubility of the native inorganic P and hydrolysis of organic phosphate during acid extraction of the nonignited soil. Recent studies on the determination of organic P in soils include those of Steward and Oades[64] and Williams et al.[65]

Because organic P is always determined by the difference between total and inorganic P, accuracy is low when the soil is low in organic P, such as in the subsoil.

Organic P Content of Soils

Although organic P in soil is somewhat biologically stable, continuous turnover occurs through mineralization–immobilization (discussed later) and the equilibrium level will depend to a considerable extent on the nature of the soil and its environment. The situation is analogous to that of soil organic matter as a whole, as discussed earlier in Chapter 2. As one might expect, the organic P content of the soil follows closely that for total organic matter. The relationship between organic P and the organic C content of some tropical soils is illustrated in Fig. 7.9.

Values for total organic P in soils are summarized in Table 7.9. In view of the diverse nature of soils, the widely different values are not surprising. As one might suspect, organic P decreases with depth in much the same way as organic C. Depth distribution patterns for organic P and organic C in several soil profiles from Iowa are shown in Fig. 7.10.

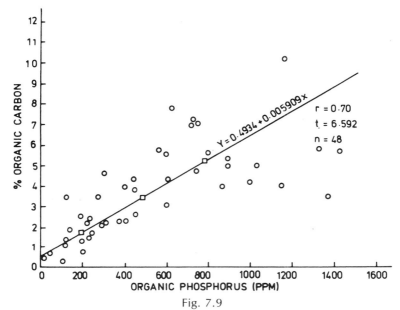

Fig. 7.9

Relationship between organic P and organic C for some tropical soils. (From A. P. Uriyo and A. Kesseba, "Organic Phosphorus in Soil Profiles," *Geoderma* **13,** 208; reproduced by permission of Elsevier Scientific Publishing Company.)

Factors influencing the proportion of P in soil organic matter include P supply, parent material, climate, drainage, cultivation, pH, and depth of soil. The effect of each is not known with certainty and contradictory results are the rule rather than the exception. Some workers have concluded that a low P content of the organic matter is a characteristic of P-deficient soils but this hypothesis has not been confirmed. A slight effect has been noted for parent material, namely, the organic matter of soils derived from granite tends to be lower in P than the organic matter of soils derived from basalt or basic igneous materials. Also, the P content of organic matter may be higher in fine-textured soils than in coarse-textured ones. Baker[68] found that organic P accumulated rapidly during the first 50 years of soil development, after which the rate declined until a steady state was reached. As one might expect, levels of organic P can be modified by manure additions.

When soils are first placed under cultivation, the content of organic C (and N) usually declines (Chapter 2). This same pattern is also followed for organic P. At present, it is not known whether the mineralization rate of organic P is greater than, equal to, or slower than the mineralization rate of organic C and N.

A combination of upward transport of P by plants and retention in the surface soil may alter the vertical distribution of P in the soil profile. For

Table 7.9
Some Values for Organic P in Soil[a]

Location	Organic P μg/g	Organic P % of Total P
Australia	40– 900	–
Canada	80– 710	9–54
Denmark	354	61
England	200– 920	22–74
New Zealand	120–1360	30–77
Nigeria	160–1160	–
Scotland	200– 920	22–74
Tanzania	5–1200	27–90
United States	4– 85	3–52

[a] From the review of Halstead and McKercher,[21] where specific references can be obtained. The range shown for the U.S. soils must be regarded as a minimum. Data for the Tanzania soils are from Uriyo and Kesseba.[66]

soils subject to moderate weathering and leaching, a minimum in the concentration of total P is often found a short distance below the surface, which is due in part to organic P enrichment of the surface soil at the expense of inorganic P.

The C/N/P/S Ratio

The amount of organic P in soils is roughly correlated to C and/or N, although C/organic P and N/organic P ratios are much more variable than C/N ratios.[69–74] The P content of soil organic matter varies from as little as 1.0% to well over 3.0%, which is reflected by the variable C/organic P ratios that have been reported for soils, as shown in Table 7.10 for the soils of several great soil groups in Canada.[69] The ratios (46 to 648) are somewhat more variable than reported for Finnish soils by Kaila,[70] who observed C/organic P ratios of from 61 to 276 for cultivated mineral soils, 141 to 526 for cultivated humus soils, and 67 to 311 for virgin soils. Relatively low C/organic P ratios have been reported for some British and New Zealand soils, with mean C/P ratios of about 60. One explanation for the extreme ratios is that N and S occur as structural components of humic and fulvic acids whereas P does not. It is noteworthy that essentially all the organic P can be recovered from the soil by alkaline extractants, and that most of the P resides in the fulvic acid fraction. Goh and Williams[75] found that C/P ratios were lower in the lower-molecular-weight components of soil organic matter.

The percentage of the total organic P that occurs in association with living microorganisms (the soil biomass) is another factor that must be taken into

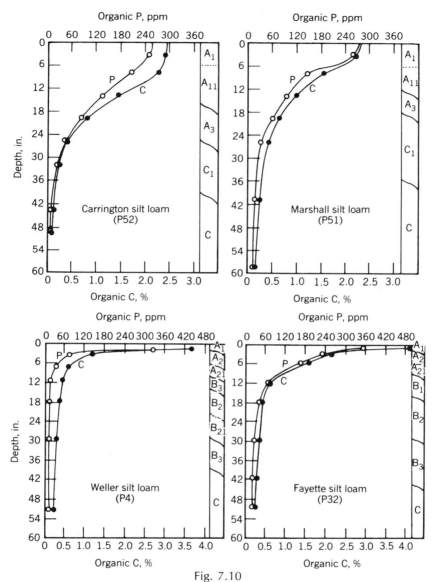

Fig. 7.10

Distribution of organic P and C in four Iowa soil profiles. The top two are Mollisols while the bottom two are Alfisols. (From Pearson and Simonson.[67])

Table 7.10
Carbon/Organic P Ratios of Soils from Several Great
Soil Groups in Canada[a]

Soil Order	Number of Soils	C/Organic P Range	C/Organic P Mean
Chernozemic	8	92–648	231
Solonetzic	5	106–177	149
Podzolic	5	103–225	172
Brunizolic	7	46–245	148
Regosolic	5	106–417	209
Gleysolic	8	62–274	124
All soils	38	46–648	172

[a] From John et al.[69]

account when comparing the C/organic P ratios of contrasting soil types. As noted later, the C/P ratio of the soil biomass is substantially lower than for the soil organic matter, and the percentage of the organic P as biomass P can vary over a broad range (>3 to <20%).

In view of the above, it is not surprising that considerable variation has also been observed in the C/N/P/S ratios of individual soils (Table 7.11). Notwithstanding, the mean for soils from different regions of the world is remarkably similar. As an average, the proportion of C/N/P/S in soil humus is 140:10:1.3:1.3.

Walker[9] and Walker and Syers[10] have concluded that the P content of the original parent material places an upper limit on organic C and N levels. They postulated that given sufficient time the supply of N comes into balance with the supply of P (assuming that S is not limiting).

Table 7.11
Organic N, Total N, Organic P, and Organic S Relationships in Soil

Location	Number of Soils	C/N/P/S	Reference
Iowa	6	110:10:1.4:1.3	Neptune et al.[71]
Brazil	6	194:10:1.2:1.6	Neptune et al.[71]
Scotland[a]			
Calcareous	10	113:10:1.3:1.3	Williams et al.[74]
Noncalcareous	40	147:10:2.5:1.4	Williams et al.[74]
New Zealand[b]	22	140:10:2.1:2.1	Walker and Adams[73]
India	9	144:10:1.9:1.8	Somani and Sarena[72]

[a] Values for S given as total S.

[b] Values for subsurface layers (35–53 cm) were 105:10:3.5:1.1.

Table 7.12
Distribution of the Forms of Organic P in Soils[a]

Source	% of Organic P		
	Inositol Phosphates	Nucleic Acids	Phospholipids
Australia	0.4–38	–	–
Bangladesh	9–83	0.2–2.3	0.5–7.0
Britain	24–58	0.6–2.4	0.6–0.9
Canada	11–23	–	0.9–2.2
New Zealand	5–43	–	0.7–3.1
Nigeria	23–30	–	–
United States	3–52	0.2–1.8	–

[a] From Dalal,[20] where specific references can be obtained. Results for the soils of Bangladesh are from Islam and Ahmed.[83]

Specific Organic P Compounds

Relatively little is known about the chemistry of organic P in soil. Principal forms are the inositol phosphates,[76–92] nucleic acids or their degradation products,[83,93,94] and phospholipids.[83,95–97] Other organic P compounds present in trace quantities include sugar phosphates and phosphoproteins. The usual ranges for percentages of organic P in the various forms is as follows:

Inositol phosphates	10–50%
Phospholipids	1–5%
Nucleic acids	0.2–2.5%
Phosphoproteins	trace
Metabolic phosphates	trace

Although up to 83% of the organic P in select soils has been accounted for as inositol phosphates (Table 7.12), recoveries of over 50% are the exception rather than the rule. The highest values recorded thus far are for the soils of Bangladesh, but, as can be seen from Table 7.13, recoveries of less than 50% were obtained for seven of the 10 soils examined. The Bangladesh study is unique in that the measurements for inositol phosphates, phospholipids, and nucleic acids were all made on the same soils. As an average, no more than one-third of the organic P in soils can be identified in known compounds.

As shown previously, recovery of soil organic P in the three principal forms follows the order: inositol phosphates ≫ phospholipids > nucleic acids. The same types of organic P compounds are found in plants, but the order is reversed. The somewhat higher abundance of inositol phosphates

Table 7.13
Recovery of Organic P in Known Compounds for 10 Bangladesh Soils[a]

Soil	Organic P (μg/g)	Recovery of Organic P (%)			
		Inositol Phosphates	Phospholipids	Nucleic Acids	Total
1	39	44	7.0	2.30	54
2	104	83	2.0	0.21	85
3	153	32	2.0	0.43	35
4	187	66	2.0	0.23	68
5	304	20	4.0	0.16	25
6	61	47	1.7	1.07	50
7	250	60	0.8	0.44	61
8	207	9	2.5	0.59	12
9	105	46	0.5	1.20	48
10	114	41	1.4	0.43	43
Avg	152	45	2.4	0.71	48

[a] From Islam and Ahmed.[83]

in soils, as contrasted to plants, may be due to their tendency to form insoluble complexes with polyvalent cations, such as Fe and Al in acid soils and Ca in calcareous soils.

Inability to account for the bulk of the organic P in most soils may be due to:[93]

1 Occurrence of polymeric phosphate-containing compounds, such as teichoic acids from bacterial cell walls and phosphorylated polysaccharides. The former consists of a polyol, ribitol, or glycerol phosphate in which adjacent polyol residues are linked by a phosphodiester bond.

2 Occurrence of stable complexes of inorganic phosphate with soil organic matter, which is recovered only after destruction of organic matter and is measured along with the organic P.

3 Presence in soil of complexes between soil organic matter and simple phosphate esters (e.g., nucleotides or inositol phosphates) that modify the properties of the phosphate esters so that they are not estimated by conventional methods.

INOSITOL PHOSPHATES. Inositol phosphates are esters of hexahydrohexahydroxy benzene, commonly referred to as inositol (I). A variety of esters are possible, the most common being the hexaphosphate. In plants, mono-, di-, and triphosphates are sometimes found in rather large quantities. The hexaphosphate ester, or phytic acid (II), occurs in cereal grains as the mixed Ca and Mg salt called phytin. For many years the inositol hexaphosphates in soil were thought to be derived from the phytin of higher plants but recent work indicates that they are probably of microbial origin.

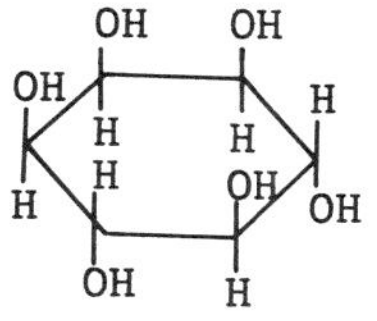

myo-Inositol

I

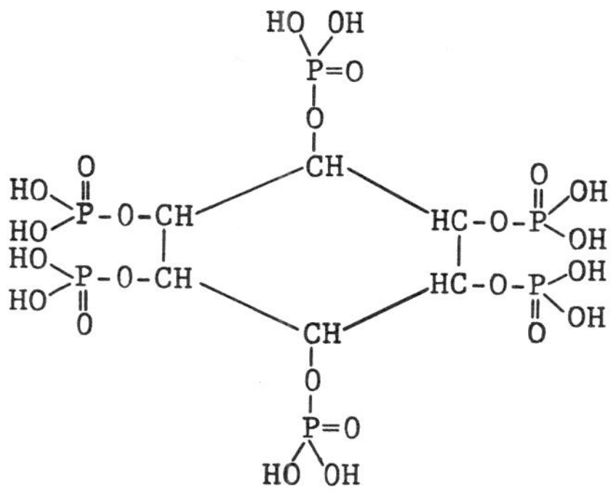

Phytic acid

II

A unique feature of inositol phosphates, particularly the hexaphosphate esters, is their high stabilities in acids and bases. Advantage is taken of this property for their extraction and preparation from soil. The steps involved have usually included the following:

1 Extraction with hot 0.5 N NaOH after pretreatment of the soil with 5% HCl.
2 Oxidation of the extracted organic matter with alkaline hypobromite.
3 Precipitation of inositol phosphates as their Fe (or Ca) salts.
4 Alkaline hypobromite oxidation of the precipitate material.
5 A second Fe (or Ca) precipitation of the inositol phosphates.
6 Decomposition of the second Fe precipitate with NaOH solution and removal of the $Fe(OH)_3$.

The P recovered in the final solution is sometimes considered to be in the form of inositol phosphates, but other P compounds are usually present (e.g., inorganic phosphate). Most workers now use chromatographic techniques to separate the inositol phosphates from inorganic phosphate and from each other.

By anion-exchange chromatography, Smith and Clark[89] demonstrated that only a part of the inositol phosphates consisted of the hexaphosphate; lower phosphate esters were also present. As can be seen from the data presented in Table 7.14, the percentage of inositol phosphates as the mono-, di-, tri-, tetra, penta-, and hexaphosphate forms varies widely from one soil to another. Caldwell and Black[77] analyzed 49 Iowa surface soils by anion-exchange chromatography and found that inositol hexaphosphate, the most abundant ester, accounted for from 3 to 52% of the organic P, the average being only 17%. The percentage was higher for soils developed under forest vegetation than under grass vegetation.

From the structural formula of inositol, it can be demonstrated that nine

Table 7.14

Range of Inositol Mono-, Di- and Tri-, Tetra-, Penta-, and
Hexaphosphates in 10 Soils of Bangladesh[a]

Form	% of Organic P	% of Inositol Phosphates (%)
Mono-	1.3– 6.7	1.8–10.9
Di- and tri-	4.0–11.0	6.5–15.8
Tetra-	5.0–18.1	8.1–29.5
Penta-	9.4–17.3	15.4–27.3
Hexa-	11.1–49.2	18.2–65.6

[a] From Mandal and Islam.[85]

positional stereoisomers are possible, depending on arrangements of H and OH groups. They include seven optically inactive forms and one pair of optically active isomers. The best-known form is myo-inositol (shown earlier), which is widely found in nature. This is the isomer from which phytin is constituted. Other naturally occurring isomers of more limited biological distribution are D-chiro and L-chiro-inositol (III and IV) and scyllo-inositol (V).

D-chiro-Inositol   L-chiro-Inositol   scyllo-Inositol   neo-Inositol
III   IV   V   VI

The advent of chromatographic techniques not only has led to more precise values for inositol phosphates in soil but a number of unusual isomers have been isolated. Smith and Clark[89] first demonstrated the presence of an unknown inositol phosphate in soil, subsequently shown by Caldwell and Black[77] to be an inositol hexaphosphate other than myo-inositol (I). The unknown isomer, which accounted for an average of 46% of the inositol hexaphosphate material, was found to be synthesized by a variety of soil microorganisms.

The unknown isomer was subsequently shown by Cosgrove[78–80] to be the hexaphosphate of scyllo-inositol (V). Cosgrove also demonstrated the occurrence in soil of the hexaphosphate of D- and L-chiro-inositol (III and IV) and of neo-inositol (VI). Penta phosphates of myo-, chiro-, and scyllo-inositol were also found.

Evidence that inositol hexaphosphates are bound to other organic components in soil has been shown by the work of Anderson and Hance[76] and

Cosgrove.[78] Other studies have shown that organic P compounds of several types are bound to high-molecular-weight organic colloids.[87,90] Unlike organic C and N, the organic P of soils can be rather easily extracted with alkaline reagents.

For additional information on inositol phosphate stereoisomers in soil, the reader is referred to the reviews of Cosgrove[80] and Halstead and McKercher.[21] Results of the various studies show that considerable differences exist in the isomeric composition of inositol phosphates in soil and that some, if not all, of the soil inositol phosphates are synthesized *in situ* by microorganisms.

NUCLEIC ACIDS.   Nucleic acids are found in all living cells, and they are synthesized by soil microorganisms during the decomposition of plant and animal residues. Two types are known, ribonucleic acid (RNA) and deoxyribonucleic acid (DNA), each consisting of a chain of nucleotides. Each nucleotide contains a pentose sugar, a purine or pyrimidine base (see Chapter 2), and a phosphoric acid residue. The latter serves as a link between adjacent pentose units. The few estimates that have been made thus far indicate that no more than 3% of the organic P in soil occurs as nucleic acids or their derivatives.[83,93,94]

Anderson[93,94] isolated two pyrimidine nucleoside diphosphates from NaOH extracts of soil. One ester contained thymine and the other uracil (see Chapter 3). The suggestion was made that the two nucleoside diphosphates were artifacts, derived from polynucleotides by acid and alkaline hydrolysis.

PHOSPHOLIPIDS.   The phospholipids represent a group of biologically important organic compounds that are insoluble in water but soluble in fat solvents, such as benzene, chloroform, or ether. Included are the glycerophatides, such as phosphatidyl inositol, phosphatidyl choline or lecithin (VII), phosphatidyl serine (VIII), and phosphatidyl ethanolamine (IX), where RC=O is a long-chain fatty acyl group.

L-α-Lecithin	Phosphatidyl serine	Phosphatidyl ethanolamine
VII	VIII	IX

The total quantity of phospholipids in soil is small, usually less than 5 μg

P/g. In terms of percentage, from 0.5 to 7.0% of the soil organic P occurs as phospholipids, with a mean value of 1% (see Table 7.12). The phospholipids in soil are undoubtedly of microbial origin.

The presence of large amounts of mineral matter may reduce the ability of organic solvents to remove phospholipids; thus most estimates must be regarded as minimal. An improved procedure for recovering phospholipids from soil using sequential extraction with ethanol–benzene and methanol–chloroform has been described by Baker.[98]

Evidence for the presence of glycerophosphates in soil has come from the detection of glycerophosphate, choline, and ethanolamine in hydrolysates of extracted phospholipids.[16] The research carried out thus far indicates that phosphatidyl choline (VII) is the predominant soil phospholipid, followed by phosphatidyl ethanolamine (IX).

MICROBIAL TRANSFORMATIONS IN THE P CYCLE

Microorganisms can affect the P supply to higher plants in three different ways: (1) by decomposition of organic P compounds, with release of available inorganic phosphate, (2) by immobilizing available phosphates into cellular material, and (3) by promoting the solubilization of fixed or insoluble mineral forms of P, such as through the production of chelating agents.

Mineralization and Immobilization

Turnover of P through mineralization–immobilization follows somewhat the same pattern as for N (Chapter 5) in that both processes occur simultaneously. Accordingly, the maintenance of soluble phosphate in the soil solution will depend to some extent on the magnitude of the two opposing processes.

$$\text{Organic P} \underset{\text{immobilization}}{\overset{\text{mineralization}}{\rightleftharpoons}} PO_4^{3-}$$

From an experimental standpoint, the turnover of organic P in soil is not as easily measured as for organic N, one reason being that the end product of P mineralization (phosphate) can be removed from solution by adsorption to colloidal surfaces or by precipitation as insoluble phosphates of Ca, Fe, or Al. Furthermore, a decrease in soluble phosphates may result in some solubilization of insoluble forms of P through reestablishment of solubility equilibria, as discussed earlier.

Factors that affect the mineralization of soil organic P include temperature, moisture, aeration, soil pH, cultivation, presence of growing plants, and fertilizer P additions.[20] Alternative wetting and drying of soil enhance organic P mineralization.[99] For a series of soils having variable P contents,

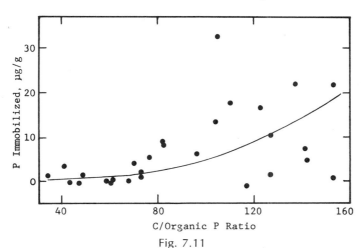

Fig. 7.11

Relationship between immobilized P and the C/organic P ratio of the soil. (Adapted from Enwezor.[101])

a positive correlation has often been observed between mineralized P and organic P content.[100] Enwezor[101] observed a net immobilization of P for 21 of the 28 soils he examined. The amount immobilized was more closely related to the percentage of the total P in the organic form than to total organic P. Although a poor correlation existed between immobilized P and the C/organic P ratio of the soil, there was a general tendency for P immobilization to be less at the lower than at the higher ratios, as can be seen by inspection of Fig. 7.11. Additions of fertilizer P can lead to increases in soil organic P through net immobilization, the net effect being a lowering of the C/organic P ratio.[102]

The microbial reduction of phosphate to phosphine (PH_3) is possible under anaerobic conditions but there is little evidence for significant evolution of this gas from waterlogged soils. Burford and Bremner[103] concluded that any phosphine formed in soils would be adsorbed by soil constituents and thereby would not escape to the atmosphere.

The P content of decomposing organic residues plays a key role in regulating the quantity of soluble P in the soil at any one time. Net immobilization of P will occur when the C/organic P ratio is 300 or more; net mineralization will result when the ratio is 200 or less. In terms of the P content of crop residues net immobilization is considered to occur when the decomposing material contains less than 0.2 to 0.3% P; net release will occur at higher P contents. The relationship between the C/P ratio of crop residues and soil phosphate levels is as follows:

	C/P Ratio	
<200	200 to 300	>300
Net gain of PO_4^{3-}	Neither a gain or loss of PO_4^{3-}	Net loss of PO_4^{3-}

Soil conditions that favor rapid decay of plant residues, such as proper aeration, moisture supply, and temperature (30° to 45°C) increase the rate of mineralization of P from added organic matter.

As was the case for N (see Chapter 5), net immobilization of P during early stages of the decay process will be followed by net mineralization of P as the C/P ratio of the decomposing residue is lowered. The recycling of P in this manner is undoubtedly of tremendous importance in the P economy of many soils, particularly those of developing countries where fertilizer P is in short supply or unavailable.

As noted earlier in Table 7.6, a significant amount of the P consumed by plants, frequently of the order of one-third, is contained in the nonharvested portion of the plant. Much of this organic P will become available to subsequent crops through mineralization. It should also be noted that availability of insoluble inorganic P may be enhanced during decay of plant residues.

With most green manure crops and animal manures, there will be an initial increase in inorganic P during decomposition. Ultimately, the microbial tissue itself will be attacked, resulting in further release of inorganic phosphate.

Because turnover of organic P is a biological process, mineralization-immobilization is strongly influenced by various physical and chemical properties of the soil, such as pH, moisture content, and energy supply. Turnover is likely to be very rapid at high soil temperatures, such as those expected for parts of the tropics.

Mineralization and immobilization of P may also be affected by the quantities of available N and S in the soil. The rate of release of P contained in crop residues is influenced by maturity, P content, application rate, time of contact of the crop residue with the soil, and kind of residue. Most plant residues probably contain sufficient inorganic P to meet microbial needs during decomposition. However, a major factor affecting P availability is the extent to which the inorganic P initially present is converted to microbial organic P and how this is further transformed as decomposition proceeds.

Phosphorus of the Soil Biomass

From 2 to 5% of the organic P in cultivated soils, to as much as 20% in grassland soils, occurs as part of the soil biomass.[104–107] The P content of microbial tissue has been reported to range from 1.5 to 2.5% for bacteria to as high as 4.8% for fungi.[104] These values are in accord with values estimated by Brookes et al.[106] for the P content of the soil biomass (1.4 to 4.7%). As

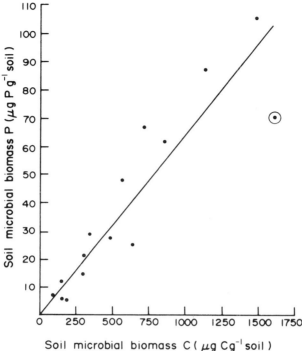

Fig. 7.12

Relationship between P and C of the soil microbial biomass. (Reprinted with permission from *Soil Biology and Biochemistry*, vol. 16, P. C. Brooks, D. S. Powlson, and D. S. Jenkinson, "Phosphorus in Soil Biomass," © 1984, Pergamon Press, Ltd.)

one might expect, the amount of biomass P in soil will follow closely the content of biomass C (Fig. 7.12).

Since the microbial biomass is relatively "labile" as compared to other organic matter fractions, the biomass P would be expected to be more available to plants than other organic P fractions. It should be noted that the P content of microbial tissue is somewhat higher than for plant tissue (0.1 to 0.5%).[28]

The following methods have been used to estimate biomass P in soils: (1) from cell weights based on microbial counts, (2) from the inorganic P released during incubation of partially sterilized soil, and (3) from ATP measurements.

Calculations Based on Cell Weights

In their review on soil organic P, Halstead and McKercher[21] concluded that as much as 5 to 10% of the organic P in many soils could be associated with living microbial tissue. Their estimate was based on a soil with 2 to 4%

organic matter, a micropopulation corresponding to 3.6 to 18.0 $\times$ 10^3 kg of fresh cell material per hectare, and a N/organic P ratio of 10:1 for the microbial population.

Incubation of Lysed Microbial Cells

In recent years the quantities of P in the microbial biomass have been estimated from the amount of inorganic P that is produced during incubation of soil fumigated with chloroform ($CHCl_3$), the extractant being 0.5 M $NaHCO_3$. Biomass P is calculated from the relationship:[104–107]

Biomass P

$$= \frac{\text{P extracted from fumigated soil} - \text{P extracted from nonfumigated soil}}{0.4}$$

The assumption is made that 40% of the P in the biomass is rendered extractable to 0.5 M $NaHCO_3$ (as inorganic P) by lysis and incubation. Also, a correction is required for phosphate fixation during extraction, as determined by adding a known amount of KH_2PO_4 during the extraction stage and measuring the recovery of added phosphate. As one might expect, measurements of biomass P in soils that contain large amounts of $NaHCO_3$-extractable phosphate will be subject to considerable error.

As noted earlier, from 2 to 5% of the soil organic P resides in the biomass, although higher and lower values are not uncommon. Some data obtained by Brookes et al.[106] are recorded in Table 7.15. In agreement with other work, the C/P ratios for the biomass (11.7 to 35.9) are substantially lower than for the soil organic matter (i.e., compare these values with the C/P ratios given earlier in Table 7.10). It appears evident, therefore, that the higher the percentage of the organic P as biomass P, the lower will be the C/organic P of the soil.

Adenosine Triphosphate (ATP)

ATP (see Fig. 7.4) is regarded as an ideal indicator of living microbial tissue, and its measurement provides the basis for assays of the soil biomass, and hence biomass P.[108–116] The inherent instability of the high-energy phosphate bond precludes ATP from remaining intact outside of living protoplasm. A survey of methods for extracting ATP from soils and sediments has been given by Webster et al.[116]

The basis for the determination of ATP in soil extracts is bioluminescence, i.e., light emission from the interaction of ATP with luciferin (LH_2), luciferase (E), and atmospheric O_2, the amount of light emitted being proportional to ATP content. The reactions are:

$$ATP + LH_2 + E \xrightarrow{Mg^{2+}} E-LH_2-AMP + P-P$$

$$E-LH_2-AMP + O_2 \longrightarrow E + \text{oxyluciferin} + AMP + \text{bioluminescence}$$

Table 7.15

Soil Biomass P of Some English Arable and Grassland Soils as a Percentage of Total
Soil Organic P[a]

Agricultural History	Total Organic P, µg/g	Biomass P		Biomass C/P
		µg/g	% of Organic P	
Arable soils				
1	180	6.0	3.3	26.3
2	210	5.3	2.5	35.9
3	242	7.0	2.9	14.1
4	810	27.5	3.4	17.9
Grassland soils				
1	330	72.3	21.9	22.5
2	352	61.7	17.5	13.7
3	190	15.0	7.9	20.0
4	210	12.0	5.7	12.3
5	500	24.8	5.0	25.6
6	200	48.6	24.3	11.7

[a] From Brookes et al.[106]

The ATP–ADP system functions in bioenergetic reactions and it occurs in all living organisms, although the amount of ATP produced is not the same in aerobic and anaerobic metabolism.

A linear relationship has been observed between ATP and biomass C in soil, a typical result being shown in Fig. 7.13. Since a linear relationship also exists between biomass C and biomass P (see Fig. 7.12), the latter can be estimated from the ATP content of the soil.

The ATP content of the soil biomass is of the order of 4.2 to 7.1 µg/mg of biomass C. Biomass C can be estimated from ATP values using the relationship:[111]

$$\text{Biomass C} = (124.3 \pm 4.2)\ \text{ATP}$$

The ATP content of the soil varies with season, storage of the soil, fertilizer P applications, and presence or absence of fresh plant residues. Hersman and Temple[117] concluded that ATP was the most satisfactory index of soil microbial activity in reclaimed coal strip mine spoils.

Phosphatase Enzymes in Soil

The final stage in the conversion of organic P to inorganic phosphate in soils occurs through the action of phosphatase enzymes, both intracellular and those released into the soil solution following lysis of microbial cells.[118–123] Two basic types exist in soils: acid phosphatases, which exhibit optimum

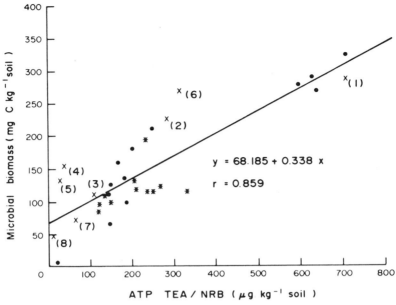

Fig. 7.13

Relationship between ATP and biomass C in soils. TEA/NRB refers to the Tris–EDTA buffer used for extraction. (Reprinted with permission from *Soil Biology and Biochemistry*, vol. 15, L. M. J. Verstraeten, K. De Coninck, and K. Vlassak, "ATP Content of Soils Estimated by Two Contrasting Extraction Methods", © 1983, Pergamon Press, Ltd.)

activity in the pH range 4 to 6, and alkaline phosphatases, with pH optimum 9 to 11. The subject of phosphatase activity in soil has been reviewed by Cosgrove[19] and Halstead and McKercher.[21]

The general equation for the reaction catalyzed by phosphatases is:[123]

$$\underset{O^-}{\overset{\overset{\displaystyle O}{\|}}{R\text{-}O\text{-}P\text{-}O^-}} + H_2O \xrightarrow[\text{enzyme}]{\text{phosphatase}} \underset{O^-}{\overset{\overset{\displaystyle O}{\|}}{HO\text{-}P\text{-}O^-}} + ROH$$

Methods for the determination of phosphatase are outlined by Tabatabai.[124] The more common procedures are based on the colorimetric determination of the *p*-nitrophenol released by phosphatase when soil is incubated with buffered (pH 6.5 for acid phosphatase and pH 11 for alkaline phosphatase) sodium *p*-nitrophenyl phosphate solution. The results, which are expressed in such units as μg P/g·x hours, can then be correlated with other measured soil parameters, or, in some cases, to soil management practices.[118–123]

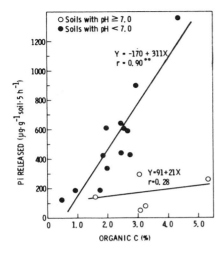

Fig. 7.14
Relationship between phosphatase activity and organic C in soils. A highly significant correlation is shown for the soils with pH values <7.0. Activity was also correlated with clay content, which itself was related to organic C. (Reprinted with permission from *Soil Biology and Biochemistry*, Vol. 11, M. A. Tabatabai and W. A. Dick, "Distribution and Stability of Pyrophosphatase in Soils," © 1979, Pergamon Press, Ltd.)

Phosphatase activity has been shown to be related to organic C content (Fig. 7.14), and to be affected by such factors as pH, moisture, storage of the soil, P fertilization, and incubation of the soil with a C substrate. Attempts to isolate phosphatase enzymes from soil have not apparently been successful as none is recorded in the literature.

Role of Microorganisms in the Solubilization of Insoluble Phosphates

As noted earlier, the availability of phosphate in soil is often limited by fixation reactions, which convert the monophosphate ($H_2PO_3^-$) to various insoluble forms. Insoluble Ca phosphates predominate in calcarerous soils while Fe and Al phosphates are formed in acidic soils. Absorption by clay minerals can affect phosphate availability under neutral or slightly acid conditions.

Several studies indicate that the availability of phosphates in soil is enhanced by additions of organic residues. Several independent, but not necessarily exclusive, reactions may be involved, including the following:[125]

1 Phosphorus tied up as insoluble Ca, Fe, and Al phosphates may be released to soluble forms through the action of organic acids and other chelates that are produced during decomposition of crops residues and excretion products from plant roots. The mechanisms involved are depicted in Fig. 7.15.

2 Humates produced during decomposition may compete with phosphate ions for adsorbing surfaces, thereby preventing fixation of phosphate.

3 Humates may form a protective surface over colloidal sesquioxides, with reduction in phosphate fixation.

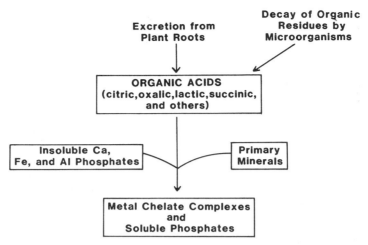

Fig. 7.15

Schematic diagram of the release of insoluble phosphates to soluble forms through the action of organic acids and other naturally occurring chelates. (From Stevenson,[125] reproduced by permission of John Wiley and Sons.)

4   The solubility of Ca and Mg phosphates may be increased as a result of the production of carbonic acid from CO_2 released during decay (CO_2 + H_2O → H_2CO_3).

5   Fresh organic matter may have a priming effect on the decomposition of native humus, with mineralization of organic P.

6   Phospho–humate complexes may be formed.

At present, it is not possible to select any given pathway from among these alternatives. Evidence that naturally occurring chelating agents enhance the availability of P to higher plants (item 1) is circumstantial and some investigators have questioned whether organic acids and other chelating agents are produced (or persist) in sufficient abundance to appreciably influence phosphate solubility.[126] The effectiveness of these compounds will undoubtedly be greatest in unfertilized soils low in natural fertility and where most of the P is tied up as insoluble Ca, Fe, or Al phosphates. The situation is somewhat analogous to the role of mycrorrhizal fungi in enhancing uptake by plants, which is restricted largely to P-deficient soils.[13] In such ecosystems the fungi are indispensable and play a primary role in providing P to plants, especially during the early stages of colonization.

The following mechanisms have been given for the increased uptake of P by mycorrhizal roots:[127]

1   There is greater exploration of soil P because of extension of hyphae into the soil.

2 Exudates are produced by the fungi (e.g., organic acids) that increase the amount of P available to plants.

3 Mycorrhizal and nonmycorrhizal roots differ in their ability to absorb phosphate from solution.

4 Differential absorption of anions and cations in the presence of mycorrhizal fungi leads to changes in pH, which thereby alter the amount of soil phosphate taken up by the plant.

Another aspect that should be mentioned is that the slow but continued solubilization of inorganic phosphates through chelation, such as may occur during soil formation, would lead to conversion of mineral P to more available organic P forms, thereby enhancing the fertility of the soil.

The action of organic chelates in solubilizing phosphates and phosphate minerals (item 2) has been attributed to the formation of complexes with Ca, Fe, or Al, thereby releasing the phosphate in water-soluble forms. The reactions are:

$$CaX_2 \cdot 3Ca(PO_4)_2 + \text{chelate} \rightarrow \text{soluble } PO_4^{-2} + \text{Ca-chelate complex}$$

where $(X = OH \text{ or } F)$.

$$Al(Fe) \cdot (H_2O)_3(OH)_2H_2PO_4 + \text{chelate} \rightarrow \text{soluble } PO_4^{2-}$$
$$+ Al(Fe) - \text{chelate complex}$$

Similar reactions may be involved in preventing the fixation of fertilizer-applied phosphate, as well as phosphate formed in situ by weathering of minerals or decay of organic matter.

Numerous laboratory studies, reviewed by Stevenson,[128] have shown that many organic acids are effective in releasing phosphates from insoluble mineral forms, as well as in reducing the precipitation of phosphate by Fe and Al. The most effective compounds are those capable of forming stable chelate complexes with metal ions, such as the di- and tricarboxylic hydroxy acids. Calcium, for example, is chelated very strongly by citric acid (a tricarboxylic compound), less strongly by dibasic acids (e.g., malic and tartaric acids), and only to a small extent by most α-hydroxy monobasic acids (such as lactic). The role of organic acids and other natural chelates in forming chelate complexes in soil is discussed in greater detail in Chapter 9.

The ability of soil microorganisms to solubilize insoluble phosphates and minerals of various types through the production of organic acids has been demonstrated in experiments carried out under laboratory conditions. This is normally done by first isolating the bacteria, actinomycetes, or fungi as per the agar-plate method for counting soil microorganisms. Individual isolates are selected and recultured, following which select purified cultures are inoculated on petri dishes containing scattered granules of insoluble mineral matter. A halo is formed around individual mineral grains when decomposition or solubilization occurs. In many instances, decomposition has been related to organic acid production, particularly those compounds that form highly stable complexes with metal ions (e.g., citric).

Conditions that promote the activities of microorganisms would be expected to enhance the solubilization of insoluble phosphates through the production of chelating agents. Thus it is not surprising that the addition of decomposable organic matter to soil has been observed to increase phosphate uptake by plants, an effect that has been confirmed in experiments with ^{32}P-labeled products.

Gerretsen[129] is usually credited as the first to draw attention to the influence of rhizosphere microorganisms on phosphate uptake by plants. He grew a variety of crop plants in sterilized and nonsterilized soil, to which various insoluble phosphates were added, and observed greater phosphate adsorption by plants growing in nonsterilized soil. When bacteria isolated from root surfaces were put into sterile soil, enhanced phosphate uptake occurred. Normal rhizosphere roots were found to bring phosphate into solution far more effectively than uninfected roots.

Several reasons can be given for the enhanced uptake of soil P, including stimulation of phosphate-solubilizing bacteria, greater turnover of P in the native humus, and production of organic acids that dissolve insoluble phosphates through chelation of Ca, Fe, and Al. A high proportion of the microorganisms in the rhizosphere have been shown to produce organic acids when cultured on laboratory growth media (see Chapter 9).

As plant roots penetrate the soil, phosphate is adsorbed at a high rate. The net result is that a zone of phosphate depletion is formed in the immediate vicinity of the root. This zone is only a few millimeters wide and coincides with the rhizosphere—the region near the root where microorganisms are particularly active. The net effect of the increased activity and solubilization of phosphate is that P is circulated at a higher order of magnitude, to the benefit of the plant.

The conclusion that rhizosphere microorganisms play an important role in the P nutrition of plants has been challenged by Tinker and Sanders.[126] Their line of reasoning is that (1) the quantities of organic acids produced or secreted by plants are too small to significantly affect phosphate solubilization and (2) any phosphate released by the bacteria will enter the normal absorption equilibria and will thereby suffer the same transport impedance as other phosphate ions. Nevertheless, the case cannot be considered closed. As Hayman[22] pointed out, plants always contain a rich population of microorganisms around their roots, and that it is unrealistic to assess plant uptake of soil P without considering the activities of these microbes.

Inoculation of Soil with "Phosphobacterin"

Inoculation of soil with *Bacillus megatherium var. phosphaticum*, for which the term "phosphobacterin" is used, has been a common practice in Russia

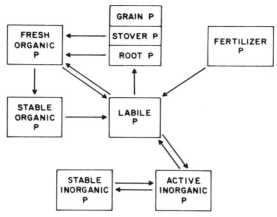

Fig. 7.16

Pools and flows of P in the soil–plant system.[28] The pool of labile inorganic P supplies the plant, in which P is divided into root, stover, and grain. Organic P is partitioned into the fresh residue P, consisting of the P in undecayed residues and microbes, and the stable organic P. (Reproduced from *Soil Sci. Soc. Amer. J.* **48,** 800–805 (1984) by permission of the Soil Science Society of America.)

for many years. At one time the phosphobacterin was said to benefit 50 to 70% of the field crops to which it is applied. The organism, which releases organic P to available mineral forms, has been reported to be particularly effective in Mollisol (Chernozem) soils, which contain much organically bound P. The early claims have not been substantiated[130–132] and the widespread adoption of bacterial inoculation for the purpose of increasing the availability of soil P cannot be recommended.

MODELING OF THE SOIL P CYCLE

Jones et al.[28] have designed a soil and plant P model (Fig. 7.16) based on pools of stable, active, and labile inorganic P; fresh organic and stable organic P; and grain, stover, and root P. The model simulates P uptake by plants in relation to soil chemical and physical properties, crop P requirements, tillage practice, fertilizer rate, soil temperature, and soil water content. Initial values of active inorganic P and stable organic P are calculated from chemical and/or adsorption properties. Estimates for initial labile P can be included, such as may be obtained from soil test data (i.e., P extracted with 0.5 M NaHCO$_3$). Promising results were obtained when the model was used to simulate long-time changes in soil organic P, soil test values, and yields of corn and wheat at several locations in the Great Plains of the United States.[28] Other P models have been devised by Cole et al.[133] and Smith et al.[134]

SUMMARY

The phosphorus cycle is complex and involves the storage of P in living organisms, dead organic matter, and inorganic forms. Phosphorus is adsorbed by plants largely as the orthophosphate ($H_2PO_4^-$), which is present in the soil solution. However, replenishment of this P by solubilization of inorganic forms, or through mineralization of organic P, is of considerable importance from the standpoint of plant nutrition. A particularly useful concept is that the P in soil can be partitioned into "pools" based on availability of varioius organic and inorganic forms to plants.

Fixation reactions of various types are of importance from the standpoint of the efficiency of fertilizer P use by plants. In acid soils, phosphate is precipitated as the highly insoluble Fe or Al phosphates, or adsorbed to oxide surfaces. A major factor affecting P availability in calcareous soils is the formation of insoluble Ca phosphates.

As much as two-thirds or more of the total P in soils occurs in organic forms, with the higher percentages being typical of peats, uncultivated forest soils, and tropical soils. An active P cycle exists in soil, in which turnover between organic and inorganic forms (mineralization–immobilization) plays a major role. Additions of decomposable organic matter to soil have been shown in laboratory and greenhouse studies to enhance the availability of insoluble Ca, Fe, and Al phosphates to plants but the importance of the process under conditions existing in the field is unknown.

Excellent progress has been made in determining the nature of organic P compounds in soils but much of the organic P has yet to be identified in known compounds. As a general rule, somewhat less than one-half of the organic P can be accounted for. Approximate recoveries of organic P are: inositol phosphates, 2 to 50%; phospholipids, 1 to 5%; nucleic acids, 0.2 to 2.5%; phosphoproteins, trace; metabolic phosphates, trace.

REFERENCES

1   J. W. B. Stewart, "The Importance of P Cycling and Organic P in Soils," in *Better Crops with Plant Food*, American Potash-Phosphate Institute, Atlanta, Georgia, Winter Issue, 1980–1981, pp. 16–19.

2   C. W. Finkl, Jr. and R. W. Simonson, "Phosphorus Cycle," in R. W. Fairbridge and C. W. Finkl, Jr., Eds., *The Encyclopedia of Soil Science*, Dowden, Hutchinson, and Ross, Stroudsburg, Pa., 1971, pp. 370–377.

3   V. Pierrou, "The Global Phosphorus Cycle," in B. H. Svensson and R. Söderlund, Eds., *Nitrogen, Phosphorus and Sulphur—Global Cycles*. SCOPE Report No. 7, Ecol. Bull. 22 (Stockholm), 1976, pp. 75–88.

4   J. E. Richey, "The Phosphorus Cycle," in B. Bolin and R. B. Cook, Eds., *The Major Biogeochemical Cycles and Their Interactions*, Wiley, New York, 1983, pp. 51–56.

5   V. Sauchelli, *Manual on Phosphates in Agriculture*, Davidson Chemical Corp., Baltimore, 1951.

6 C. V. Cole and R. D. Heil, "Phosphorus Effects on Terrestrial Nitrogen Cycling," in F. E. Clark and T. Rosswall, Eds., *Terrestrial Nitrogen Cycles*, Ecol. Bull. 33 (Stockholm), 1981, pp. 363–374.

7 G. E. Likens, F. H. Bormann, and N. B. Johnson, "Interactions Between Major Biogeochemical Cycles: Terrestrial Cycles," in G. E. Likens, Ed., *Perspectives of the Major Biogeochemical Cycles*, SCOPE Report No. 17, Chichester, Wiley, 1981, pp. 93–112.

8 W. B. McGill and C. V. Cole, *Geoderma* **26**, 267 (1981).

9 T. W. Walker, "The Significance of Phosphorus in Pedogenesis," in *Experimental Pedology*, Easter School in Agricultural Sciences, University of Nottingham, 1964, pp. 295–316.

10 T. W. Walker and J. K. Syers, *Geoderma* **15**, 1 (1976).

11 R. L. Bieleski, *Ann. Rev. Plant Physiol.* **24**, 225 (1973).

12 F. E. Khasawneh, E. C. Sample, and E. J. Kamprath, Eds., *The Role of Phosphorus in Agriculture*, American Society of Agronomy, Madison, Wis., 1980.

13 K. Mengel and E. A. Kirkby, *Principles of Plant Nutrition*, 2nd ed., International Potash Institute, Bern, Switzerland, 1982.

14 W. H. Pierre, *J. Amer. Soc. Agron.* **40**, 1 (1948).

15 W. H. Pierre and A. G. Norman, Eds., *Soil and Fertilizer Phosphorus*, Academic Press, New York, 1953.

16 G. Anderson, "Other Organic Phosphorus Compounds," in J. E. Gieseking, Ed., *Soil Components, Vol. 1, Organic Compounds*, Springer-Verlag, Berlin, 1975, pp. 305–331.

17 G. Anderson, "Assessing Organic Phosphorus in Soils," in F. E. Khasawneh, E. C. Sample, and E. J. Kamprath, Eds., *The Role of Phosphorus in Agriculture*, American Society of Agronomy, Madison, Wis., 1980, pp. 411–431.

18 D. J. Cosgrove, "Metabolism of Organic Phosphates in Soil," in A. D. McLaren and G. H. Peterson, Eds., *Soil Biochemistry*, Vol. 1, Dekker, New York, 1967, pp. 216–228.

19 D. J. Cosgrove, *Adv. Microbial Ecol.* **1**, 95 (1977).

20 R. C. Dalal, *Adv. Agron.* **29**, 83 (1977).

21 R. L. Halstead and R. B. McKercher, "Biochemistry and Cycling of Phosphorus," in E. A. Paul and A. D. McLaren, Eds., *Soil Biochemistry*, Vol. 4, Dekker, New York, 1975, pp. 31–63.

22 D. S. Hayman, "Phosphorus Cycling by Soil Micro-organisms and Plant Roots," in N. Walker, Ed., *Soil Microbiology*, Butterworth, London, 1975, pp. 67–91.

23 C. G. Kowalenko, "Organic Nitrogen, Phosphorus and Sulfur in Soils," in M. Schnitzer and S. U. Khan, Eds., *Soil Organic Matter*, Elsevier, New York, 1978, pp. 95–136.

24 S. Larsen, *Adv. Agron.* **19**, 151 (1967).

25 J. R. Ramirez Martinez, *Folia Microbiol.* **13**, 161 (1968).

26 J. Emsley, "The Phosphorus Cycle," in P. J. Craig et al., Eds. *The Natural Environment and the Biogeochemical Cycles*, Springer-Verlag, New York, 1980, pp. 147–167.

27 E. J. Griffith, A. Beeton, J. M. Spencer, and O. T. Mitchell, Eds., *Environmental Phosphorus Handbook*, Wiley, New York, 1973.

28 C. A. Jones, C. V. Cole, A. N. Sharpley, and J. R. Williams, *Soil Sci. Soc. Amer. J.* **48**, 800, 805, 810 (1984).

29 A. Finck, *Fertilizers and Fertilization*, Verlag Chemeim, Deerfield Beach, Fla., 1982.

30 R. H. Olson, T. J. Army, J. J. Hanway, and V. J. Kilmer, Eds., *Fertilizer Technology and Use*, Soil Science Society of America, Madison, Wis., 1971.

31 C. W. Finkl, Jr., and R. W. Simonson, "Phosphorus Cycle," in R. W. Fairbridge and C. W. Finkl, Jr., Eds., *Encyclopedia of Soil Science*, Part 1, Dowden, Hutchinson and Ross, Stroudsburg, Pa., 1979, pp. 370–377.

32   R. W. Simonson, "Loss of Nutrient Elements During Soil Formation," in O. P. Engelstad, Ed., *Nutrient Mobility in Soils: Accumulations and Losses*, Soil Science Society of America, Madison, Wis., 1970, pp. 21–45.

33   J. C. Ryden, J. K. Syers, and R. F. Harris, *Adv. Agron.* **25**, 1 (1973).

34   R. Thomas and J. P. Law, "Properties of Waste Waters," in L. F. Elliott and F. J. Stevenson, Eds., *Soils for Management of Organic Wastes and Waste Waters*, American Society of Agronomy, Madison, Wis., 1977, pp. 47–72.

35   H. J. Haas, D. L. Grunes, and G. A. Reichman, *Soil Sci. Soc. Amer. Proc.* **25**, 214 (1961).

36   H. Tiessen, J. W. B. Stewart, and J. R. Bettany, *Agron. J.* **74**, 831 (1982).

37   S. A. Barber, *Commun. Soil Sci. Plant Anal.* **10**, 1459 (1979).

38   O. G. Oniani, M. Chater, and G. E. G. Mattingly, *J. Soil Sci.* **24**, 1 (1973).

39   R. A. Bowman and C. V. Cole, *Soil Sci.* **125**, 95 (1978).

40   J. M. Sadler and J. W. B. Stewart, *Can. J. Soil Sci.* **55**, 149 (1975).

41   A. N. Sharpley and S. J. Smith, *Soil Sci. Soc. Amer. J.* **47**, 581 (1983).

42   S. C. Chang and M. L. Jackson, *Soil Sci.* **84**, 133 (1957).

43   J. D. H. Williams, J. K. Syers, and T. W. Walker, *Soil Sci. Soc. Amer. Proc.* **31**, 736 (1967).

44   J. K. Syers, G. W. Smillie, and J. D. H. Williams, *Soil Sci. Soc. Amer. Proc.* **36**, 20 (1972).

45   B. S. Chauhan, J. W. B. Stewart, and E. A. Paul, *Can. J. Soil Sci.* **39**, 387 (1979).

46   B. S. Chauhan, J. W. B. Stewart, and E. A. Paul, *Can. J. Soil Sci.* **61**, 373 (1981).

47   M. J. Hedley, R. E. White, and P. H. Nye, *New Phytol.* **91**, 45 (1982).

48   M. J. Hedley, J. W. B. Stewart, and B. S. Chauhan, *Soil Sci. Soc. Amer. J.* **46**, 970 (1982).

49   R. A. Bowman and C. V. Cole, *Soil Sci.* **125**, 49 (1978).

50   R. J. Atkinson, A. M. Posner, and J. P. Quirk, *J. Inorg. Nucl. Chem.* **34**, 2201 (1972).

51   F. J. Hingston, A. M. Posner, and J. P. Quirk, *J. Soil Sci.* **25**, 16 (1974).

52   J. H. Kyle, A. M. Posner, and J. P. Quirk, *J. Soil Sci.* **26**, 32 (1975).

53   D. Muljadi, A. M. Posner, and J. P. Quirk, *J. Soil Sci.* **17**, 212 (1966).

54   R. L. Parfitt, R. J. Atkinson, and R. S. C. Smart, *Soil Sci. Soc. Amer. Proc.* **39**, 837 (1975).

55   R. L. Parfitt, A. R. Fraser, J. D. Russell, and V. C. Farmer, *J. Soil Sci.* **28**, 40 (1977).

56   J. C. Ryden, J. R. McLaughlin, and J. K. Syers, *J. Soil Sci.* **28**, 62, 72, 585 (1977).

57   F. Borie and H. Zunino, *Soil Biol. Biochem.* **15**, 599 (1983).

58   W. A. Stoop, *Geoderma* **31**, 57 (1983).

59   U. Kafkafi, A. M. Posner, and J. P. Quirk, *Soil Sci. Soc. Amer. Proc.* **31**, 348 (1967).

60   R. L. Fox, "External Phosphorus Requirements of Crops," in R. H. Dowdy et al., Eds., *Chemistry in the Soil Environment*, ASA Special Publ. No. 40, American Society of Agronomy, Madison, Wis., 1981, pp. 223–239.

61   W. H. Pierre and F. W. Parker, *Soil Sci.* **24**, 119 (1927).

62   R. C. Dalal and E. G. Hallsworth, *Soil Sci. Soc. Amer. J.* **40**, 541 (1976).

63   S. R. Olsen and L. E. Sommers, "Phosphorus," in A. L. Page, R. H. Miller, and D. R. Keeney, Eds., *Methods of Soil Analysis*, Part 2, 2nd ed., American Society of Agronomy, Madison, Wis., 1982, pp. 403–430.

64   J. H. Steward and J. M. Oades, *J. Soil Sci.* **23**, 38 (1972).

65   J. D. H. Williams, J. K. Syers, T. W. Walker, and R. W. Rex, *Soil Sci.* **110**, 13 (1970).

66   A. P. Uriyo and A. Kesseba, *Geoderma* **13**, 201 (1975).

67   R. W. Pearson and R. W. Simonson, *Soil Sci. Soc. Amer. Proc.* **4**, 162 (1939).

68   R. T. Baker, *J. Soil Sci.* **27,** 504 (1976).
69   M. K. John, P. N. Sprout, and C. C. Kelley, *Can. J. Soil Sci.* **45,** 87 (1964).
70   A. Kaila, *Soil Sci.* **95,** 38 (1963).
71   A. M. L. Neptune, M. A. Tabatabai, and J. J. Hanway, *Soil Sci. Soc. Amer. Proc.* **39,** 51 (1975).
72   L. L. Somani and S. W. Sarena, *Anal. Edafol. Agrobiol.* **37,** 809 (1978).
73   T. W. Walker and A. F. R. Adams, *Soil Sci.* **85,** 307 (1958).
74   C. H. Williams, E. G. Williams, and N. M. Scott, *J. Soil Sci.* **11,** 334 (1960).
75   K. M. Goh and M. R. William, *J. Soil Sci.* **33,** 73 (1982).
76   G. Anderson and R. J. Hance, *Plant Soil* **19,** 296 (1963).
77   A. G. Caldwell and C. A. Black, *Soil Sci. Soc. Amer. Proc.* **22,** 290, 293, 296 (1958).
78   D. J. Cosgrove, *Aust. J. Soil Res.* **1,** 203 (1963).
79   D. J. Cosgrove, *Soil Biol. Biochem.* **1,** 325 (1969).
80   D. J. Cosgrove, *Inositol Phosphates*, Elsevier, New York, 1980.
81   J. F. Dormaar, *Soil Sci.* **104,** 17 (1967).
82   R. L. Halstead and G. Anderson, *Can. J. Soil Sci.* **50,** 111 (1970).
83   A. Islam and B. Ahmed, *J. Soil Sci.* **24,** 193 (1973).
84   M. F. L'Annunziata, *Soil Sci. Soc. Amer. Proc.* **39,** 377 (1975).
85   R. Mandal and A. Islam, *Geoderma* **22,** 315 (1979).
86   R. B. McKercher and G. Anderson, *J. Soil Sci.* **19,** 47, 302 (1968).
87   J. R. Moyer and R. L. Thomas, *Soil Sci. Soc. Amer. Proc.* **34,** 80 (1970).
88   T. I. Omotoso and A. Wild, *J. Soil Sci.* **21,** 216, 224 (1970).
89   D. H. Smith and F. E. Clark, *Soil Sci. Soc. Amer. Proc.* **16,** 170 (1952).
90   R. L. Veinot and R. L. Thomas, *Soil Sci. Soc. Amer. Proc.* **36,** 71 (1972).
91   C. H. Williams and G. Anderson, *Aust. J. Soil Res.* **6,** 121 (1968).
92   J. K. Martin, *New Zealand J. Agric. Res.* **7,** 723, 736, 750 (1964).
93   G. Anderson, "Nucleic Acids, Derivatives, and Organic Phosphates," in A. D. McLaren and G. H. Peterson, Eds., *Soil Biochemistry*, Dekker, New York, 1967, pp. 67–90.
94   G. Anderson, *J. Soil Sci.* **21,** 96 (1970).
95   J. F. Dormaar, *Soil Sci.* **110,** 136 (1970).
96   C. G. Kowalenko and R. B. McKercher, *Soil Biol. Biochem.* **3,** 243 (1971).
97   C. G. Kowalenko and R. B. Mckercher, *Can. J. Soil Sci.* **51,** 19 (1971).
98   R. T. Baker, *J. Soil Sci.* **26,** 432 (1975).
99   H. F. Birch and M. T. Friend, *Nature* (*London*) **191,** 731 (1961).
100 D. N. Chakravarty and S. N. Tewari, *Indian J. Agric. Chem.* **3,** 21 (1970).
101 W. O. Enwezor, *Soil Sci.* **103,**62 (1967).
102 A. van Diest, *Plant Soil*, **29,** 241, 248 (1968).
103 J. R. Burford and J. M. Bremner, *Soil Biol. Biochem.* **4,** 489 (1972).
104 J. P. E. Anderson and K. H. Domsch, *Soil Sci.* **130,** 211 (1980).
105 P. C. Brookes, D. S. Powlson, and D. S. Jenkinson, *Soil Biol. Biochem.* **14,** 319 (1982).
106 P. C. Brookes, D. S. Powlson, and D. S. Jenkinson, *Soil Biol. Biochem.* **16,** 169 (1984).
107 M. J. Hedley and J. W. B. Stewart, *Soil Biol. Biochem.* **14,** 377, (1982).
108 M. Ahmed, J. M. Oades, and J. N. Ladd, *Soil Biol. Biochem.* **14,** 273 (1982).
109 F. Eiland, *Soil Biol Biochem.* **11,** 31 (1979).
110 D. S. Jenkinson and J. M. Oades, *Soil Biol. Biochem.* **11,** 193 (1979).

111 D. S. Jenkinson, S. A. Davidson, and D. S. Powlson, *Soil Biol. Biochem.* **11,** 521 (1979).

112 J. M. Oades and D. S. Jenkinson, *Soil Biol. Biochem.* **11,** 201 (1979).

113 G. P. Sparling and F. Eiland, *Soil Biol. Biochem.* **4,** 227 (1983).

114 K. R. Tate and D. S. Jenkinson, *Soil Biol. Biochem.* **14,** 331 (1982).

115 L. M. J. Verstraeten, K. De Coninck, and K. Valassak, *Soil Biol. Biochem.* **15,** 397 (1983).

116 J. J. Webster, G. J. Hampton, and F. R. Leach, *Soil Biol. Biochem.* **16,** 335 (1984).

117 L. E. Hersman and K. L. Temple, *Soil Sci.* **127,** 70 (1979).

118 F. Eivazi and M. A. Tabatabai, *Soil Biol. Biochem.* **9,** 167 (1977).

119 A. F. Harrison, *Soil Biol. Biochem.* **15,** 93 (1983).

120 A. F. Harrision and T. Pearce, *Soil Biol. Biochem.* **11,** 405 (1979).

121 R. E. Malcolm, *Soil Biol. Biochem.* **15,** 403 (1983).

122 G. A. Spiers and W. B. McGill, *Soil Biol. Biochem.* **11,** 3 (1979).

123 M. A. Tabatabai and W. A. Dick, *Soil Biol. Biochem.* **11,** 655 (1979).

124 M. A. Tabatabai, "Soil Enzymes," in A. L. Page, R. H. Miller, and D. R. Keeney, *Methods of Soil Analysis, Part 2*, 2nd ed., American Society of Agronomy, Madison, Wis., 1982, pp. 903–947.

125 F. J. Stevenson, *Humus Chemistry: Genesis, Composition, Reactions*, Wiley, New York, 1982.

126 P. B. H. Tinker and F. E. Sanders, *Soil Sci.* **119,** 363 (1975).

127 N. S. Bolan, A. D. Robson, N. J. Barrow, and L. A. G. Aylmore, *Soil Biol. Biochem.* **16,** 299 (1984).

128 F. J. Stevenson, "Organic Acids," in A. D. McLaren and G. H. Peterson, Eds., *Soil Biochemistry*, Dekker, New York, 1967, pp. 119–146.

129 F. C. Gerretsen, *Plant and Soil*, **1,** 51 (1948).

130 R. Cooper, *Soils Fert.* **22,** 327 (1959).

131 R. Azcon, J. M. Barea, and D. S. Hayman, *Soil Biol. Biochem.* **8,** 135 (1976).

132 J. Raf, D. J. Bagyaraj, and A. Manjunath, *Soil Biol. Biochem.* **13,** 105 (1981).

133 C. V. Cole, G. S. Innis, and J. W. B. Stewart, *Ecology* **58,** 1 (1977).

134 O. L. Smith, *Soil Biol. Biochem.* **11,** 585 (1979).

EIGHT

THE SULFUR CYCLE

Sulfur, the tenth most abundant element in the universe (about fifth by weight), occurs naturally as gypsum ($CaSO_4 \cdot 2H_2O$); as pyrite (FeS_2) in shale, limestone, and sandstone; as elemental $S°$ in bituminous shales and various sedimentary traps, such as salt domes; and as part of soil and marine humus. Sulfur is a constituent of all living organisms, where it ranks in importance with N as a constituent of proteins. Sulfur is also an integral component of vitamins and hormones.

Despite its essentiality, S has been described as the neglected plant nutrient. The sporadic attention that has been given to this nutrient is due in part to the fact that crop responses to applied S have been restricted to a few geographical areas and to only a few crops. Sulfur deficiencies are now being reported with increasing frequency and on a wider variety of crops, a result that has been attributed to increased use of high-analysis S-free fertilizers, to a reduction in the amount of S used as a pesticide, and to higher crop yields, which means that requirements for all essential plant nutrients will be greater. Another contributing factor is a reduction in the amount of S that reaches the soil in rainwater or dry deposition. As effective emission control systems are adopted for reducing the sulfur dioxide (SO_2) content of the atmosphere through burning of fossil fuels, the size of S-deficient areas may increase and new deficient areas may develop.

A pictorial representation of the S cycle is shown in Fig. 8.1. During weathering processes, the S of primary minerals is converted to sulfate (SO_4^{2-}), which, in turn, is consumed by plants and converted into organic forms, such as the cystine and methionine of proteins. When plant and animal residues are returned to the soil and subjected to decay by microorganisms, part of the organic S reappears as SO_4^{2-}; part is incorporated into microbial tissue, and hence into humus. Sulfur is added to soil in fertilizers, in pesticides of various types, in irrigation water, and through absorption of S gases (e.g., SO_2) from the atmosphere. Losses of soil S result from runoff and leaching. Under certain circumstances the soil may serve as a source of hydrogen sulfide (H_2S) and other S gases to the atmosphere.

The importance of organic matter in providing S to growing plants will,

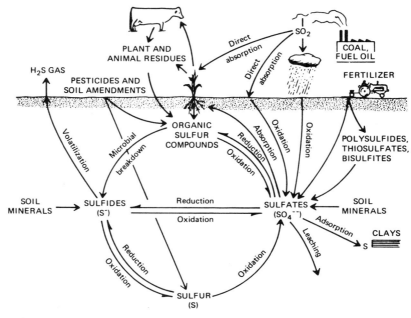

Fig. 8.1

Pictorial representation of the S cycle in soils. (Reprinted with permission of Macmillan Publishing Company from *The Nature and Properties of Soils* by N.C. Brady. Copyright © 1974 by Macmillan Publishing Company.)

of course, depend on the adequacies of extraneous additions in meeting crop requirements. As far as atmospheric S is concerned, rainfall deposition ranges from less than 5 kg/ha·yr in rural areas remote from industrial activity to as much as 200 kg/ha·yr in areas of congested industrial activity.

In this chapter a cursory examination is given of pertinent facts regarding the S cycle in soil. Additional data and documentation can be found in numerous monographs and reviews, covering such topics as biogeochemistry,[2-7] plant nutrition,[8,9] forms and reactions in soil,[10-18] and microbial oxidations and reductions.[19-21]

BIOGEOCHEMISTRY OF THE GLOBAL S CYCLE

Major reservoirs of S in the earth (Table 8.1) are the lithosphere (24.3×10^{18} kg) and the hydrosphere (1.3×10^{18} kg). Moderate amounts of S occur in the pedosphere (2.7×10^{14} kg); small amounts are found in the atmosphere (4.8×10^9 kg). Fluxes of S between some of the more important reservoirs are approximately as follows:[4]

	10^{12} kg/yr
Atmosphere to the ocean through fallout	258
Pedosphere to the ocean in runoff	208
Ocean to the atmosphere in sea spray	140
Lithosphere to the hydrosphere by weathering	114
Lithosphere to the atmosphere by fuel combustion and metal processing	111

Principal geologic sources of S for industrial uses include elemental $S°$ deposits associated with salt domes and sedimentary rocks. Total annual production of $S°$ in the United States exceeds 5×10^9 kg, most of which is converted to H_2SO_4 for use in agricultural, chemical, and related industries.[22] Known reserves of mineable S in the United States exceed 200×10^9 kg.

Chemical Properties of S

Sulfur, with an atomic weight of 32.064, exists in several valence or oxidation states, the lowest being -2 in sulfide (gaseous H_2S, ferrous sulfide, FeS) and the highest being $+6$ in sulfate (SO_4^{2-}). Important inorganic S species in the environment are given in Table 8.2. Typical oxidation states are 0 for elemental $S°$, -1 for disulfide (e.g, pyrite, FeS_2), and $+4$ for sulfur dioxide gas (SO_2) and the sulfite ion (SO_3^{2-}). Oxidation and reduction of S occur rather easily under proper conditions, thereby accounting for the great diversity of reactions that S undergoes in the environment.

Sulfur exists in four stable isotopes—^{32}S, ^{33}S, ^{34}S, and ^{36}S—for which

Table 8.1

Major S Reserves in the Earth[a]

Reservoir	Amount of S, kg
Atmosphere	4.8×10^9
Lithosphere	24.3×10^{18}
Hydrosphere	
Sea	1.3×10^{18}
Fresh water	3.0×10^{12}
Marine organisms	2.4×10^{11}
Pedosphere	
Soil	2.6×10^{14}
Soil organic matter	0.1×10^{14}
Land plants	7.6×10^{11}

[a] Adapted from Freney et al.[4] and Trudinger.[6] The lithosphere refers to the crust of the earth.

Table 8.2
Important Inorganic S Species in the Environment[a]

Sulfides, −2		Thiosulfate, −2 and +6	
Sulfide ion	S^{2-}	Thiosulfate ion	$S_2O_3^{2-}$
Bisulfide ion	HS^-	Bithiosulfate ion	$HS_2O_3^-$
Hydrogen sulfide	H_2S	Thiosulfuric acid	$H_2S_2O_3$
Polysulfide, −1		Sulfites, +4	
Disulfide ion	S_2^{2-}	Sulfite ion	SO_3^{2-}
Pyrite	FeS_2	Hydrogen sulfite	HSO_3^-
		Sulfur dioxide	SO_2
Elemental, 0			
Sulfur	$S°$	Sulfate, +6	
		Sulfate ion	SO_4^{2-}
		Bisulfate ion	HSO_4^-
		Sulfuric acid	H_2SO_4

[a] Sulfur often occurs in more than one oxidation state in the same species. Thus thiosulfate ($S–SO_3^{2-}$) has one S at −2 and one at +6, for a mean value of +2.

natural abundances are approximately 95.1, 0.74, 4.2, and 0.017%, respectively. The $^{32}S/^{34}S$ ratio of naturally occurring S compounds is highly variable ($\pm 4.5\%$ from a mean value).[22] Sedimentary SO_4^{2-} is normally enriched with ^{34}S as compared to other deposits.[22] Radioactive isotopes are ^{31}S, ^{35}S, and ^{37}S.

Origin of S in Soils

The principal original source of S in soils is the pyrite (FeS_2) of igneous rocks. During weathering and soil formation, the S of pyrite undergoes oxidation to the SO_4^{2-} form, which is ultimately assimilated by plants and microorganisms and incorporated into the soil organic matter. In some soils part of the S is retained as gypsum ($CaSO_4\cdot2H_2O$) and epsomite ($MgSO_4\cdot7H_2O$), or leached. In dry regions, where rainfall is insufficient to leach SO_4^{2-} from the soil profile, gypsum often accumulates in a horizon below a zone of $CaCO_3$ accumulation.

Sulfur Content of Soils

The total S content of soils varies over a broad range, from as little as 20 μg/g (0.002%) in highly leached and weathered soils of humid regions (especially sands) to well over 50,000 μg/g (5%) in calcareous and saline soils of arid and semiarid regions. While the S content of soils within a geographical region varies widely, total S tends to be lower in weathered and leached soils. Jordan and Reisenauer[23] reported a mean S content of 540 μg/g in nonleached soils of the United States; for moderately leached soil, mean total S content was 210 µg/g. Tropical soils generally contain low amounts of S, which can be explained by their low organic matter content. High levels of S (35,000 µg/g or more) are found in soils developed in tidal areas, where sulfides have accumulated.

The normal range of S in agricultural soils of humid and semihumid regions is 100 to 500 µg/g, or 0.01 to 0.05% S. This amounts to from 224 to 1,120 kg/ha within the plow layer (200 to 1,000 lb/acre). However, as noted later, much of the S is tied up in organic forms and will not be directly available to plants.

The long-term effect of cropping on S levels (without additions of agricultural chemicals) has been to decrease the total amounts present, largely because of mineralization of organic matter. As noted in Chapter 2, losses of organic matter through cultivation have ranged from as little as 15% in well-managed soils to well over 50% in soils subject to continuous and intensive cropping. Losses of organic S undoubtedly have been of the same order of magnitude. At steady state, losses of soil S through cropping and leaching must equal gains of S through atmospheric deposition, return in crop residues, or additions in S-containing fertilizers and pesticides. Agricultural practices that result in increases in the organic matter content of the soil, such as long-time applications of organic residues, can lead to increases in S content.[24,25] Larson et al.[24] found that the S (and C) content of the soil increased in proportion to the amount of plant residues added; a 15% increase in S content was obtained by the annual application of 16 tons of plant residues per hectare (40 tons/acre) over an 11-year period.

Sulfur in the form of $CaSO_4$ (gypsum) is a constituent of "normal" superphosphate, and small amounts are found in concentrated "superphosphates" (see Chapter 7). For this reason S deficiencies seldom occur on soils well fertilized with P. Several "controlled release" fertilizers contain elemental $S°$ as a coating, the best-known example being S-coated urea. A coating of about 20% elemental $S°$ is normally used, thereby providing a source of SO_4^{2-} to plants through microbial oxidation (discussed later).

Atmospheric S

Much interest has been shown in recent years in the flux of S gases between soil and the atmosphere.[26–33] These constituents are undesirable as atmospheric components because they have the potential for adversely affecting climate and the environment through their aerosol-forming properties.[7]

Sulfur dioxide (SO_2) and other miscellaneous S compounds are emitted into the atmosphere by power and industrial plants, as well as by natural sources, such as volcanic activity. The atmospheric SO_2 may be absorbed directly from the air or washed into the soil in rainwater. The gas readily combines with water to form sulfurous acid ($SO_2 + H_2O \rightarrow H_2SO_3$), but this acid is soon oxidized to H_2SO_4 by S-oxidizing microorganisms.

Anthropogenic input of S into the atmosphere (e.g, by combustion of fossil fuels) is now about one-half as much as from natural sources.[7] In highly industrial regions (e.g., Europe, parts of eastern Asia, and the eastern portion of North America) emissions of S gases into the atmosphere have probably increased by at least one order of magnitude in recent years because of burning of fossil fuels. The net effect of this increase is that because of fallout of atmospheric S, the amount of S in many lakes has nearly doubled.

Biologically produced S compounds represent a significant contribution to S entering the atmosphere each year. Essentially all the biologically produced H_2S arises from coastal areas or land surfaces, and this H_2S accounts for about 2×10^9 kg/yr of the estimated 106×10^9 kg/yr of S believed to be emitted into the atmosphere by biological sources. An alternate biological source of atmospheric S is dimethyl sulfide.[7,30] In considering soils contribution to S gases in the atmosphere, one must keep in mind that most soils are able to adsorb S gases of various types.[34,35]

The amount of S returned to the soil in rainwater varies with location and season of the year. Since the S arises largely from the combustion of coal and other fossil fuels, larger quantities are brought down near industrial areas and during winter months. In rural regions remote from industrial sites, 5 kg or less of S/ha·yr are returned to the soil in this way. Higher amounts of S (to 200 kg/ha·yr) are added to the soils of highly industrialized areas, much of which will be leached below the rooting depth, the exception being arid regions of low rainfall

Losses of Soil S

Sulfur is lost from agricultural soils mainly through leaching. The extent to which SO_4^{2-} is lost depends on rainfall, SO_4^{2-} retention capacity of the soil, drainage characteristics, and immobilization by microorganisms during decay of plant residues. As one might expect, losses are greatest on coarse-textured soils under high rainfall. Because of leaching, SO_4^{2-} seldom accumulates in soils of humid and semihumid regions.

In addition to leaching, S can be lost through the production of H_2S and volatile organic S compounds by microorganisms (see previous section). Losses of S by this mechanism are particularly serious in submerged soils, as will be noted later.

SULFUR IN PLANT NUTRITION

Plants contain about the same amounts of S as P, the usual range being 0.2 to 0.5% on a dry weight basis. Among the S compounds in plants are glutathionine, thiamine, vitamin B, biotin, ferredoxin, coenzyme A, and the S amino acids (cysteine, cystine, and methionine). The latter accounts for the bulk of the S in most plants (90%). Small but variable amounts of the S occur

Table 8.3

Chemical Structures of Some Biogenic Organic S Compounds. Structures of the
Common S-containing Amino Acids are Given in the Text.

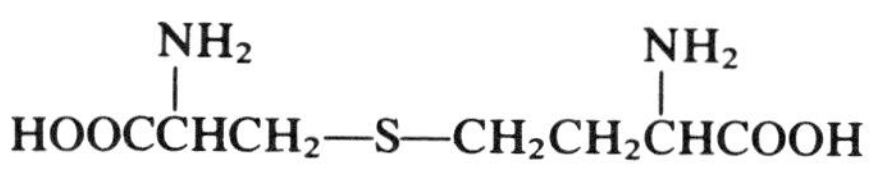

$$\underset{\text{Cystathionine}}{HOOC\overset{\overset{\displaystyle NH_2}{|}}{C}HCH_2-S-CH_2CH_2\overset{\overset{\displaystyle NH_2}{|}}{C}HCOOH}$$

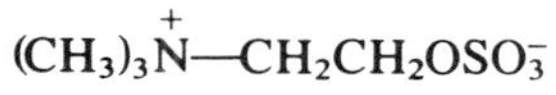

$$(CH_3)_3\overset{+}{N}-CH_2CH_2OSO_3^-$$
Choline sulfate

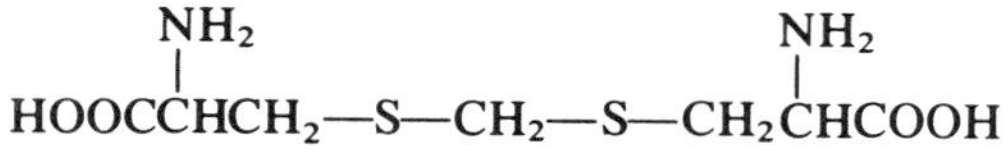

$$\underset{\text{Djenkolic acid}}{HOOC\overset{\overset{\displaystyle NH_2}{|}}{C}HCH_2-S-CH_2-S-CH_2\overset{\overset{\displaystyle NH_2}{|}}{C}HCOOH}$$

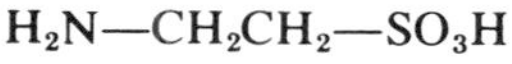

$$H_2N-CH_2CH_2-SO_3H$$
Taurine

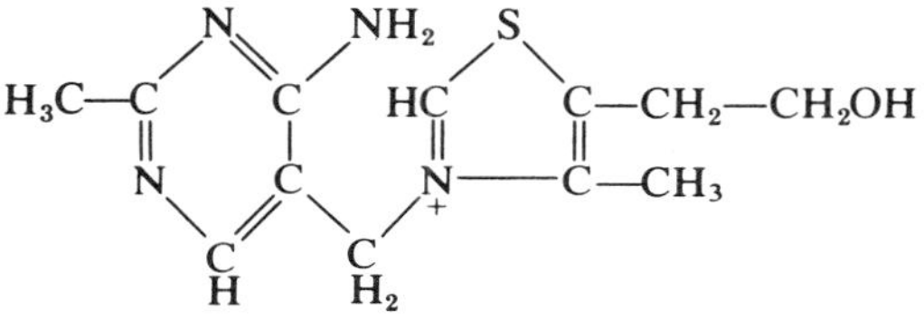

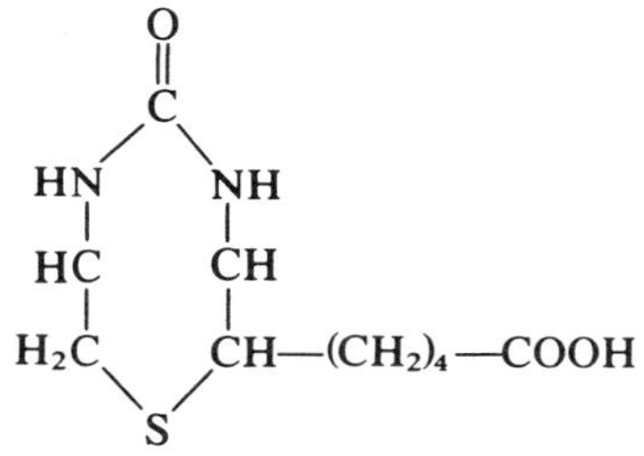

Biotin

Thiamine

$$\underset{\text{α-Lipoic acid}}{\overset{\displaystyle CH_2CH_2CH(CH_2)_4COOH}{\underset{\displaystyle S\text{———}S}{|\qquad\qquad|}}}$$

Glutathione

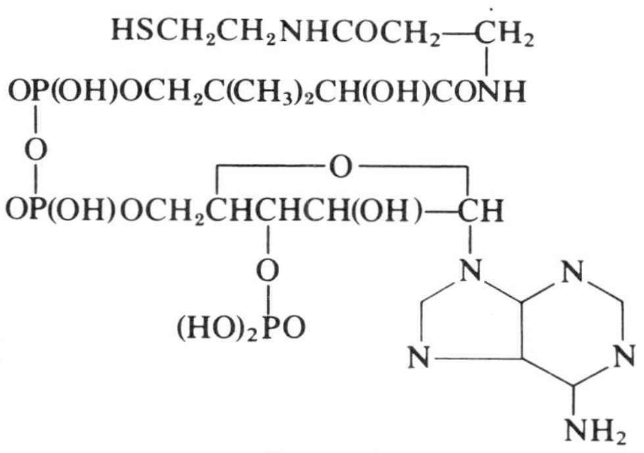

Coenzyme A

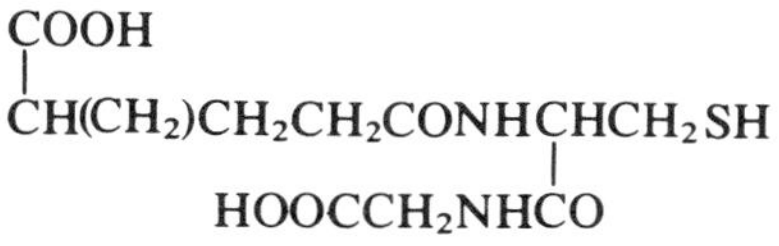

in inorganic forms. Chemical structures of some S-containing biochemicals
are given Table 8.3.

Sulfur has numerous functions in plant, many of which are related to
transformations mediated by enzymes. Ferredoxin, an Fe–S protein, is the
first stable redox compound of the photosynthetic electron chain. The disulfide (—S—S—) bond serves as a linkage between the polypeptide chains
in a protein molecule; the sulfhydral (—S—H) group provides sites for metal
cation binding and for attachment of prosthetic groups to enzymes. The
methyl group of methionine is involved in the biosynthesis of lignin, pectin,
chlorophyll, and flavonoids.

Plants suffering from lack of S show deficiency symptoms characteristic of reduced photosynthetic activity. Growth is retarded and maturity is often delayed. In most plants, the younger leaves are light green in color, which is sometimes confused with rather similar symptoms of N deficiency (see Chapter 5). Sulfur deficiencies are widely found in leguminous crops, both grain and forage legumes.

Sulfate taken up by the plant must first be reduced before incorporation into organic forms. The first step of S incorporation is a reaction between sulfate and ATP (adenosine triphosphate) to form adenosine phosphosulfate, as follows:

$$ATP + SO_4^{2-} + 2H^+ \longrightarrow \text{Adenosine-O-} \underset{\overset{|}{OH}}{\overset{\overset{O}{\|}}{P}} \text{-O-} \underset{\overset{\|}{O}}{\overset{\overset{O}{\|}}{S}} \text{-OH} + HO\text{-}\underset{\overset{|}{OH}}{\overset{\overset{O}{\|}}{P}}\text{-O}\sim\underset{\overset{|}{OH}}{\overset{\overset{O}{\|}}{P}}\text{-OH}$$

Adenosine phosphosulfate   Pyrophosphate

Further transformations occur according to the sequence shown in Fig. 8.2. The sulfate group of adenosine phosphosulfate (APS) is first transferred to a carrier–SH complex through the action of APS transferase, with the H of the carrier being replaced by the sulfural group. Following reduction (carrier—S—SO_3H → carrier —S—SH), the SH of the carrier complex is transferred to acetylserine, with formation of cysteine. Ferredoxin serves as an electron source for the transformations (see Fig. 8.2). Cysteine, the first stable product in which S is present in organically bound form, is the precursor of cystine, methionine, and other S-containing biochemical compounds. The overall reduction process leading to the formation of cysteine is:[9,36]

$$H_2SO_4 + ATP + 8H^+ + \text{acetylserine} \longrightarrow$$

$$\text{cysteine} + \text{acetate} + 3H_2O + AMP + P{\sim}P$$

The total amount of S consumed by plants is highly variable, as illustrated in Table 8.4. Cruciferous crops (e.g., cabbage and turnips) require considerably more S than most crops, because of their tendency to synthesize unusual quantities of mercaptans and glucosides. For the common field crops, total S taken up by the plant (grain plus straw or stover) amounts to from 10 to 30 kg/ha. At a medium value, a typical agricultural soil (100 to 500 μg S/g) will contain sufficient S to plow depth for only a few decades of cropping without any further S additions. However, as noted later, the S in many soils is tied up in organic forms and is not directly available to plants. It should also be noted that part of the plant S is recycled when crop residues are returned to the soil following harvest.

Sulfate is the main form of S taken up by the plant. The amount consumed depends on the concentration of SO_4^{2-} at the root surface, which, in turn, depends on the rate at which SO_4^{2-} is replenished through mass flow and

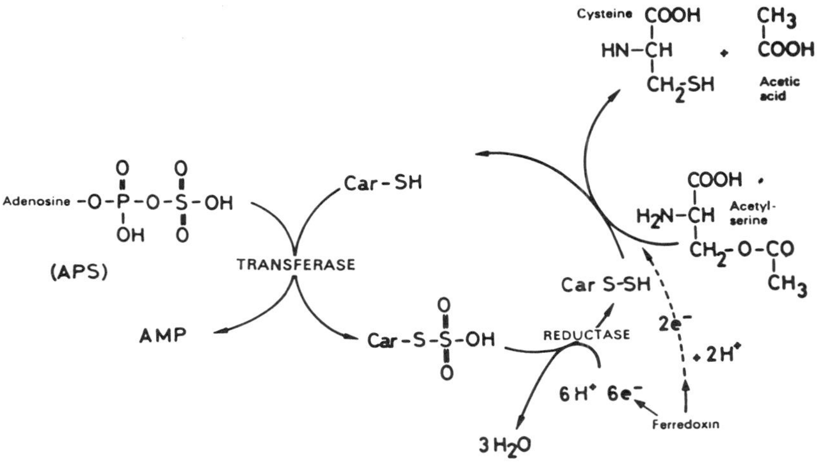

Fig. 8.2

Simplified scheme for the transfer of S from adenosine phosphosulfate to cysteine. Additional transformations lead to the formation of cysteine, methionine, and other S-containing biochemicals. (From Mengle and Kirkby,[9] reproduced by permission of the International Potash Institute, Bern, Switzerland.)

diffusion. In most soils the reserve of SO_4^{2-} in the solution phase is inadequate for optimum crop yields, and S deficiencies would develop within a short time were it not for additions in rainfall and through organic matter decomposition.

Table 8.4
Amounts of S Removed from the Soil in Specific Crops[a]

Crop	S Removed, kg/ha
Corn	9–11
Wheat	10–13
Potatoes	8–11
Grasses	9–11
Cotton	13–17
Clovers	17–25
Alfalfa hay	22–27
Sugar beet	21–31
Onions	20–22
Cabbage	21–43
Turnip	28–39

[a] From Whitehead[17] and Beaton.[37]

Table 8.5
Inorganic and Organic S in 37 Iowa Surface Soils[a]

Form of S	as μg/g of Soil		as % of Total S	
	Range	Average	Range	Average
Total	55–618	292		100
Organic	55–604	283	95–99	97
Inorganic	1–26	8	1–5	3
Sulfate	1–26	8	1–5	3
Nonsulfate	0	0	0	0

[a] From Tabatabai and Bremner.[38]

CHEMISTRY OF SOIL S

Sulfur occurs in soils in both organic and inorganic forms. The amounts of S in the two forms vary widely, depending on the nature of the soil (pH, drainage status, organic matter content, mineralogical composition) and depth in the profile.

Unlike phosphate, SO_4^{2-} is subject to leaching; thus in highly leached soils, inorganic forms of S have been removed and only the S in organic forms remains. Table 8.5 summarizes data obtained by Tabatabai and Bremner[38] for total and organic S in 37 Iowa surface soils (0–15 cm depth). Organic S, obtained by subtracting inorganic S from total S, accounted for 95 to 99% of the total S and was significantly correlated with organic C and total N. The percentage of the soil S in organic forms is similar to that recorded by Freney[12] for Australian soils. Evans and Rost[39] found that from 46 to 73% of the S in Minnesota soils was organic, being lowest in some Spodosols and highest in some prairie soils.

Inorganic Forms

Inorganic S occurs in soil largely as SO_4^{2-}, although compounds of lower oxidation states are often found, such as sulfide (e.g., FeS), sulfite, thiosulfate, and elemental $S°$. Under water-logged conditions, inorganic S occurs in reduced forms, such as H_2S, FeS, and FeS_2 (pyrites). Pyrite is often the main inorganic form of S in wetland and submerged soils. In some instances, elemental $S°$ can be formed. A major fraction of the S in calcareous and saline soils occurs as gypsum ($CaSO_4 \cdot 2H_2O$).

Some soils have the capacity to retain SO_4^{2-} in an adsorbed form.[40-46] Sorption is restricted to those soils that are acidic and is due primarily to anion exchange by positive charges on Fe and Al oxides and clay minerals. The lower soil horizons adsorb more SO_4^{2-} than the surface layer; kaolinite adsorbs more than montmorillonite; soils rich in Fe and Al oxides adsorb even more SO_4^{2-}; and the amount adsorbed increases with decreasing soil pH. Organic matter exerts a negative effect on SO_4^{2-} adsorption,[40,45,46] which may be the reason for the low sorption capacities of surface soils. Accumulations of adsorbed SO_4^{2-} have been observed in the B horizons of forest soils due to inputs from atmospheric deposition.[42,47] The reactions involved in SO_4^{2-} sorption are similar to those for phosphate adsorption (see Chapter 7).

Organic Forms

As noted earlier, essentially all the S in soils of humid and semihumid regions occurs in organic forms.[48–51] For these soils the absolute amount of both total and organic S will be influenced to a large extent by those factors that affect the organic matter content of the soil, as documented in Chapter 2.

Determination of Organic S

Organic S is usually taken as the difference between total soil S as obtained by wet ashing and inorganic S (SO_4^{2-} and sulfide) extracted by reagents such as dilute HCl or $NaHCO_3$. In lieu of extraction, analyses for SO_4^{2-} can be made on ignited soil with and without addition of $NaHCO_3$, in which case organic S is calculated from the difference between the two values.[52] A direct method based on ignition of acid-leached soil has also been described.[53] Extractants used for removing inorganic S compounds undoubtedly remove small amounts of organic S as well.

The C/N/S Ratio

Evidence that most of the S in soils of humid and semihumid regions occurs in organic combinations has come from studies indicating that a close relationship exists between the amounts of C, N, and S that are present.[54–61] While considerable variation is found in the C/N/S ratio of individual soils, the mean ratio for soils from different regions of the world is remarkably similar, as can be seen from results given in Table 8.6. As an average, the proportion of C/N/S in soils is approximately 140:10:1.3. Data presented in Table 8.6 also show that the sulfur content of the soils examined were highly correlated with both C and N.

Difference in the C/S and N/S ratios between soil groups has been attributed to such factors as parent material and type of vegetative cover. Bettany et al.[56] found that the C/S and N/S ratios of some grassland soils of Saskatchewan, Canada, were lower than those for some comparable forest soils of the region; mean C/N/S ratios for the two groups were 91:10:1.56 and 122:10:0.94, respectively.

Fractionation of Organic S

The organic S in soils is often divided into two main groups of compounds, namely, S directly bound to C (C-bonded S) and as sulfate esters (R—O—SO_3H linkage).[51,56,61–68]

Table 8.6

Relationships Between C, N, and S in Soils from Different Regions

Soils and Locations[a]	Correlation Coefficients		C/N/S
	C and S	N and S	
Scotland (10 ea. group)			
Granite	0.690	0.942	169:10:1.4
Slate	0.868	0.817	148:10:1.4
Old red sandstone	0.971	0.984	130:10:1.4
Basic igneous	0.926	0.896	140:10:1.4
Calcareous	0.898	0.927	113:10:1.3
All soils	0.866	0.938	140:10:1.4
Australia			
Agricultural	0.981	0.982	150:10:1.3
Grassland	0.879	0.906	113:10:1.2
New Zealand			
Grassland		0.970	130:10:1.3
Aridic Haploboroll (6)	—	—	93:10:1.6
Typic Haploboroll (13)	—	—	94:10:1.5
Udic Haploboroll (9)	—	—	113:10:1.4
Transitional (14)	—	—	125:10:1.3
Typic Cryoboralf (12)	—	—	119:10:0.9
Iowa (USA)			
Cultivated (6)	0.980	0.970	110:10:1.3
Brazil			
Agricultural (6)	0.850	0.890	194:10:1.6

[a] For specific references see Freney and Stevenson[13] Results for Canada, Iowa, and Brazil are more recent.[56,58] Figures in parentheses signify number of samples.

C-BONDED S.   Compounds in this group include the S-containing amino acids, such as cysteine and methionine.

```
CH2-SH              CH2-CH2-S-CH3
|                   |
CH-NH2              CH-NH2
|                   |
COOH                COOH

   Cysteine            Methionine
```

NON-C-BONDED S.   Non-C-bonded S is readily reduced to H_2S by hydriodic acid (HI) and is assumed to occur as ester sulfates, such as phenolic sulfates and sulfated polysaccharides. The term *ester sulfate-S* will be used to describe this fraction even though positive proof is lacking that all of the S recovered as H_2S by HI reduction occurs in this form.

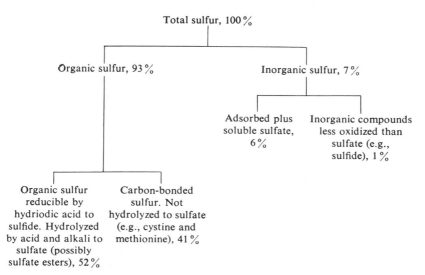

Fig. 8.3

Average distribution of the forms of S in 24 Australian soils. (Reprinted from Freney,[12] p. 235, by courtesy of Marcel Dekker, Inc.)

Methods of fractionating soil S have been described by Freney et al.,[61] Lowe and DeLong,[64] and Tabatabai and Bremner.[51] Carbon-bonded S can be calculated from the difference between total organic S and ester sulfate S, but a direct method based on reduction to H_2S by Raney nickel (Ni-Al alloy) has been described. However, the Raney nickel method does not reduce all C-bonded S compounds to H_2S, including the S of cysteic acid and methionine sulfone. Also, S combined with humic and fulvic acids may not be reduced. Thus it is not surprising that all of the C-bonded S cannot be accounted for by reduction with Raney nickel. Iron and other soil constituents can lead to low results.[62]

The average distribution of the forms of S in 24 Australian soils is given in Fig. 8.3. The percentage of the total S as ester sulfates (52%) is similar to that reported in more recent studies by Bettany et al.[56] for the Ap layers of 54 Saskatchewan soils (36 to 50%) and by Tabatabai and Bremner[51] for 37 Iowa surface soils (35 to 61%).

The fraction of the soil S as ester sulfates is higher in grassland soils than in forest soils, and the percentage may increase with depth in the soil profile, as indicated in Fig. 8.4. The S that is reducible with HI occurs in both high- and low-molecular-weight compounds, but mostly in the former.[62]

McLaren and Swift[66] found that a high proportion (75%) of the S that is lost when soils are subject to long-time cultivation consists of C-bonded forms. However, ester sulfate-S was believed to be more transitory and of greater importance in the short-term mineralization of organic S. McLachan

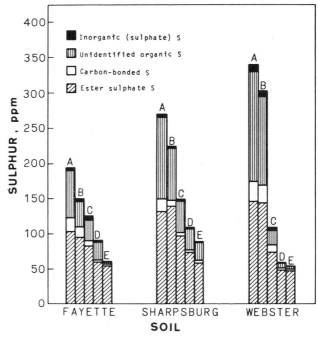

Fig. 8.4
Amounts of different forms of S in three Iowa soils. A = 0–15 cm, B = 15–30 cm, C = 30–60 cm, D = 60–90 cm, E = 90–120 cm. The unidentified fraction is the C-bonded S not reducible with Raney nickel. (From Tabatabai and Bremner,[50] reproduced by permission of the Sulphur Institute, Washington, D.C.)

and Demarco[65] also found that, on cropping, more S was withdrawn from C-bonded forms than from sulfate ester forms.

Sulfur-Containing Amino Acids

The proportion of the soil S that occurs as amino acids is not known with certainty. On the basis of published data for amino acids in soil hydrolysates, Whitehead[17] estimated that from 11 to 16% of the soil S occurs in this form. This estimate may be low, for the reason that cystine and methionine, the two main S-containing amino acids, are both known to undergo extensive destruction during acid hydrolysis. In an attempt to avoid this problem, Freney et al.[67] oxidized cystine and methionine to stable forms (cysteic acid and methionine sulfone) prior to acid hydrolysis. For two soils, an average of 26% of the soil S, equivalent to 46% of the C-bonded S, was estimated to be present as amino acids. The results are summarized in Table 8.7. Scott et al.[68] used a somewhat similar procedure to determine S-containing amino acids in some Scottish soils. From 11 to 15% of the total organic S (19 to

Table 8.7
Distribution of Amino Acid-S in Two Soils[a]

Sulfur Fraction	Virgin Soil, μg/g	Improved Soil, μg/g
Total S	145	266
Hydriodic acid-reducible	78	94
C-bonded	67	172
Cysteine and cystine[b]	19	48
Methionine[c]	11	33
Total amino acids	30	81
Amino Acid-S		
As % of total S	20.7	30.5
As % of C-bonded S	44.8	47.0

[a] From Freney et al.[67]
[b] Determined as cysteic acid.
[c] Determined as methionine sulfone.

31% of the C-bonded S) was accounted for as amino acids. Other results, tabulated by Anderson,[10] have shown a range of 27 to 125 μg/g amino acid-S/g of soil.

Lipid S

The existence of sulfolipids in soil has been demonstrated from the occurrence of S in lipid extracts of soil.[69–71] Chae and Tabatabai[71] found that the amount of lipid S in 10 Iowa surface soils ranged from 0.87 to 2.63 μg/g, equivalent to from 0.29 to 0.45% (average = 0.37%) of the total soil S. From 20 to 40% of the sulfolipid S was HI-reducible, indicating that most of the S occurred in C-bonded forms. The technique of column chromatography was used by Chae and Lowe[70] to fractionate soil lipid S into three main fractions: polar lipids, 7 to 37%; glycolipids, 35 to 50%; less polar lipids, 13 to 53%.

Complex Forms of Organic S

Only a small fraction of the soil organic S can be accounted for in known compounds. This has led to the belief that part of the organic S, particularly C-bonded forms, occurs as complex condensation products resulting from the reaction of thiol compounds with quinones and reducing sugars. For example, cysteine, thiourea, and glutathione all react with o- and p-quinones to form characteristic brown pigments. The basic mechanisms are similar to those discussed in Chapter 5 for the reaction of amino compounds with quinones. Mason[72] pointed out that when there is a possibility for competition between sulfhydral (SH) and NH_2 groups for the quinone, such as

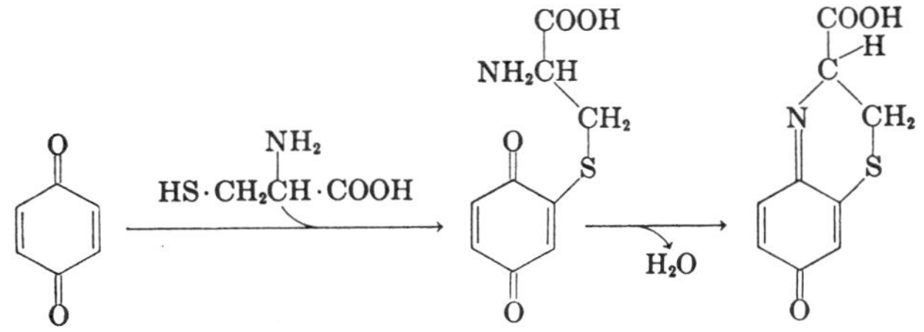

Fig. 8.5

Reaction between cysteine and *p*-quinone to form complex structures containing C–S linkages.

in the case of cysteine, the sulfhydral groups will react first. For example, the first product of the reaction between cysteine and *p*-benzoquinone is 1, 4-benzoquinone-2-cysteine, which subsequently undergoes an inner condensation between the quinone C=O group and the free NH_2 group to form a cyclic product. These reactions are shown in Fig. 8.5. The resistance of soil organic S to attack by microorganisms may be due in part to the existence of such C-S linkage in high-molecular-weight humic substances.

Soluble Forms

In view of the broad synthetic activities of microorganisms, numerous S-containing biochemicals would be expected to be produced in soil. However, because these same compounds are themselves susceptible to further decomposition, they do not persist in the *free* form, and the low amounts found at any time represent a balance between synthesis and destruction by microorganisms. In accordance with this concept, the concentration of *uncombined* organic S compounds in soil has been found to be low and variable. For example, only trace quantities of S-containing amino acids (cysteine, cystine, methionine, methionine sulfoxide, and cysteic acid) have been recovered from soil by extraction with water or such solvents as neutral ammonium acetate (see Chapter 5). The concentration of S-containing organics may be higher in rhizosphere soil than in nonrhizosphere soil, since they are known to be exuded from plant roots.

DYNAMICS OF ORGANIC S TRANSFORMATIONS

The conversion of S in soil organic matter and organic residues to plant available forms is strictly a microbiological process. When the soil is well aerated, organic S is oxidized (mineralized) to SO_4^{2-}, which is the form taken up by most plants. Concurrently, SO_4^{2-} is assimilated by microorganisms and incorporated into microbial tissue (biomass), a process called *immobilization*. Increases in the organic S content of the soil will occur only when conditions are suitable for the accumulation of organic matter, such as by frequent additions of organic residues.[24,25] Williams and Donald[73] found that essentially all the fertilizer S added in superphosphate to soils over a 15- to 25 year period could be recovered in the organic matter.

Turnover of S through mineralization–immobilization follows somewhat the same pattern as for N (Chapter 5) and P (Chapter 7) in that both processes occur simultaneously. Accordingly, the quantitity of plant available SO_4^{2-} in the soil solution at any one time represents the difference between the magnitude of the two opposing processes.

$$\text{Organic S} \underset{\text{immobilization}}{\overset{\text{mineralization}}{\rightleftharpoons}} SO_4^{2-}$$

The relative rates at which the two processes occur will, of course, be influenced by those factors affecting the activities of microorganisms, including additions of crop residues. Observations concerning the mineralization of nutrients from soil organic matter were outlined in Chapter 2. Those aspects that pertain to organic S can be summarized as follows:

1   The amount of S mineralized from unamended soils does not appear to be directly related to soil type, total amount of C, N, or S; C/S, N/S, or C/N ratios; soil pH; or mineralizable N.

2   Sulfur mineralization follows a number of patterns, namely, (a) initial immobilization followed by net mineralization in later stages; (b) a steady, linear release of SO_4^{2-} with time; (c) an initial rapid release followed by a slower linear release; and (d) a rate of release that decreases with time. The pattern of SO_4^{2-} release may be related to the chemical nature of the decomposing fraction, a subject that requires further study.

3   Mineralization of S in the presence of plants is greater than that in fallow soil, a result that may be due to greater proliferation of microorganisms under plants.

4   Mineralization of S is affected by those factors that influence the growth of microorganisms, such as temperature, moisture, pH, and availability of food supply.

   a.   *Temperature.* Mineralization is suppressed markedly at 10°C but increases with increasing temperature from 20 to 40°C and decreases thereafter.

   b.   *Moisture.* Mineralization is considerably retarded at low (<15%) and high (>80%) moisture levels. Optimum moisture content for mineralization is at 60% of the maximum water-holding capacity.

   c.   *pH.* Mineralization is usually directly proportional to pH up to a value of about 7.5.

5   More SO_4^{2-} is released when soils are dried and remoistened prior to incubation than when they are incubated without prior drying.

Reasons for differences in the mineralization rate of organic S in soils are not fully understood. Errors involved in estimating mineralization in the presence of growing plants include atmospheric contamination by SO_2, incomplete recovery of roots from soil, additions of S during watering, and failure to account for all the S in plant material due to inadequacies of the analytical method.

While some workers have shown a direct relationship between the amount of SO_4^{2-} released from soil during incubation and the S content of the soil, others have failed to show such a relationship. A plausible explanation for the divergent results is that the S content of recently added plant residues regulates the amount of SO_4^{2-} produced, rather than total organic S in the soil. It has long been known that the C/S ratio of organic residues provides a rough guide to the amount of SO_4^{2-} that accumulates during decay. As shown in the following table, when the C/S ratio of added plant residues is below 200, there is a net gain of SO_4^{2-}; when the C/S ratio exceeds 400, there is a net loss. For C/S ratios between 200 and 400, there is neither a gain nor loss of SO_4^{2-}. Carbon/S ratios of 200 and 400 correspond to S contents of about 0.25 and 0.5%, respectively.

	C/S Ratio	
< 200	200 to 400	> 400
Net gain of SO_4^{2-}	Neither a gain or loss of SO_4^{2-}	Net loss of SO_4^{2-}

One might expect that the relative rates of mineralization of S and N from soil organic matter would be similar and that the two would be released in the same ratio in which they occur in soil organic matter.[74-78] This has not always been the case, for the following possible reasons:

1 Nitrogen and S may not exist in the same organic compounds or fractions of the soil organic matter; consequently, they will not be released at the same time.

2 The inclusion of plant or animal residues with large N/S ratios would cause greater immobilization of S relative to N.

3 The presence of Ca^{2+} may obscure the release of SO_4^{2-} by formation of insoluble $CaSO_4$.

4 Air-drying the soil before treatment may affect N and S release differently.

The addition of $CaCO_3$ to soil has been found to lead to an increase in soluble SO_4^{2-} on incubation. This increase may be due to:

1 Enhanced mineralization of soil organic matter because of better growth of bacteria.

2 Release of adsorbed SO_4^{2-} because of the increase in pH.

3 Sulfate added in the $CaCO_3$.

Several studies have shown that slightly more N is lost relative to S when soils are cropped. Accordingly, the N/S ratio of the organic matter of cultivated soils is generally lower than that for virgin soils. Data obtained by Stewart and Whitfield[79] for 10 paired soils from the Great Plains region of the United States are as follows:

	Total N, $\mu g/g$	Organic S, $\mu g/g$	N/S Ratio
Virgin soils	1600	183	8.7
Cultivated soils	934	117	8.0
% loss	42	36	

The microbial assimilation of SO_4^{2-}, with conversion to organic forms, is believed to be a major process influencing the behavior of SO_2 and other S compounds introduced into forest soils by acid precipitation.[80,81]

Biochemical Transformations of S-Containing Amino Acids

Little information is available concerning the pathways of decomposition of organic S compounds in soil. A few workers have followed the decomposition of known substances added to soil, and others have studied their decomposition by pure cultures of microorganisms isolated from soil. Pathways for the decomposition of S-containing amino acids in soil have been reviewed by Freney[12] and Freney and Stevenson.[13]

Cystine, cystine "disulfoxide," cysteine sulfinic acid, and cysteic acid have been shown to be involved in the breakdown of cysteine, yielding SO_4^{2-} as the end product. Oxidation can proceed in several ways, as illustrated in Fig. 8.6. Because of the diversity of microorganisms affecting organic S transformations, several pathways are undoubtedly involved. It should be noted that decomposition in mixed microbial populations may be different than that in single-species populations because of interactions between species.[12]

A reductive scheme yielding H_2S has been shown to exist for the decomposition of cysteine in cultures of microorganisms. Again, several equally valid pathways have been proposed for this transformation. Type reactions formulated for the formation of H_2S from cysteine during anaerobic decomposition are:

$$\underset{\text{Cysteine}}{\text{HS-CH}_2\text{-CH(NH}_2)\text{-COOH}} \xrightarrow{\text{H}_2\text{O}} \underset{\text{Pyruvic acid}}{\text{CH}_3\text{-CO-COOH}} + \text{NH}_3 + \text{H}_2\text{S}$$

$$\xrightarrow{\text{H}_2\text{O}} \underset{\text{Serine}}{\text{HO-CH}_2\text{-CH(NH}_2)\text{-COOH}} + \text{H}_2\text{S}$$

A number of pathways have also been proposed for methionine decomposition in soils. In the studies recorded by Freney,[12] methyl mercaptan (CH₃SH) and NH₃ were common end products of metabolism. The overall reaction is:

$$\underset{\text{Methionine}}{\text{CH}_3\text{-S-CH(NH}_2)\text{-COOH}} \longrightarrow \underset{\substack{\alpha\text{-Ketobutyric}\\\text{acid}}}{\text{CH}_3\text{-CH}_2\text{-CO-COOH}} + \text{NH}_3 + \underset{\substack{\text{Methyl}\\\text{mercaptan}}}{\text{CH}_3\text{SH}}$$

Sulfur in Living Tissue (Soil Biomass)

From 1 to 3% of the organic S in soils can be accounted for as part of the soil biomass.[82–84] The C/S ratio of microbial tissue is of the order of 57 to

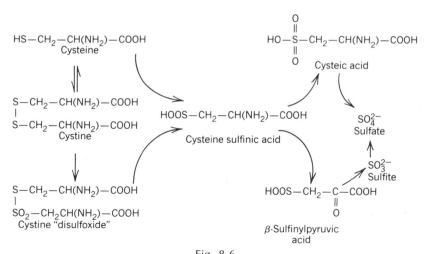

Fig. 8.6
Generalized pathway for the biological oxidation of cysteine. (Adapted from Freney.[12])

85 for bacteria and 180 to 230 for fungi. The ratios for bacteria are generally lower than those for soils (see Table 8.6). The S content of most microorganisms lies between 0.1 and 1.0% on a dry weight basis.

The quantity of S in the microbial biomass has been estimated from the amount of inorganic SO_4^{2-} that is produced during incubation of soil fumigated with chloroform ($CHCl_3$), the extractant being 0.1 M $NaHCO_3$ or 10 mM $CaCl_2$. The equation is:

$$B = F/K_s$$

where B is the μg of biomass S, F is the amount of S (in μg) released from $CHCl_3$-fumigated soil minus that released from nonfumigated soil, and K_s is the percentage of the biomass S released by the $CHCl_3$ treatment and recovered by extraction (K_s = 0.41 for 0.1 M $NaHCO_3$ as extractant and 0.35 for 10 mM $CaCl_2$).

Although only a small fraction of the soil organic S resides in the biomass at any one time, this fraction is extremely labile and is the main driving force for S turnover in soil. The higher the amount of organic S in the biomass, the greater will be the availability of S to higher plants.

Arylsulfatase Enzymes in Soil

Since much of the soil organic S is present in the form of sulfate esters, arylsulfatase (arylsulfate sulfohydrolase) may play a key role in S mineralization.[85-89] The overall reaction is:

$$R \cdot OSO_3^- + H_2O \rightarrow R \cdot OH + H^+ + SO_4^{2-}$$

Arylsulfatase was first reported in Iowa soils by Tabatabai and Bremner.[88] The enzyme has since been detected in soils from other geographical regions. The method used to determine arylsulfatase activity is based on the colorimetric determination of the *p*-phenol released when the soil is incubated at pH 5 with potassium *p*-nitrophenyl sulfate and toluene.

Arylsulfatase activity in soil has been shown to vary according to soil type, depth, season, and climate. Data obtained by Tabatabai and Bremner[89] for six Iowa soil profiles show that arylsulfatase activity decreased markedly with depth (Fig. 8.7) and was closely correlated with the decrease in organic matter content (r = 0.783). However, as can be seen from Fig. 8.8, considerable deviation in arylsulfatase activity occurred at the higher C contents, corresponding to the upper soil layers (0- to 15- and 15- to 30-cm sampling depths). This suggests that additional factors (other than organic matter content) were involved in arylsulfatase activity. Cooper[85] found a significant correlation between arylsulfatase activity in 20 Nigerian soils and total C (r = 0.691), organic S (r = 0.565), and HI-reducible S (r = 0.878).

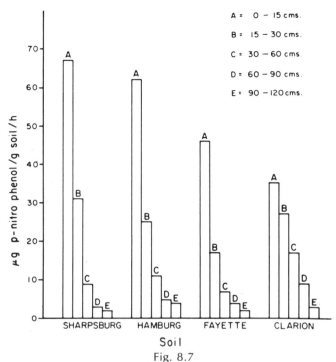

Fig. 8.7

Arylsulfatase activity in some Iowa soil profiles. (From Biederbeck[11] as prepared from data of Tabatabai and Bremner,[89] reproduced by permission of Elsevier Scientific Publishing Company.)

Use of ^{35}S for Following Organic S Transformations

As noted earlier, when fertilizer SO_4^{2-} is applied to soil, some of the S is converted to organic forms; some is consumed by plants. As was the case for C and N (Chapters 1 and 5), the use of isotopes (i.e, ^{35}S) has provided valuable information on S transformations in soil.[90-95]

Freney et al.[91] applied ^{35}S-labeled SO_4^{2-} to soil and observed a steady incorporation of S into organic forms over a 24-week period. A maximum of 50% of the added SO_4^{2-} was immobilized during this period, but the percentage was increased to 82% by addition of an energy source (glucose). The labeled S was incorporated into both C-bonded and ester sulfate forms, but the latter had the greatest specific activity.

In continuing these studies, Freney et al.[92] found that plants utilized S from both the C-bonded and ester sulfate forms. The greater part of the indigenous soil S (60 to 65%) came from the C-bonded fraction, a result in agreement with the studies described previously. In the case of recently immobilized ^{35}S, the greatest part (60 to 70%) was converted into ester sul-

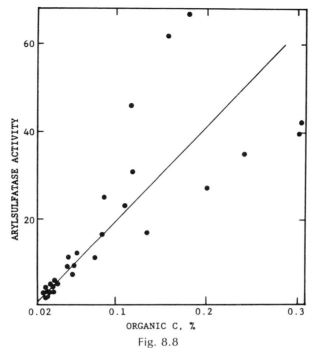

Fig. 8.8

Relationship between arylsulfatase and organic C content of the soil profiles shown in Fig. 8.7. Samples were taken at 15 cm intervals to a depth of 120 cm. (Adapted from data reported by Tabatabai and Bremner.[89])

fates, and it was from this fraction that all the labeled S was taken up. Since both S fractions contributed available S for plant uptake, the conclusion was reached that neither is likely to be of value for predicting the S requirements of plants.

In other work, Bettany et al.[90] found that mineralization of soil organic S was unaffected by addition of inorganic S. The amount of S mineralized could not be related to the quantity of total S in the soil or to the percentage of the S present as ester sulfates. The largest amount of S mineralized occurred from the soil with the lowest C/S ratio.

Formation of Volatile S Compounds

The decomposition of organic S compounds leads to the formation of trace amounts of mercaptans, alkyl sulfides, and other volatile organic S compounds, as well as H_2S.[26-33] Hydrogen sulfide is a common constituent of poorly drained soils, of swamp gases, and of putrifying organic remains. Included with the volatile organic S compounds produced by microorganisms are the following:

CS_2	COS	CH_3-SH
Carbon disulfide	Carbonyl sulfide	Methyl mercaptan

$CH_3-CH_2-S-CH_2-CH_3$	CH_3-S-CH_3	$CH_3-S-S-CH_3$
Diethyl sulfide	Dimethyl sulfide	Dimethyldisulfide

Anaerobic conditions are especially favorable for the synthesis of volatile organic S compounds by microorganisms. Volatile S compounds in soil may be of importance as they may stimulate or suppress the growth of certain pathogenic fungi.[29] Also, they may inhibit nitrification and other biochemical processes.

Much interest has been shown in recent years to the flux of S gases between soil and the atmosphere. These constituents are undesirable as atmospheric components because they have potential for adversely affecting climate and the environment through their aerosol-forming properties.[30]

Elliott and Travis[28] concluded that the major organic S compound produced by the anaerobic decomposition of beef cattle manure was carbonyl sulfide (COS). In other work, Nicolson[96] obtained incomplete recoveries of S from the soil–plant systems he examined and concluded that the unaccounted for S resulted from emission of volatile S compounds. Pathways for the formation of S gases in aerobic soils were postulated by Nicolson[96] to be as follows:

Determination of Plant-Available S

The status of soil tests to determine available S is similar to that noted earlier for N (Chapter 5) in that only limited success has been obtained and no one procedure has been consistently superior to all others. Results of soil tests are particularly difficult to interpret in those soils where essentially all the S occurs in organic forms.

Procedures for evaluating plant-available S in soil include (1) extraction with water, salt solutions (e.g, 0.15% $CaCl_2$ and 5 mM $MgCl_2$), dilute acids (0.5 M NH_4OAc + 0.25 M acetic acid), and weak bases (0.5 M $NaHCO_3$ at pH 8.5), (2) release of SO_4^{2-} upon incubation, and (3) S uptake by the plant. Extractants of soil S can be categorized as those that remove readily soluble SO_4^{2-}, those that remove soluble SO_4^{2-} plus a portion of the absorbed SO_4^{2-}, and those that also remove a portion of the soil organic S. The use of soil tests for assessing plant available S has been discussed by Biederbeck[11] and Tabatabai.[52]

Selection of an extractant is dependent to some extent on the nature of the soils encountered. Acid–phosphate solutions appear best for acid soils containing variable amounts of organic S and where part of the SO_4^{2-} occurs in adsorbed forms; neutral salt solutions are preferred for near neutral soils of semiarid regions. Factors that affect the extraction of SO_4^{2-} include sample preparation, soil/extractant ratio, and shaking time. The effect of air drying is to increase the amount of SO_4^{2-} extracted.

Difficulties in predicting S-fertilizer needs of a particular crop are due to one or more of the following:

1 The rate of release of S from the soil organic matter is highly variable and affected by the activities of microorganisms.
2 The presence of undecomposed organic residues low in S content can result in a decline in SO_4^{2-} levels because of immobilization.
3 The contribution of SO_4^{2-} in rainwater, or of SO_2 in dry deposition, cannot be estimated accurately.
4 Sulfate becomes unavailable through sorption, leaching, or reduction to volatile gases during waterlogging.
5 Environmental factors (e.g., temperature, moisture) influence the uptake of SO_4^{2-} by plants. Also, the amount taken up from the subsoil is highly variable and depends on the rooting characteristics of the plant.

OXIDATION AND REDUCTION OF INORGANIC S COMPOUNDS

In addition to assimilation by living organisms (discussed previously), inorganic S compounds are subject to a variety of oxidation–reduction reactions, nearly all of which are mediated by microorganisms.[19–21,97–101] An outline of the main reactions is given in Fig. 8.9.

A full account of inorganic S transformations is beyond the scope of the present paper and the reader is referred to reviews on the subject.[19–21] Oxidation–reduction reactions of S have special significance to the genesis and management of acid sulfate soils, a topic reviewed by Bloomfield and Coulter.[98] For the purpose of the present discussion, it can be said that inorganic S transformations affect soil color, soil reaction, and availability of plant nutrients. However, under most agricultural conditions, the effect of these S transformations is relatively obscure and it is the completely oxidized (SO_4^{2-}) and completely reduced (H_2S) forms of S that are of greatest concern.

Reduction of S

Reduction of SO_4^{2-} by microorganisms occurs in two different ways. In one case the S is incorporated into cellular organic constituents, such as S-amino acids. This process is referred to as *assimilatory* SO_4^{2-} *reduction*, and/or

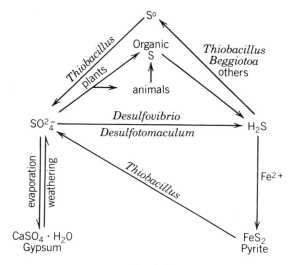

Fig. 8.9
Pathways of S transformations. The list of organisms involved in oxidation-reduction is incomplte (see text). (Adapted from Kaplan.[99])

"immobilization." A second type of reduction is carried out by a special group of bacteria and leads to the formation of equivalent quantities of sulfide (e.g., H_2S) as the end product. This process is known as *dissimilatory or respiratory* SO_4^{2-} *reduction* and occurs when the environment is favorable for the growth of anaerobic bacteria. A typical metabolic reaction is as follows:

$$2CH_3\text{-}CHOH\text{-}COOH + SO_4^{2-} \longrightarrow 2CH_3COOH + 2CO_2 + S^{2-} + H_2O$$

   Lactic acid   Sulfate   Acetic acid   Sulfide

Sulfate reducing microorganisms have important effects in causing the precipitation of metal sulfides, notably of ferrous Fe^{2+}, in causing pollution of natural waters, and in bringing about the corrosion of steel cables buried in poorly drained or wet soils. Such deposits are usually stained black through the accumulation of ferrous sulfides. Microbial SO_4^{2-} reduction also accounts for the occurrence of sulfides in shales, strip-mine refuge, and the H_2S in natural gas.

The ability to reduce SO_4^{2-} to H_2S is a property of a small number of strict anaerobes belonging to two genera of bacteria, *Desulfovibrio*, consisting of curved, rod-shaped cells, and *Desulfotomaculum*, which are straight or somewhat curved, sporulating rods. These organisms are particularly abundant in stagnant water basins and marine muds. Essentially, SO_4^{2-} serves as a substitute for O_2 during C metabolism in much the same way that NO_3^- is used by denitrifying bacteria (see Chapter 4). Most of the

Table 8.8
Amounts of Energy Generated by Oxidation of Some More
Reduced S Species

Oxidation Reaction	Free Energy Change $(F°)$, kcal/mole S
$H_2S + \frac{1}{2} O_2 \rightarrow S° + H_2O$	-50
$S° + \frac{3}{4} O_2 \rightarrow \frac{1}{2} S_2O_3^{2-}$	-64
$\frac{1}{2} S_2O_3^{2-} + \frac{3}{4} O_2 \rightarrow SO_3^{2-}$	-55
$SO_3^{2-} + \frac{1}{2} O_2 \rightarrow SO_4^{2-}$	-59

S that accumulates as iron sulfide (FeS) in waterlogged soils, bogs, ditches, and marine sediments is produced by SO_4^{2-}-reducing bacteria.

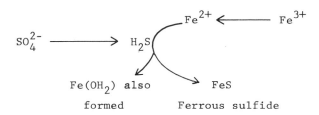

Further transformations of the sulfide in anaerobic soils lead to the formation of such products as thiosulfate and elemental S°, compounds that can themselves be oxidized and/or reduced by microorganisms.

Sulfate-reducing bacteria require an E_h of less than $+100$ mV for growth, although they can survive in a dormant state for long periods in aerobic environments. Typical habitats for SO_4^{2-}-reducing bacteria include soils and waters, sewage, polluted waters, deep-sea sediments, muds, and estuarine sand. Their presence is often indicated by a black color and a smell of H_2S. Since the organisms that reduce SO_4^{2-} require an organic substrate, their activity is favored in aqueous sediments or submerged soils containing abundant amounts of decomposable organic matter.

Sulfur-Oxidizing Bacteria

Reduced inorganic S compounds (e.g., elemental S°, H_2S, FeS_2, etc.) are readily oxidized in soils under suitable conditions by a group of bacteria that utilize the energy thus released to carry out their life processes. The amounts of energy produced by some typical oxidations are listed in Table 8.8. Many S-oxidizing bacteria are obligate autotrophs in that the C required for synthesis of carbohydrates, proteins, and other cellular products is derived from inorganic CO_2.

Practical Importance in Soils

Elemental $S°$ is sometimes applied to soil, such as for reclamation of saline and alkaline soils but also in combination with various fertilizers (e.g, S-coated urea). When applied to well-aerated, moist soils, elemental $S°$ is attacked by microorganisms to form SO_4^{2-}. The net effect of $S°$ oxidation is to lower the soil pH, which is often desirable in calcareous and saline soils. For neutral or acid soils, however, the acidifying effect is undesirable; accordingly, the use of elemental $S°$ as a soil amendment can only be recommended under special conditions.

At one time, attempts were made to utilize S-oxidizing bacteria for the formation of soluble phosphate from rock phosphate, both in composts and by additions of $S°$ with rock phosphate fertilizer. The H_2SO_4 thus formed reacts with the phosphate rock to produce the more soluble mono- and dicalcium phosphates. The practice has not been found to be feasible under most soil conditions.

Drainage of tidal marshes, as well as exposure of acid-forming underclays of coal beds in outcrops, causes H_2SO_4 to form from FeS_2, which results in extremely acid soils (pH's as low as 3.5 are not uncommon). Thus when wetland areas (e.g., marshes and swamps) are drained and reclaimed for agricultural use, serious problems arise due to the exceptional acidity that results when FeS_2 is oxidized to H_2SO_4. Such soils, sometimes known as "catclays," are called acid sulfate soils, referring to soils that have been drained, that contain free and adsorbed SO_4^{2-}, that show pale yellow mottles of jarosite $[KFe_3(SO_4)_2(OH)_6]$, and that usually have a pH below 4 in water.[98] The oxidation of pyrite is partly chemical and partly biological, as suggested by the following sequence:

$$2FeS_2 + 2H_2O + 7O_2 \xrightarrow{\text{Chemical}} 2FeSO_4 + 2H_2SO_4$$

$$4FeSO_4 + O_2 + 2H_2SO_4 \xrightarrow{\textit{T. thiooxidans}} 2Fe_2(SO_4)_3 + 2H_2O$$

$$Fe_2(SO_4)_3 + FeS_2 \xrightarrow{\text{Chemical}} 3FeSO_4 + 2S°$$

$$2S° + 3O_2 + 2H_2O \xrightarrow{\textit{T. thiooxidans}} 2H_2SO_4$$

Sulfates from seawater are the original source of sulfides in acid sulfate soils of deltaic and coastal regions. In inland swamps, the sulfides are derived from the FeS_2 of surrounding rocks, namely, through oxidation and movement of soluble SO_4^{2+} salts to the lower-lying areas in drainage waters. In each case, bacterial reduction of SO_4^{2-} (see previous section) produced the sulfides from which SO_4^{2-} is regenerated.

Coastal deposits of sulfide-bearing muds occur over broad areas, particularly in the tropics, a typical example being Malaysia, where deposits up to 60 km (40 miles) wide and 105 m (450 ft) deep are known. Documentation of the occurrence of acid sulfate soils is available.[98]

Species of S-Oxidizing Bacteria in Soil

Included among the S-oxidizing bacteria are the chemolithotrophic bacteria of the genera *Thiobacillus* (family Nitrobacteriaceae), colorless filamentous bacteria of the family Beggiatoaceae (e.g., *Beggiatoa* and *Thiothrix*), and the photosynthetic S bacteria (e.g., *Thiospirillum*, and *Thiocystis*. In general, only organisms of the first group are important in soils; bacteria of the other two groups are confined mostly to aquatic environments where H_2S is formed. A number of heterotrophic microorganisms, including bacteria, actinomycetes and fungi, are also capable of oxidizing reduced forms of inorganic S, but their importance in soil is unknown.[20]

Sulfur-oxidizing bacteria of the genus *Thiobacillus* are non-spore-forming, gram negative rods about 0.3 μm in diameter and 1 to 3 μm long. Most species are motile by polar flagella. Typical *Thiobacillus* oxidations are:

$$SH^- \longrightarrow S^\circ \longrightarrow S_2O_3^{-2} \longrightarrow S_4O_6^{2-} \longrightarrow SO_4^{2-}$$

-2	0	$-2, +6$	$+1.66$ to $+3$	$+6$
Sulfide	Elemental S°	Thiosulfate	Tetrathionate	Sulfate

A number of *Thiobacillus* species have been identified, five of which are well defined and of some importance in soils. Descriptions of the more important oxidizers follow.

T. THIOOXIDANS.   This obligate autotroph, first isolated by Waksman and Joffe,[102] is remarkable in that it is tolerant to extreme acidity. Optimum pH for oxidation is pH 2.0 to 3.5; growth essentially ceases above pH 5.5. The organism is frequently encountered in environments where very acid conditions exist because of oxidation of sulfides. Oxidation reactions of *T. thiooxidans* include the following:

$$\underset{\substack{\text{Elemental}\\S^\circ}}{2S^\circ} + 3O_2 + 2H_2O \longrightarrow \underset{\text{Sulfuric acid}}{2H_2SO_4}$$

$$\underset{\text{Thiosulfate}}{Na_2S_2O_3} + 2O_2 + H_2O \longrightarrow \underset{\text{Sulfuric acid}}{H_2SO_4} + \underset{\text{Sodium sulfate}}{Na_2SO_4}$$

$$\underset{\text{Tetrathionate}}{2Na_2S_4O_6} + 7O_2 + 6H_2O \longrightarrow \underset{\text{Sulfuric acid}}{6H_2SO_4} + \underset{\text{Sodium sulfate}}{2Na_2SO_4}$$

T. thiooxidans has been implicated in the corrosion of iron pipes and concrete, and it is the organism largely responsible for the production of H_2SO_4 in coal mines and strip mine spoil banks. Leachate waters from such environments often contain high amounts of H_2SO_4 and ferrous Fe^{2+}, thereby leading to the pollution of streams and rivers. At one time, the amount of H_2SO_4 entering the Ohio river and its tributaries in leachate waters from coal-bearing areas was estimated to be equivalent to 5×10^9 kg of H_2SO_4 per year.

The acidification of mine waters has been the subject of intense research (see review of Bloomfield and Coulter[98]). The main source of the H_2SO_4 is by oxidation of elemental $S°$, which is formed from FeS_2 by oxidation, such as by ferric Fe^{3+}.

$$FeS_2 + 2Fe^{3+} \longrightarrow 3Fe^{2+} + 2S°$$

The subsequent oxidation of $S°$ by *T. thiooxidans* decreases the pH and thereby brings more Fe^{3+} into solution, thereby facilitating further oxidation of FeS_2. However, a main source of Fe^{3+} is believed to be by oxidation of Fe^{2+} by *T. ferrooxidans* (described later).

T. FERROOXIDANS.   The most distinctive feature of this organism is its ability to oxidize Fe^{2+}, as well as $S°$ and $S_2O_3^{2-}$. The organism is an obligate autotroph, obligate aerobe that develops best under highly acidic conditions (optimum pH, 2.0 to 3.5). It is a common inhabitant, along with *T. thiooxidans*, of acid sulfate soils and coal mine drainage waters. Oxidation reactions of *T. ferrooxidans* for $S°$ and $S_2O_3^{2-}$ are similar to those shown earlier for *T. thiooxidans.* The reaction for Fe^{2+} oxidation is:

$$12FeSO_4 + 3O_2 + 6H_2O \longrightarrow 4Fe_2(SO_4)_3 + 4Fe(OH)_3$$

T. THIOPARUS.   This aerobic, obligate autotroph is widely distributed in soils, primarily those with higher pH values than for *T. thiooxidans*. A unique feature of the organism is the deposition of globules of elemental $S°$ outside the cells. The main substrate is $S_2O_3^{2-}$, but sulfide, elemental $S°$ and tetrathionate can also be oxidized to SO_4^{2-}. Accumulations of H_2SO_4 can curtail activity of the organism; growth ceases below pH 4.

T. NOVELLUS.   Unlike other *Thiobacillus* sp. this organism is a facultative autotroph and can grow on organic substrates. Thiosulfate is oxidized to SO_4^{2-} at reactions near neutrality; elemental $S°$ cannot be oxidized. The optimum pH for growth is in the neutral or slightly alkaline range.

T. DENITRIFICANS.   This organism is unique in that, in addition to aerobic oxidation of reduced S compounds, it can grow anaerobically using NO_3^- as the hydrogen acceptor for oxidation of $S°$. The overall reaction leading to the production of molecular N_2 is as follows:

$$5S° + 6KNO_3 + 2H_2O \longrightarrow K_2SO_4 + 4KHSO_4 + 3N_2$$

An oxidizing organism of some importance in thermal zones is *Sulfolobus*. This organism resembles *T. thiooxidans*, the main difference being that *Sulfolobus* is thermophilic (optimum temperature for growth is 70 to 75°C).

General Observations of S Oxidation

The main facts concerning the oxidation of S in soils can be summarized as follows:

1 The oxidation of reduced inorganic forms of S in soils is largely a biological process and occurs chiefly through bacteria of the genus *Thiobacillus*.

2 Practically all soils contain S-oxidizing bacteria, although their numbers in many arable soils is low because of lack of reduced S to be oxidized. When reduced forms of S are introduced into the soil, either from the atmosphere or in fertilizer products (e.g., S-coated urea), a rapid increase in numbers of S-oxidizing organisms occurs. Because most soils contain S-oxidizers, artificial inoculation is generally not considered necessary.

3 Thiosulfates, certain sulfides, SO_2, and elemental $S°$ of fertilizer formulations (S-coated urea, ammonia-S solutions) are generally oxidized quite readily. Some of the S in sulfides is probably autoxidized to elemental $S°$; however, oxidation of $S°$ to SO_4^{2-} is largely biological.

4 Environmental factors favoring the growth of soil microorganisms in general also favor the activities of S-oxidizing organisms. The following statements summarize the effect of these factors on the biological oxidation of S in soils.

    a. *Temperature.* Oxidation of S takes place from 4° to 55°C, but the most favorable range is 27° to 40°C.

    b. *Moisture and aeration.* The moisture content for most rapid S oxidation is near field capacity, although other factors such as texture can affect this level.

    c. *Soil reaction.* Microbial oxidation of S can take place between pH 2 and 9. Most soils have pH values between 3.5 and 8.5, well within this range. Oxidation generally increases with increasing pH, although the process is not critically limited by this soil property. Addition of lime to acid soils usually increases the rate of S oxidation.

    d. *Microbial population.* Inoculation of arable soils with S-oxidizing organisms usually increases the rate of oxidation, because of the low number of indigenous species. Additions of elemental $S°$ or some other reduced form stimulates the rapid multiplication of S oxidizers.

5 The rate of oxidation of applied elemental $S°$, such as for reclamation of saline and alkaline soils, is affected by several factors, some of which are:

    a. *Particle size.* The rate of oxidation of elemental $S°$ increases with a decrease in particle size. Large (6–12 mesh) $S°$ granules can be oxidized fairly rapidly, provided they disintegrate into small particles after application to the soil.

    b. *Placement.* Mixing elemental $S°$ with the soil usually results in the most rapid oxidation. Nevertheless, band placement and top dressing can be quite effective under certain conditions.

    c. *Rate of Application.* Increasing the amount of elemental $S°$ applied to a soil generally does not affect the percentage that is oxidized within a specified period of time. To a certain degree, slow oxidation of large $S°$ particles can be compensated for by increased rates of application.

SUMMARY

Although the role of S in soil fertility has been ignored, sufficient information has accumulated to indicate that, in agricultural soils of the humid and semi-humid regions, transformations involving organic forms are of utmost importance in the S nutrition of plants. Sulfur deficiencies have been limited to a few geographical areas but may be increasing because of more wide-spread use of S-free fertilizers, to a reduction in the amount of S used as a pesticide, and to lowering of atmospheric levels of S. Pertinent information regarding the S cycle can be summarized as follows

1 Sulfur is a constituent of the amino acids cysteine, cystine, and me-thionine, and it is an integral component of several vitamins and hor-mones. Two important plant growth regulators, thiamine and biotin, contain S.

2 Until recently, extraneous additions of S have been sufficient to meet the S requirements of most agricultural crops. Many fertilizers have contained S; some S has been added in insecticides and fungicides. Irrigation water used in arid regions is an important source of plant available S. Prolonged use of S-free fertilizers may eventually induce a S deficiency in crops.

3 The amount of S brought down in rainwater varies with location and season of the year. This S arises largely from the combustion of coal and other fossil fuels, which means that larger quantities are brought down near industrial areas and during winter months. In rural regions remote from industrial sites, no more than 5 to 6 kg/ha of S will be returned each year to the soil in this way. Leaching losses of S are greatest on coarse-textured soils under high rainfall.

4 Plant roots absorb S almost entirely in the SO_4^{2-} form. The concen-tration of this ion in the soil solution is governed to some extent by those factors affecting adsorption, as well as by mineralization–im-mobilization relationships.

5 As far as plant residues are concerned, the same type of mineraliza-tion–immobilization relationship exists for S as is known for N. When the C/S ratio of added residues is below 200, there will be a net gain in inorganic SO_4^{2+}; when the ratio is over 400, there is a net loss.

6 Very little is known about the forms of organic S in soil. About 50% of the total S in soils of humid and semiarid regions occurs in C-bonded

forms, only a fraction of which can be accounted for as amino acids. Another 40% of the total S occurs as unknown ester sulfates. The C/N/S ratio of the soil varies widely within any given location, but the mean ratio for soils from different locations is remarkably constant at about 140:10:1.3.

7 Some soils, notably those that are acidic and that contain kaolinite and/or Fe and Al oxides, are able to retain SO_4^{2-} in an adsorbed form.

REFERENCES

1 N. C. Brady, *The Nature and Properties of Soils*, Macmillan, New York, 1974.

2 M. V. Ivanova and J. R. Freney, Eds., *The Global Biogeochemical Sulfur Cycle*, Wiley, New York, 1983.

3 L. Granat, H. Rodhe, and R. D. Hallberg, "The Global Sulfur Cycle," in B. A. Svensson and R. Soderlund, Eds., *Nitrogen, Phosphorus and Sulfur-Global Cycles*. SCOPE Report 7. Ecol. Bull. (Stockholm) **22**, 89 (1976).

4 J. R. Freney, M. V. Ivanova, and H. Rodhe, "The Sulfur Cycle," in B. Bolin and R. B. Cook, Eds., *The Major Biogeochemical Cycles and Their Interactions*, Wiley, New York, 1983, pp. 56–65.

5 J. R. Freney and C. H. Williams, "The Sulphur Cycle in Soil," in M. V. Ivanova and J. R. Freney, Eds., *The Global Biogeochemical Sulphur Cycle*, Wiley, New York, 1983, pp. 129–201.

6 P. R. Trudinger, "The Biogeochemistry of Sulphur," in K. D. McLachlan, Ed., *Sulphur in Australasian Agriculture*, Sydney University Press, Sydney, 1975, pp. 11–20.

7 W. W. Kellogg, R. D. Cadle, E. R. Allen, A. L. Lazrus, and E. A. Martell, *Science* **175,** 587 (1972).

8 J. W. Anderson, "The Function of Sulphur in Plant Growth and Metabolism," in K. D. McLachlan, Ed., *Sulphur in Australasian Agriculture*, Sydney University Press, Sydney, 1975, pp. 87–97.

9 K. Mengel and E. A. Kirkby, *Principles of Plant Nutrition*, 3rd ed., International Potash Institute, Bern, Switzerland, 1982.

10 G. Anderson, "Sulfur in Soil Organic Substances," in J. E. Gieseking, Ed., *Soil Components*, Vol. 1, Springer-Verlag, New York, 1976, pp. 333–341.

11 V. O. Biederbeck, "Soil Organic Sulfur and Fertility," in M. Schnitzer and S. U. Khan, Eds., *Soil Organic Matter*, Elsevier, New York, 1978, pp. 273–310.

12 J. R. Freney, "Sulfur-Containing Organics," in A. D. McLaren and G. H. Peterson, Eds., *Soil Biochemistry*, Dekker, New York, 1967, pp. 229–259.

13 J. R. Freney and F. J. Stevenson, *Soil Sci.* **101**, 307 (1966).

14 C. G. Kowalenko, "Organic Nitrogen, Phosphorus and Sulfur in Soils," in M. Schnitzer and S. U. Khan, Eds., *Soil Organic Matter*, Elsevier, New York, 1978, pp. 95–136.

15 J. W. Fitzgerald, "Naturally Occurring Organosulfur Compounds in Soil," in J. O. Nriagu, Ed., *Sulfur in the Environment, Part II. Ecological Impacts*, Wiley, New York, 1978, pp. 391–443.

16 W. B. McGill and C. V. Cole, *Geoderma* **26**, 267 (1981).

17 D. C. Whitehead, *Soils Fertilizers* **27**, 1 (1964).

18 C. H. Williams, "The Chemical Nature of Sulphur Compounds in Soil," in K. D.

McLachlan, Ed., *Sulphur in Australasian Agriculture*, Sydney University Press, Sydney, 1975, pp. 21–30.

19 G. R. Burns, *Oxidation of Sulphur in Soils*. Tech. Bull. 13, The Sulphur Institute, Washington, D.C., 1967.

20 M. Wainwright, *Sci. Progr.* **260,** 459 (1978).

21 R. L. Starkey, *Soil Sci.* **111,** 297 (1966).

22 C. W. Field, "Sulfur: Element and Geochemistry," in R. W. Fairbridge, Ed., *Encylopedia of Geochemistry and Environmental Sciences*, Vol. IVA, Van Nostrand Reinhold, New York, 1972, pp. 1142–1148.

23 H. V. Jordan and H. M. Reisenauer, "Sulfur and Soil Fertility," in *The Yearbook of Agriculture, USDA*, U.S. Government Printing Office, Washington, D.C., 1957, pp. 101–111.

24 W. E. Larson, C. E. Clapp, W. H. Pierre, and Y. B. Morachan, *Agron. J.* **64,** 204 (1972).

25 J. K. Syers, J. A. Adams, and T. W. Walker, *J. Soil Sci.* **21,** 146 (1970).

26 W. L. Banwart and J. M. Bremner, *J. Environ. Qual.* **4,** 363 (1975).

27 W. L. Banwart and J. M. Bremner, *Soil Biol. Biochem.* **8,** 19, 439 (1976).

28 L. F. Elliott and T. A. Travis, *Soil Sci. Soc. Amer. Proc.* **37,** 700 (1973).

29 J. A. Lewis and G. C. Papavizas, *Soil Biol. Biochem.* **2,** 239 (1970).

30 P. J. Maroulis and A. R. Bandy, *Science* **196,** 647 (1977).

31 J. M. Bremner, "Role of Organic Matter in Volatilization of Sulphur and Nitrogen from Soils," in *Soil Organic Matter Studies*, International Atomic Energy Agency, Vienna, 1977, pp. 229–240.

32 J. M. Bremner and C. G. Steele, *Adv. Microbial Ecol.* **2,** 155 (1978).

33 D. C. Grey and M. L. Jensen, *Science* **177,** 1099 (1972).

34 J. M. Bremner and W. L. Banwart, *Soil Biol. Biochem.* **8,** 79 (1975).

35 K. A. Smith, J. M. Bremner, and M. A. Tabatabai, *Soil Sci.* **116,** 313 (1973).

36 W. H. Allaway and J. F. Thompson, *Soil Sci.* **101,** 240 (1966).

37 J. D. Beaton, *Soil Sci.* **101,** 267 (1966).

38 M. A. Tabatabai and J. M. Bremner, *Sulphur Inst. J.* **8,** 1 (1972).

39 C. A. Evans and C. O. Rost, *Soil Sci.* **59,** 125 (1945).

40 W. Couto, D. J. Lathwell, and D. R. Bouldin, *Soil Sci.* **127,** 108 (1979).

41 M. E. Harward and H. M. Reisennauer, *Soil Sci.* **101,** 326 (1966).

42 D. W. Johnson and G. S. Henderson, *Soil Sci.* **128,** 34 (1979).

43 S. S. S. Rajan, *Soil Sci. Soc. Amer. J.* **43,** 65 (1979).

44 D. S. Shriner and G. S. Henderson, *J. Environ. Qual.* **7,** 392 (1978).

45 B. R. Singh, *Acta Agr. Scand.* **28,** 313 (1980).

46 D. W. Johnson and D. E. Todd, *Soil Sci. Soc. Amer. J.* **47,** 792 (1983).

47 D. W. Johnson, G. S. Henderson, and D. E. Todd, *Soil Sci.* **132,** 422 (1981).

48 N. M. Scott and G. Anderson, *J. Sci. Fd. Agric.* **27,** 358 (1976).

49 M. B. David, M. J. Mitchell, and J. A. Nakas, *Soil Sci. Soc. Amer. J.* **46,** 847 (1982).

50 M. A. Tabatabai and J. M. Bremner, *Sulphur Inst. J.* **8,** 1 (1972).

51 M. A. Tabatabai and J. M. Bremner, *Soil Sci.* **114,** 380 (1972).

52 M. A. Tabatabai, "Sulfur," in A. L. Page, R. H. Miller, and D. R. Keeney, Eds., *Methods of Soil Analysis*, Part 2, 2nd ed., American Society of Agronomy, Madison, Wis., 1982, pp. 501–538.

53 C. E. Bardsley and J. D. Landcaster, "Sulfur," in C. A. Black et al., Eds., *Methods of Soil Analysis*, American Society of Agronomy, Madison, Wis., 1965, pp. 1102–1116.

54 J. R. Bettany, J. W. B. Stewart, and S. Saggar, *Soil Sci. Soc. Amer. J.* **43**, 981 (1979).

55 J. R. Bettany, S. Saggar, and J. W. B. Stewart, *Soil Sci. Soc. Amer. J.* **44**, 70 (1980).

56 J. R. Bettany, J. W. B. Stewart, and E. H. Halstead, *Soil Sci. Soc. Amer. Proc.* **37**, 915 (1973).

57 K. M. Goh and M. R. Williams, *J. Soil Sci.* **33**, 73 (1982).

58 A. M. L. Neptune, M. A. Tabatabai, and J. J. Hanway, *Soil Sci. Soc. Amer. Proc.* **39**, 51 (1975).

59 N. M. Scott and G. Anderson, *J. Soil Sci.* **27**, 324 (1976).

60 C. H. Williams, E. G. Williams, and N. M. Scott, *J. Soil Sci.* **11**, 334 (1960).

61 J. R. Freney, G. E. Melville, and C. H. Williams, *Soil Sci.* **109**, 310 (1970).

62 J. R. Freney, G. E. Melville, and C. H. Williams, *J. Sci. Food Agric.* **20**, 440 (1969).

63 L. E. Lowe, *Can. J. Soil Sci.* **45**, 297 (1965).

64 L. E. Lowe and W. A. DeLong, *Can. J. Soil Sci.* **43**, 151 (1963).

65 K. D. McLachan and D. G. DeMarco, *Aust. J. Soil Res.* **13**, 169 (1975).

66 R. G. McLaren and R. S. Swift, *J. Soil Sci.* **28**, 445 (1977).

67 J. R. Freney, F. J. Stevenson, and A. H. Beavers, *Soil Sci.* **114**, 468 (1972).

68 N. M. Scott, W. Bick, and H. A. Anderson, *J. Sci. Food Agr.* **32**, 21 (1981).

69 Y. M. Chae and L. E. Lowe, *Can. J. Soil Sci.* **60**, 633 (1980).

70 Y. M. Chae and L. E. Lowe, *Soil Biol. Biochem.* **13**, 257 (1981).

71 Y. M. Chae and M. A. Tabatabai, *Soil Sci. Soc. Amer. J.* **45**, 20 (1981).

72 H. S. Mason, *Adv. Enzymol.* **16**, 105 (1955).

73 C. H. Williams and C. M. Donald, *Aust. J. Agr. Res.* **8**, 179 (1957).

74 I. Hague and D. Walmsley, *Plant and Soil* **37**, 255 (1972).

75 C. G. Kowalenko and L. E. Lowe, *Can. J. Soil Sci.* **55**, 9 (1975).

76 M. A. Tabatabai and J. M. Bremner, *Agron. J.* **64**, 40 (1972).

77 M. A. Tabatabai and A. A. Al-Khafaji, *Soil Sci. Soc. Amer. J.* **44**, 1000 (1980).

78 C. H. Williams, *Plant and Soil* **26**, 205 (1967).

79 B. A. Stewart and C. J. Whitfield, *Soil Sci. Soc. Amer. Proc.* **29**, 752 (1965).

80 J. W. Fitzgerald, T. C. Strickland, and W. T. Swank, *Soil Biol. Biochem.* **14**, 529 (1982).

81 W. T. Swank, J. W. Fitzgerald, and J. T. Ash, *Science* **223**, 182 (1984).

82 S. Saggar, J. R. Bettany, and J. W. B. Stewart, *Soil Biol. Biochem.* **13**, 493 (1981).

83 S. Saggar, J. R. Bettany, and J. W. B. Stewart, *Soil Biol. Biochem.* **13**, 499 (1981).

84 J. E. Strick and J. P. Nakas, *Soil Biol. Biochem.* **16**, 289 (1984).

85 P. J. M. Cooper, *Soil Biol. Biochem.* **4**, 333 (1972).

86 R. Lee and T. W. Speir, *Plant and Soil* **53**, 407 (1979).

87 T. W. Speir, R. Lee, E. A. Pansier, and A. Cairns, *Soil Biol. Biochem.* **12**, 281 (1980).

88 M. A. Tabatabai and J. M. Bremner, *Soil Sci., Soc. Amer. Proc.* **34**, 225 (1970).

89 M. A. Tabatabai and J. M. Bremner, *Soil Sci. Soc. Amer. Proc.* **34**, 427 (1970).

90 J. R. Bettany, J. W. B. Stewart, and E. H. Halstead, *Can. J. Soil Sci.* **54**, 309 (1974).

91 J. R. Freney, G. E. Melville, and C. H. Williams, *Soil Biol. Biochem.* **3**, 133 (1971).

92 J. R. Freney, G. E. Melville, and C. H. Wiilliams, *Soil Biol. Biochem.* **7**, 217 (1975).

93 K. M. Goh and P. E. H. Gregg, *New Zealand J. Sci.* **25**, 135 (1982).

94 P. E. H. Gregg and K. M. Goh, *New Zealand J. Agr. Res.* **21**, 593 (1978).

95 P. E. H. Gregg and K. M. Goh, *New Zealand J. Agr. Res.* **22**, 425 (1979).

96 A. J. Nicolson, *Soil Sci.* **109**, 345 (1970).

97 C. C. Ainsworth and R. W. Blanchar, *J. Environ. Qual.* **13,** 193 (1984).

98 C. Bloomfield and J. K. Coulter, *Adv. Agron.* **25,** 265 (1973).

99 I. R. Kaplan, "Sulfur Cycle," in R. W. Fairbridge, Ed., *Encyclopedia of Geochemistry and Environmental Sciences,* Vol. IVA, Van Nostrand Reinhold, New York, 1972, pp. 1148–1152.

100 Y. M. Nor and M. A. Tabatabai, *Soil Sci. Soc. Amer. J.* **41,** 736 (1977).

101 M. Wainwright, *Soil Biol. Biochem.* **11,** 95 (1979).

102 S. A. Waksman and J. S. Joffe, *J. Bacteriol.* **7,** 239 (1922).

NINE

THE MICRONUTRIENT CYCLE

The term *micronutrient* refers to an element that is required in small amounts by plants or animals and that is necessary in order for the organism to complete its life cycle. The trace metals regarded as being essential for plants are iron (Fe), zinc (Zn), manganese (Mn), copper (Cu), boron (B), and molybdenum (Mo). Recent evidence indicates that nickel (Ni) is also required. Three other elements, cobalt (Co), chromium (Cr), and tin (Sn), are not required by plants but are essential for animals. Other heavy metals, notably cadmium (Cd), mercury (Hg), and lead (Pb), are of interest because they represent potential hazards to the environment, including plant and animal health.

An outline of the micronutrient cycle in soil is depicted in Fig. 9.1. The quantity of any given micronutrient that occurs in the solution phase as free ions, or as soluble metal–chelate complexes, is influenced by chemical reactions involving the formation of insoluble precipitates, as well as by transformations carried out by microorganisms. The return of plant residues to the soil leads to recycling of micronutrients, a factor of considerable importance in micronutrient-deficient soils. Chelating agents produced by microorganisms or excreted by plant roots are shown to be involved in the weathering of rocks and minerals, and they function as vehicles for the movement of micronutrients to roots.

The conversion of micronutrients to unavailable forms can occur through complexation by insoluble humic substances and by fixation reactions involving clay minerals and oxide surfaces. Microorganisms can act as solubilizers of mineral matter, thereby increasing the availability of micronutrients to plants.

Reviews are available concerning all aspects of the micronutrient cycle, including geochemistry,[1–3] forms and reactions in soils,[3–10] role in plant nutrition,[11–14] and environmental cycling.[15–21]

DISTRIBUTION OF TRACE ELEMENTS IN SOILS

The trace elements in soil include those derived from: (1) the parent rocks and minerals from which the soil was formed, (2) impurities in fertilizers

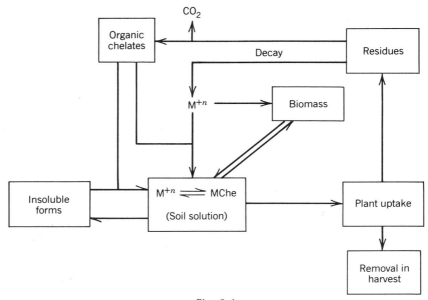

Fig. 9.1
Micronutrient cycle in soil.

and lime, as constituents of pesticides and manures, and as contaminants of sewage sludge, and (3) debris from industrial and mining wastes, fossil fuel combustion products, wind-eroded soil particles, and meteoric and volcanic material that settles out or is added via rainfall.[18]

The main source of trace elements in most soils is the parent material from which the soil was formed, where the elements existed in a host of naturally occurring minerals that were formed in the lavas from which igneous rocks were formed.[1-6] The quantities of trace elements added to soil in the ordinary amendments (item 2) are generally too low to significantly affect the total amounts present, but they may have an influence on the amounts taken up by the plant.

The quantities of six micronutrient cations in igneous rocks, along with the main minerals in which they occur, are listed in Table 9.1. Among the elements, Fe is by far the most abundant, which can be explained by the fact that it is a major constituent of ferromagnesian minerals. Important original sources of Zn, Cu, and Mo are the sulfides of igneous rocks. Zinc and Cu, as well as Mn, also occur in ferromagnesian minerals, where they substitute for Fe and Mg. Boron is found largely as the borosilicate mineral, tourmaline.

The two groups of igneous rocks (granite and basalt) contain their own characteristic array of trace elements. Iron, Zn, Mn, and Cu are somewhat more abundant in basalt; B and Mo are more concentrated in granite. Prin-

Table 9.1

Sources, Ionic Forms, and Abundance of Micronutrients (μg/g) in Igneous and Sedimentary Rocks[a]

Nutrient	Common Soluble Ionic Forms	Principal Sources	Igneous Rocks		Sedimentary Rocks		
			Granite	Basalt	Limestone	Sandstone	Shale
Iron	Fe^{2+}, Fe^{3+}	Ferromagnesian minerals	27,000	86,000	3,800	9,800	47,000
Zinc	Zn^{2+}	Sulfides; substituted ion in common minerals	40	100	20	16	95
Manganese	Mn^{2+}	As substituted ion in ferromagnesian minerals	400	1,500	1,100	<100	850
Copper	Cu^{2+}	Sulfides; substituted ion in common minerals	10	100	4	30	45
Boron	BO_3^{3-}	Tourmaline; soluble borates	15	5	20	35	100
Molybdenum	MoO_4^{2-}	Sulfides; substituted ion in common minerals	2	1	0.4	0.2	2.6

[a] From Krauskopf[1] and Hodgson.[3]

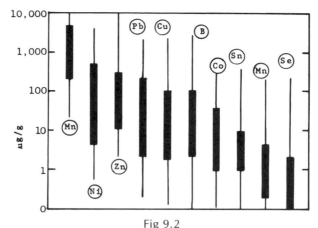

Fig 9.2
Trace element content of soils. (Adapted from Mitchell.[7])

cipals governing the accumulation and distribution of trace elements in igneous rocks have been discussed by Krauskopf.[1]

Included in Table 9.1 are data for the concentrations of trace elements in sedimentary rocks. Sediments represent debris from many sources and they thereby contain all the trace elements found in igneous rocks, although not necessarily in the same proportion. Processes of disintegration, weathering, mixing, and deposition, together with other complex factors, lead to redistribution of the elements, and, in some cases, changes in their chemical forms. Fine-grained sediments (e.g., shales) become enriched in trace elements such as Zn, Cu, Co, B, and Mo (see Table 9.1). Shales that contain high amounts of organic matter are often enriched in Cu and Mo.

The range of trace elements in soils is given in Fig. 9.2, where the thin lines indicate the more unusual values. The amount of any given trace element can vary as much as a thousandfold, depending on the soil and its origin. As a general rule, the trace element content follows closely the clay and/or organic matter content.[3]

Except for Fe and Mn, the total amount of any given trace element per unit weight of soil is very low, usually amounting to no more than a few kilograms per hectare (lb/acre) within the plow depth. Numerous exceptions can be found, however. For example, some volcanic ash soils of Hawaii have high Mn contents (>10%), which is far above the usual range shown in Fig. 9.2.

The distribution of trace elements in soil reflects to some extent the amounts contained in the parent material from which the soil was formed. However, the elements exist in somewhat different chemical forms. During the various stages of soil formation, trace elements are released from the primary minerals and incorporated into other forms, such as structural components of secondary silicate (clay) minerals, as complexes with organic matter, and occlusion in Fe and Mn oxides, as discussed later.

Geochemistry of Trace Element Deficiencies and Excesses

Micronutrient deficiencies are common in soil, although to a lesser extent than for the macronutrients (e.g., N, P, K).

A deficiency or excess of a particular trace element can often be traced to the parent material from which the soil was formed. Deficiencies are particularly serious on soils derived from coarse-grained materials (e.g., sands and sandstones), which characteristically contain very low concentrations of one or more of the micironutrient cations. Highly weathered soils, even those containing high amounts of clay, often have low micronutrient contents. Organic soils may contain low or high amounts of trace elements, depending on the type of plant material from which the peat was formed and the amount of micronutrients carried into the peat in seepage waters.

Carbonate rocks per se are low in micronutrients but loss of carbonate during weathering leads to their enrichment in the altered debris. Accordingly, soils derived from carbonate rocks are not necessarily deficient in micronutrients. Certain bauxite and kaolin-rich sediments, as well as carbonates, are usually low in B; marine shales and glauconitic sandstones are usually high.

In no country has more widespread occurrences of micronutrient deficiencies been observed than in Australia.[6,9] Deficiencies occur on a wide variety of soil types, but mostly in extensive areas of calcareous sands. Vast expanses of the southwest of western Australia (8 million hectares) and of the Ninety Mile Plain of south Australia (over 1 million hectares) have been reclaimed by application of one or more micronutrients, notably Zn, Cu, and Mo. In England, Cu deficiencies are common on soils formed from shallow chalk. Peats and mucks deficient in one or more of the micronutrients occur throughout northern and western Europe, New Zealand, the eastern United States, and elsewhere. Many soils in New Zealand and Tasmania are deficient in one or more of the trace elements; acreage of usable land that require additions of micronutrient fertilizers has been estimated at 200,000 ha.

National maps for micronutrient deficiencies have been prepared for the United States, but, unlike the Australian experience, generalizations can seldom be made as to the land area affected. As Kubota and Allaway[2] pointed out, such summaries fail to depict actual areas of deficiencies, and they can give a distorted picture of the severity of a given trace element problem. Whereas Zn deficiencies are common in soils of the western half of the United States, the number of deficient fields is small compared to the total land area. The indiscriminate use of micronutrient fertilizers on nondeficient soils cannot be recommended; in some cases, undesirable accumulations of the element in the soil can lead to plant toxicities.

Lucas and Knezek[22] pose the following questions regarding programs that encourage additions of micronutrients in N–P–K fertilizers. Why continuously apply Fe and Mn when the average soil contains about 50,000 μg/g Fe and 1,000 μg/g Mn? Would one be satisfied with trace amounts of Zn, B, and Mo in the fertilizer if he has no responsive crops on deficient soils? Would one want to use such elements as Mn, B, and Mo in the fertilizer when existing soil conditions could cause toxicity to plants and animals?

Correction of a trace element deficiency is not an easy task, partly because of the small amount required and because addition of an excess may have a deleterious effect on the plant. For any given micronutrient, no general rule can be given for the proper amount to apply because each plant species has its own requirement and tolerance.

The six elements regarded as being essential to plants (Fe, Zn, Mn, Cu, B, Mo), along with Co (required by animals but not plants), are discussed briefly below. Detailed information can be found in several reviews.[3-10]

Iron

Of the various trace elements, Fe is by far the most abundant. It ranks fourth (about 5%) among the elements in the lithosphere, after oxygen, Si, and Al. The Fe content of soil varies from about 200 μg/g in coarse textured soils to well over 10% (100,000 μg/g) in ferruginous latosols (tropical soils). Leached sands contain the lowest amount of Fe.

The ferrous form (Fe^{2+}) is the most available to plants. Under alkaline conditions, Fe^{2+} becomes oxidized to the ferric (Fe^{3+}) state, which is relatively unavailable to plants. The reaction responsible for reduced availability is precipitation of $Fe(OH)_3$ as the concentration of OH^- ions is increased. The solubility product of $Fe(OH)_3$ is 10^{-38}, indicating that very little Fe will be in solution on the alkaline side of neutrality.

In calcareous or heavily limed soils, plants frequently suffer from lack of Fe, a condition referred to as lime-induced Fe deficiency. In some environments, availability of Fe under stress conditions is enhanced through the production of organic chelating agents by microorganisms, a typical case being the synthesis of hydroxamic siderophores by mycorrhizal fungi, a subject discussed later.

Zinc

The normal level of Zn in soils is of the order of 2 to 50 μg/g, with some soils having contents up to 200 μg/g. As is the case for other micronutrient cations, sandy soils usually contain very low amounts of Zn. Soils formed from limestone generally contain more Zn than those formed from gneiss or quartzite.

Zinc deficiencies are particularly common in the western half of the United States, although deficient areas are also found elsewhere. The crops most often affected are corn, field beans, sorghum, and potatoes. Deficiencies have also been observed in sugarcane fields of Florida. Cool, wet periods during the growing season favor Zn deficiencies. Leached sandy soils, eroded soils, and peat and muck soils are generally low in available Zn.

Certain soil conditions reduce the availability of Zn, notably high pH. The solubility of Zn decreases with an increase in pH. Thus a high incidence of Zn deficiency often occurs on calcareous or limed soils. High levels of soil phosphates also tend to enhance Zn deficiency.

Manganese

The concentration of Mn in soil covers a particularly wide range, from as little as 20 $\mu g/g$ to well over 6,000 $\mu g/g$. The lower values are typical of severely leached acid soils. In contrast, Mn excesses or toxicities often occur in unleached acid soils, as well as in waterlogged soils.

The main factors that determine Mn availability in soil are pH and the oxidation-reduction potential. A pH value below 6.0 to 6.5 favors reduction of Mn and the formation of the more available divalent form (Mn^{2+}); higher pH values favor oxidation to the Mn^{4+} ion, from which insoluble oxides are formed (MnO_2, Mn_2O_3, and Mn_3O_4). In this respect, Mn resembles Fe in its behavior in soil. Both are involved in oxidation-reduction reactions and both occur as insoluble oxides at high pH values.

As one might expect, deficiencies of Mn are most often found in calcareous soils. One of the more sensitive crops is oats, some varieties being so sensitive that the deficient plant fails to mature and produce grain.

Copper

The Cu content of most agricultural soils ranges from 5 to 60 $\mu g/g$, although both lower (<2 $\mu g/g$) and higher values (up to 250 $\mu g/g$) are not uncommon. Soils derived from coarse-grained sediments (sands and sandstones), as well as acid igneous rocks, are usually low in Cu. Many highly weathered soils, such as those of the coastal plains of the southeastern United States, contain low amounts of Cu.

Factors affecting the soils ability to provide Cu to plants include pH, humus content, nature of the previous crop, drainage status, and proportion of sand to clay. Crops grown on mineral soils with Cu contents of less than about 4 $\mu g/g$, or on organic soils with less than 20 to 30 $\mu g/g$, are likely to suffer from Cu deficiency.

Boron

Total B in soils ranges from 2 to 100 $\mu g/g$, but is generally in the 7 to 80 $\mu g/g$ range. The element occurs as tourmaline, a very insoluble fluorine-containing borosilicate, as Ca and Mg borates, as complexes with Fe and Al, and in association with organic matter.

Light-colored acid soils of humid regions (e.g., Atlantic Coastal Plain and the Mississippi Valley) are often deficient in B because of the ease with which the borate ion ($H_2BO_3^-$) is leached. Peats usually contain high amounts of B, possibly as borate complexes with organic matter.

Deficiencies of B are rather common on light-colored sands and silt loams in humid regions. The crops most often affected are alfalfa and root and cruciferous crops, such as cabbage, cauliflower, and turnips. More B is applied to alfalfa than any other field crop. Liming of acid soils may trigger a temporary B deficiency in specific crops.

As suggested previously, crop species differ markedly in their requirements for B, and any geographic pattern of B deficiency may be due to the nature of the crop than to B levels in the soil.

Molybdenum

Total Mo in soils varies from 0.2 to 5 $\mu g/g$, the average being about 2 $\mu g/g$. Deficiencies are geographically widespread and have been recorded for a large number of crops despite the very small amounts generally required by plants. Molybdenum is essential for N_2 fixation by *Rhizobium;* accordingly, herbage legumes growing on Mo-deficient soils will often show symptoms of N deficiency. Care must be taken to apply Mo at the recommended fertilizer rate because excess Mo in feeds and forages is toxic to cattle, as noted later.

Main forms of Mo in soils are as follows: (1) water-soluble, (2) combined with organic matter, (3) as the exchangeable anion MoO_4^{2-}, and (4) held in the lattice structures of minerals.

Molybdenum is the only micronutrient cation that increases in availability with increasing pH. Unlike Fe and Mn, availability is lowest under acidic soil conditions. Application of lime to the soil will often sufficiently increase the availability of native soil Mo to correct Mo deficiency.

Cobalt

Cobalt is not required by plants but it is essential for animals. The amount contained in soils varies over a wide range (1 to 300 $\mu g/g$), but the usual range is of the order of 10 to 15 $\mu g/g$. In the United States, soils low in Co (often 1 $\mu g/g$ or less) are found in sandy soils (e.g., Spodosols) of glaciated regions of the Northeast. Low amounts are also found in soils of the coastal plains of the southeastern United States.

Plants growing on soils containing less than about 5 $\mu g/g$ will often contain insufficient Co to meet the dietary requirements of cattle and sheep. The availability of Co varies considerably from soil to soil and a direct correlation does not exist between total content and plant uptake.

Trace Element Toxicities

Many soils contain such high concentrations of one or more of the trace elements that they constitute a health hazard to plants or animals. A few examples are given below.

Toxic concentrations of Cu are common in soils near cupriferous outcrops, such as occur in Europe and southern Africa. Affected areas are readily recognized from the distinctive species and anomalous growth and color of the plants associated with them. In the Cu belt of southern Africa, tolerant herbaceous species colonize the most contaminated soils and these are surrounded by an intermediate zone of stunted woody plants that form a transition to the prevailing forest.

Toxicities due to excess Cu have also been observed on sandy soils that have been treated with $CuSO_4$, at one time used extensively as a pesticide, such as for the control of fungal growth on grapes. Other examples of toxic concentrations of Cu are waste heaps from Cu mining and smelting operations.

Boron toxicities are also common and arise in arid-zone soils where Na- and Ca-borate salts accumulate at the soil surface. In the western United States, toxicities can result when waters rich in B are used for irrigation.

Abnormalities in the concentration of trace elements in plants can cause nutritional disorders in animals, for two reasons: (1) a higher amount is required by the animal, as compared to the plant, Co being a typical example, and (2) the animal has a lower tolerance level for the trace element (e.g, Mo and Se).

Selenium is one of the least abundant of the elements found in soil-forming rocks and minerals. Notwithstanding, soils formed from sedimentary rocks of the Cretaceous age in the western United States contain such high concentrations of Se that a condition called selenosis is often induced in cattle grazing in the region.

Molybdenum is a micronutrient that can occur in plants in concentrations of several hundred μg/g without apparent harm. Yet concentrations in forages as low as 20 μg/g can cause toxic effects in some animals, especially ruminants. Molybdenum toxicities due to consumption of forages by livestock have been experienced in Florida and in several western states, as well as other countries. Toxicity problems in the western United States characteristically occur on poorly drained neutral or alkaline soils formed from Mo-rich granitic alluvium. Liming of acid soils inherently high in Mo can lead to accumulations of Mo in forages.

MICRONUTRIENT REQUIREMENTS OF PLANTS

Data for the micronutrient concentrations of mature leaf tissue on the basis of deficiencies, sufficiencies, and excesses or toxicities are given in Table 9.2. Of the micronutrients, Fe is required in the largest amount, followed by Zn, Mn, and B. Molybdenum is required in the smallest amount, with Cu being next to the lowest.

The range of values for any given micronutrient in the plant is considerable, because of luxury consumption by the plant and differences among plants in their requirements or ability to adsorb micronutrients from the soil. The micronutrient content of various parts of the same plant also varies over

Table 9.2
Approximate Micronutrient Concentrations in Mature Leaf Tissue That May be
Classified as Deficient, Sufficient, or Excessive (in μg/g)[a]

Micronutrient	Deficient	Sufficient	Excessive or Toxic
Fe	<50	50–250	Unknown
Zn	<20	25–150	400+
Mn	<20	20–50	500+
Cu	<4	5–20	20+
B	<15	20–100	200+
Mo	<0.1	0.5–?	Unknown

[a] From Jones.[23] Average values are for recently matured leaves. Values will vary for specific crops and time of sampling.

a broad range. The relationship between plant growth or response and the concentration or uptake of a micronutrient is illustrated in Fig. 9.3. An initial increase in growth is followed by a plateau, representing first optimum uptake and then luxury consumption. With increasing amounts of the micronutrient, growth is reduced because of toxicity effects. Each micronutrient has its own characteristic response curve (e.g., amount needed for optimum growth, width of plateau, etc.). All micronutrients are potentially toxic when the range of safe and adequate exposure is exceeded. Except for the initial growth response, a nonessential trace element has a similar effect, notably a plateau at low concentrations followed by reduced growth thereafter.

It is apparent, therefore, that the concentration of any given micronutrient

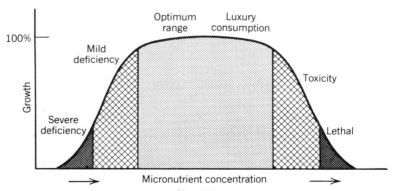

Fig. 9.3
Relationship between plant growth or response and the concentration or uptake of a micronutrient cation. Except for the initial growth response, a nonessential trace element has a similar effect. (Adapted from Mertz.[12])

Table 9.3
Incomplete List of the Function of Micronutrients in Plants

Micronutrient	Function in Plants
Fe	An essential component of many heme and nonheme Fe enzymes and carriers, including the cytochromes (respiratory electron carriers) and the ferredoxins. The latter are involved in key metabolic functions, such as N_2 fixation, photosynthesis, and electron transfer.
Zn	An essential component of several dehydrogenases, proteinases, and peptidases, including carbonic anhydrase, alcohol dehydrogenase, glutamic dehydrogenase, and malic dehydrogenase, among others.
Mn	Involved in the O_2-evolving system of photosynthesis and is a component of the enzymes arginase and phosphotransferase.
Cu	A constituent of a member of important enzymes, including cytochrome oxidase, ascorbic acid oxidase, and laccase.
B	The specific biochemical function of B is unknown but it may be involved in carbohydrate metabolism and synthesis of cell wall components.
Mo	Required for the normal assimilation of N in plants. An essential component of nitrate reductase, as well as nitrogenase (N_2 fixation enzyme).

in the plant is influenced by a wide variety of factors that include availability of the element in the soil, cultural practice, the proportion of plant parts (leaves and stems), and environmental conditions for plant growth (e.g., climate).

Function of Micronutrients

The low requirement of plants for trace elements can be accounted for by their participation in enzymatic reactions and as constituents of growth hormones, rather than as components of major plant products (e.g., structural or protoplasmic tissue). As can be seen from Table 9.3, a multitude of enzymatic processes are affected. They include enzymes that play a key function in carbohydrate metabolism (photosynthesis, respiration), N metabolism (protein synthesis and biological N_2 fixation), cell wall metabolism (e.g., lignin synthesis), water relations, ion uptake, seed production, metabolism of secondary plant substances, and disease resistance. Many transformations involve oxidation-reduction reactions, in which case the element assumes a different oxidation state.

Deficiency Symptoms

A deficiency of any given trace element is nearly always manifested by definitive symptoms that occur in the leaves or stems of the growing plant.

Chlorosis is an early and typical symptom. Also, in many plants the stems fail to elongate. Symptoms vary between plant species and are often masked by other nutritional stresses.

With chlorosis, the veins of the leaf usually remain green whereas the tissue does not. Often the specific color of the chlorotic area is an aid in identifying the micronutrient that is deficient, namely, bright yellow to yellow-green for Fe, reddish-yellow or orange for B, and white for Zn or Mn. The distribution of chlorotic areas over the leaf or the degree to which the leaf is affected is also useful in identifying the nutrient that is deficient.

A deficiency that restricts stem elongation results in a leaf deformation condition referred to as rosetting. This symptom is most often manifested by inadequate B and Zn in the plant.

Levels and deficiency patterns for specific trace elements in plants follow.

Iron

Plants differ in their contents of Fe, as well as in their abilities to extract Fe from the soil. Typical Fe contents are as follows: pasture grasses, 100 to 200 μg/g; pasture legumes, 200 to 300 μg/g; alfalfa, to 1,000 μg/g.

Iron deficiency is by far the most prevalent of the trace element deficiencies, and typically occurs with plants growing on calcareous soils, and more frequently in fruit crops than field crops. As one might expect, Fe deficiencies in the United States pose more of a problem in arid and semiarid regions of the west than in other zones. Farmyard manures and other organic wastes, either enriched with Fe or unenriched, have been found to be good sources of Fe for sensitive crops in deficient soils.

The most characteristic plant deficiency symptom is inability of the young leaves to produce chlorophyll, leading to chlorotic mottling or yellow striping. Deficiencies in fruit trees are often corrected with foliar spray applications.

Zinc

Levels of Zn in plants range from 10 to 100 μg/g for most agricultural crops (15 to 60 μg/g in grasses and legumes). As is the case for other micronutrients, plants differ widely in the level of Zn for optimum growth.

Zinc deficiency in agronomic crops is second only to Fe as the most common nutrient deficiency. Typical symptoms are stunted growth and changes in leaf morphology. Some crops, however, fail to show leaf symptoms even though deficient. Several well-known disorders attributed to Zn deficiency include "white bud" of corn, so-called because the bud may turn white or light yellow while the leaves show bleached bands or stripes. Zinc fertilizers have long been in common use on citrus in Florida.

Manganese

The Mn content of plants (to 500 μg/g) varies much more than the other micronutrients. For many plants, deficiencies occur when the plant tissue contains less than 20 μg/g. The best-known species that is susceptible to Mn deficiency is the common oat.

Visible effects of Mn deficiency are more diverse than for most other micronutrients. The symptoms may occur first in young or old leaves and may consist of a variety of chlorotic patterns and necrotic spotting. Field disorders such as "gray speck" of oats, "marsh spot" of peas, and "speckled yellows" of beets are typical examples. The most characteristic symptom of "gray speck" of oats is kinking of the leaf near the base and brownish-black specks on the leaves. Soybeans are also sensitive to Mn deficiency. A deficiency of Mn is manifested by chocolate-brown lesions in barley and white necrotic streaks in wheat or rye.

Manganese toxicities are common where sensitive plants are grown on strongly acid soils, a condition that can be corrected through pH adjustment with lime.

Copper

Plant tissues normally contain 3 to 40 μg/g of Cu, depending on plant species and soil factors. Deficiencies are well known, particularly in deciduous fruit trees, cereals, and herbage legumes. The concentration in the plant below which deficiencies occur varies with the species and the plant part that is sampled but an adverse effect on plant growth is usually noted at concentrations less than 2 to 3 μg/g in the dry matter. Symptoms of Cu deficiency are stunted growth, shortened internodes, and yellowing, withering, and curling of leaves. In the cereals the leaves may become coiled into a tight spiral.

Copper toxicities are also known, such as where Cu has accumulated in the surface soil through prolonged use of Cu-containing fungicides. The more common symptoms of Cu toxicity include reduced shoot vigor, poorly developed and discolored roots, and leaf chlorosis. For many plants, growth is restricted whenever Cu content in whole shoots exceeds 20 μg/g.

Boron

The normal level of B in mature leaves is of the order of 20 to 100 μg/g. Deficiency symptoms often occur at less than 15 μg/g; toxicities occur at contents greater than 200 μg/g.

A deficiency of B in plants produces a variety of symptoms which are given descriptive terms such as "heart rot" of sugar beets, "yellows" of alfalfa, and "top-sickness" of tobacco. With wheat, the younger leaves are white in color, are rolled, and are frequently trapped at the apex within the rolled subtending leaf.

Molybdenum

The Mo content of plant tissues varies from a trace (<0.1 μg/g) to more than 300 μg/g dry matter, with legumes containing higher amounts than nonleguminous plants. Deficiency symptoms occur in most plants when the concentration is less than 0.1 μg/g.

Typical deficiency symptoms are interveinal mottling and cupping of the older leaves, followed by necrotic spots at leaf tips and margins. In alfalfa and other legumes, N deficiency symptoms can arise because the rhizobia do not contain the Mo they require for biological N_2 fixation.

As noted earlier, high concentrations of Mo in pasture plants (e.g., >15 to 20 μg/g) can be toxic to cattle. The condition typically occurs on calcareous soils high in Mo.

Antagonistic Effects

A number of trace elements are known to reduce or limit the utilization of one or more micronutrients by the plant. The effect can be both external and internal to the plant. Some of the more widespread interactions include: (1) B, Fe, and Zn deficiencies induced by liming acid soils, (2) Fe and Zn deficiencies associated with application of P fertilizers, (3) Fe deficiencies caused by the accumulation of excess Cu in the soil, and (4) Fe deficiencies that result when Mn is present in either abnormally high or low concentrations. A discussion of micronutrient interactions has been given by Olsen.[13]

Reduction in the uptake of Fe and Mn by plants as a result of P fertilization or lime additions is believed to be caused by reactions occurring outside of the root; interactions of Fe with Cu, Mn, and Zn are thought to be physiological in nature. There is some evidence that lime-induced B deficiency may reflect a condition of nutrient inbalance in the plant, possibly caused by the uptake of excess Ca supplied by the lime. Both Ca and B are concentrated in the cell wall of plant tissue, and normal growth may occur only when the ratio of the two elements in the plant falls within a certain limited range.

Factors Affecting the Availability of Micronutrients to Plants

A number of factors affect the availability of micronutrients to plants, including soil reserves, organic matter content, pH, oxidation and reduction, soil moisture and aeration, and seasonal variation.

Soil Reserves

Deficiencies and toxicities of trace elements can often be traced to the nature of the soil and its content of micronutrients, a subject discussed earlier. Specifically, alluvial sands and certain organic soils often have low reserves of micronutrients.

Organic Matter

Organic matter plays a key role in the availability of micronutrients in soil, a subject discussed in a later section. Only a few salient points follow.

The formation of metal–organic complexes affects the behavior of micronutrient cations in two opposing ways. Complexation by insoluble organic matter reduces availability whereas the formation of soluble complexes enhances availability. Those micronutrients that form strong complexes (e.g., Cu^{2+}) are influenced to a greater extent than those that form weak complexes. Hodgson[3] concluded that the low incidence of Cu deficiencies in mineral soils was due to the formation of soluble complexes by natural chelating agents in the soil solution.

The effect of organic matter on the availability of Mn appears to be particularly pronounced. Organic matter can influence Mn transformations in at least three ways: (1) formation of complexes that effectively reduce the activity of the free ion in solution; (2) a decrease in the oxidation-reduction potential of the soil, either directly or indirectly through increased microbial activity; and (3) stimulation in microbial activity that results in incorporation of Mn in biological tissue.[3] Iron undoubtedly responds in much the same way.

pH and Oxidation State

The pH of the soil profoundly affects the solubilities of trace elements and, consequently, their availabilities to plants. For the pH range 5 to 8, Co, Cu, Fe, Mn, and Zn are more available at the lower extreme than at the higher; availability of Mo is in the reverse direction.

The pH of the soil has a particularly pronounced effect on the solubilities of Fe and Mn; both are converted to highly insoluble oxides as the pH is increased. The reaction for Fe is as follows:

$$Fe^{3+} + 3OH^- \rightleftharpoons Fe(OH)_3 \tag{1}$$

where $Fe(OH)_3$ is equivalent to the hydrated oxide, $Fe_2O_3 \cdot 3H_2O$. Its solubility product is extremely low (10^{-38}). Iron deficiencies are particularly serious on calcareous soils, which typically have pH's of the order of 8.0.

The relationship between Mn solubility and pH is complicated by changes in oxidation state from Mn^{2+} to Mn^{4+} as the pH is increased. The two opposing reactions are:

$$Mn^{2+} + 2OH^- + (0) \rightleftharpoons MnO_2 + H_2O \tag{2}$$

$$MnO_2 + 4H^+ + 2e^- \rightleftharpoons Mn^{2+} + 2H_2O \tag{3}$$

Oxidation in neutral soils (reaction 2) appears to be due to biological activity and leads to precipitation of Mn as the highly insoluble MnO_2. A number of soil microorganisms have the ability to oxidize Mn^{2+} to Mn^{4+}, but only in soils above pH 5; little if any oxidation occurs under acidic conditions. Thus, in highly acidic soils, Mn occurs as the divalent cation (see reaction 3) and may accumulate in toxic concentrations. Toxicities due to excess Mn usually disappear as the pH is elevated above 5; a further increase in pH above the neutral point may induce Mn deficiency.

The solubilities of Cu and Zn are also reduced as the pH of the soil increases, but the effect is less pronounced than that for Fe and Mn. As is the case with Fe, increases in soil pH by application of lime can induce Zn deficiency. In contrast, additions of lime can reduce the uptake of B by plants, a result that may be due to the influence of increasing Ca on the physiological availability of B.[3]

Molybdenum is the main micronutrient that increases in availability with increasing pH. Deficiencies to plants are common in acid soils, whereas toxicities to animals are sometimes encountered in neutral or alkaline soils. The chemistry of Mo in acid soils is similar to that of P in that both exist as anions (MoO_4^{2-} and $H_2PO_4^-$) and both can be strongly adsorbed to Fe and Al oxide surfaces.

Rhizosphere

Plant roots are known to exude a wide variety of organic compounds in amounts sufficient to alter the availability of trace elements. Root exudates can affect nutrient availability through their influence on microbial activity or through chelation of one or more trace element. A more general discussion as to how micronutrients are affected by the rhizosphere is given later.

Seasonal Variations

The availability of micronutrients in soil varies considerably over the growing season, which is not surprising in view of seasonal changes in temperature, moisture, microbial activity, and so on. Considerable variation also occurs from one season to another. The subject of seasonal variation has been discussed in detail by Hodgson.[3]

DYNAMICS OF MICRONUTRIENT CYCLING BY PLANTS AND MICROORGANISMS

An important aspect of the micronutrient balance in soil is the recycling of trace elements through the return of crop residues, or animal manures in the case of grazed pastures. The amount of any given trace elemnet that is removed by harvesting will depend upon its distribution within the plant (roots, stems, leaves, grain, etc.), the nature of the plant and its intended use, and whether or not the nonharvested residues are returned to the soil. In some plants, and under certain circumstances, the micronutrient becomes concentrated in the roots and is thereby returned to the soil even though the top growth is removed.

Depletion of Soil Resources Through Cropping

For many crops high percentages of the micronutrients taken up by the plant are not lost from the soil–plant system but are recycled through the return of organic residues. For example, trace elements in the cereals (e.g., corn, wheat, rice) are concentrated in the straw, roots, and stubble, which are normally left in the soil after harvesting of the grain. Data given for corn in Table 9.4 show that only about 20% of the plant Cu or Zn, representing less than 0.2% of the soil reserves, will occur in the shelled corn (assumed yield of 7,270 kg/ha, or 100 bushels per acre). Values for the percentage of soil Cu or Zn accounted for in the shelled corn were calculated by comparing the content of each trace element in the grain with the amount of that element in a hectare of soil to the rooting depth (2.240 million kg to a depth of 15 cm and an assumed average content of 20 µg/g for both Cu and Zn). In this hypothetical example, the soil will contain sufficient amounts of the two micronutrients to supply comparable crops for over 1,000 years of cropping, provided the residues are returned to the soil following harvest. It should be noted, however, that significant amounts of the soil Cu or Zn may occur in forms that are not available to plants (see next section).

As with Cu and Zn, the amount of Fe and Mn removed from the soil by cropping will be trivial relative to total reserves. In contrast, reserves of B and Mo can be depleted in a relatively short time. Both of these micronutrients are often present in low amounts in the soil (<2 µg/g) and both are required in rather high amounts by certain plants. A forage legume will remove as much as 100 g of Mo per hectare, and, for a soil containing 1 µg/g, the supply will not last very long.

Role of Microorganisms

Microorganisms affect the availability of micronutrients in several ways, as summarized:[24]

1 Through release of trace elements during decay of plant and animal residues.
2 Through immobilization of trace elements into microbial tissue.
3 Through synthesis of biochemical chelating agents that immobilize insoluble forms of the micronutrients (see subsequent section).
4 By oxidation of a trace element (Fe, Mn) into a less available form.
5 By reduction of the oxidized form of an element (e.g., Fe^{3+} to Fe^{2+} in an O_2-deficient environment).
6 Through indirect transformations resulting from pH or oxidation-reduction potential changes.

Items 1 and 2 represent the opposing processes of mineralization and immobilization. Bacteria, actinomycetes, and fungi require the same micronutrients as higher plants and will compete with plants for available micronutrients when levels are suboptimum for growth. Accordingly, some of the trace elements released from plant residues by mineralization, or solu-

Table 9.4

Distribution of Cu and Zn in Various Components of the Corn Plant

Plant Part	Dry Weight kg/ha	Cu				Zn			
		µg/g	Total Cu in plant, g	% of Plant Cu	% of Soil Cu[a]	µg/g	Total Zn in plant, g	% of Plant Zn	% of Soil Zn[a]
Shelled corn	7,270	5	36.4	20.3	0.08	10	73	19.7	0.16
Stover & cobs	6,050	12	72.6	40.6	0.16	25	151.0	40.8	0.34
Roots & stubble	5,825	12	69.9	39.1	0.16	25	146.0	39.5	0.33

[a] Estimated amount of Cu and Zn to the rooting depth (15 cm) is 44.8 kg/ha for an assumed content of 20 µg/g of Cu and Zn.

Table 9.5

Micronutrient Content of Manures and Herbaceous Plants (Grasses and Legumes)

Element	Manures,[a] μg/g	Herbaceous plants,[b] μg/g
B	4.5–52.0	1–5
Mn	75.0–549.0	25–200
Co	0.25–4.70	0.05–0.30
Cu	7.6–40.8	2–15
Zn	43.0–247.0	15–60
Mo	0.84–4.18	0.1–4.0

[a] From Atkinson et al.[25] Results represent the range for 44 samples of diverse types. One exceptionally high value for Mo is not included.

[b] From Whitehead.[26] The values represent the usual ranges found in grasses and legumes. Legumes generally contain higher amounts of B than grasses. For Mo, values as high as 200 μg/g have been reported.

bilized from the soil, can be immobilized and incorporated into the biomass, ultimately to be released to inorganic forms by cell lysis. The relationship is analogous to N immobilization when crop residues with wide C/N ratios are applied to soil.

Several studies have shown that addition of straw residues to a micronutrient-deficient soil can reduce the availability and relative efficiency with which the applied micronutrient is used by plants. A plausible explanation for this effect is that available forms of the micronutrient were immobilized by microorganisms active in the decay process.

As can be seen from Table 9.5, the micronutrient content of farmyard manure is significantly higher than that for herbage (grasses and legumes). The application of 5,000 kg farmyard manure/hectare (2.23 tons/acre) results in the addition of the following approximate quantities of micronutrients (in kg): B, 0.1; Mn, 1.0; Co, 0.005; Cu, 0.08; Zn, 0.48; Mo, 0.01. The rate at which these micronutrients are released will depend upon conditions affecting microbial activity and will be highest in warm, moist, well-aerated soils that have a near neutral reaction.

Still another factor to consider in micronutrient cycling is an apparent enrichment in the organic fraction of the soil due to long-time upward translocation by plant roots and the subsequent incorporation of micronutrients into humic and fulvic acids. The review of Hodgson[3] shows that the micronutrient content of some, but certainly not all, soils is related to organic matter content. Unfortunately, very little is known concerning the relative amounts of the micronutrients in the various soil types that occur as insoluble metal–organic matter complexes or the factors affecting the availability of the organically bound nutrients to plants and microorganisms.

Finally, downward movement of micronutrients as soluble metal–organic matter complexes can be of considerable importance. An interesting ecological relationship has been described by Fraser,[27] who concluded that the accumulation of toxic amounts of Cu in a forest peat was due to a sequence of events that included removal of Cu from a large volume of surrounding soil by plant roots, upward translocation into the leaf tissue, incorporation into the humus layer of the soil, and transport to the swamp in seepage water as a soluble organic complex.

A significant effect that microorganisms have on the trace element cycle is oxidation–reduction of Fe and Mn. As drainage becomes impeded, and the oxidation potential (E_h) approaches 0.2 V, Fe^{3+} and Mn^{4+} can be reduced to Fe^{2+} and Mn^{2+}, respectively. In its oxidized state, Mn is essentially unavailable to plants.

FORMS OF TRACE ELEMENTS IN SOIL

The micronutrient cations in soil occur in the following forms, or pools:

1   Water-soluble.
    a.   As the free cation.
    b.   As complexes with organic and inorganic ligands.
2   On exchange sites of clay minerals. Cations held in this manner are defined as those that can be extracted with a weak exchanger, such as NH_4^+
3   Specifically adsorbed. Some trace elements (e.g., Cu^{2+}) are retained by clay minerals and/or Fe and Mn oxides in the presence of a large excess of Ca^{2+} or some other electrostatically bonded cation. Trace elements bound in this manner are said to be "specifically adsorbed."
4   Adsorbed or complexed by organic matter, including plant residues, native humus, and living organisms (soil biomass).
5   As insoluble precipitates, including occlusion by Fe and Mn oxides.
6   As the primary minerals and as cations that have undergone isomorphorus substitution for Fe and Al in octahedral positions of silicate clays.

The distinction between the various forms is not well defined. For example, sorption reactions cannot be discerned easily from precipitation. Also, organic matter and clay provide sites for cation exchange; both retain cations in difficultly available forms.

In most soils, only very small amounts of micronutrient cations are found in plant available forms, consisting of the water-soluble (1) and exchangeable (2) pools. However, the amounts in these pools may be in reversible equilibrium with the specifically adsorbed (3) and organically bound pools (4).[28,29] Thus when the concentration of micronutrients in the soil solution is reduced by plant consumption, or diluted by excess moisture from rainfall, release of micronutrients from insoluble to soluble forms may occur. In many soils, the organic pool is of considerable significance because of its large size.

The percentage of trace elements in soil that occurs in association with organic matter will be particularly high in soils rich in organic matter (e.g., Histosols and Mollisols) and will be influenced by such properties as pH and kind and amount of clay.

Clay and organic colloids (e.g., humic acids) are major soil components involved in retention of trace elements. However, individual effects are not as easily ascertained as might be supposed, for the reason that, in most mineral soils, organic matter is intimately bound to the clay, probably as a clay–metal–organic complex. Accordingly, clay and organic matter function more as a unit than as separate entities and the relative contribution of organic and inorganic surfaces to adsorption will depend upon the extent to which the clay is coated with organic substances. The amount of organic matter required to coat the clay will vary from one soil to another and will depend on both the kind and amount of clay in the soil. For soils having similar clay and organic matter contents, the contribution of organic matter to the binding of trace elements will be highest when the predominant clay mineral is kaolinite and lowest when montmorillonite is the main clay mineral.

Another factor of importance in the binding of divalent transition metal cations is the extent to which binding sites on organic surfaces are occupied by trivalent cations, notably Fe and Al. These cations generally form stronger complexes with organic molecules than divalent cations. The difficulty of removing Fe and Al from extracted organic matter is well known.

The mechanism whereby trace elements are specifically adsorbed by oxides and layer silicate minerals (3) is unknown, but hydrolysis of the metal ion and adsorption of weakly hydrated species (e.g., MOH^+) may be involved. Leeper[6] pointed out that the monovalent form of the cation, being less hydrated than the divalent cation (M^{2+}), would be more tightly bound to colloidal surfaces. Another possibility is the formation of covalent linkages with surface hydroxyl groups, such as a Cu–O–Fe bond. Mechanisms for specific adsorption of trace elements by soils have been discussed by McBride.[30]

The transition metals also have the potential for being "fixed" by clays through displacement of Mg^{2+} from octahedral sites of the clay. This is possible because such ions as Co^{2+} and Zn^{2+} have sizes similar to that of Mg^{2+}.

Many noncalcareous soils contain trace elements that are solubilized by chemical treatments that dissolve Fe and Mn oxides. The elements thus bound are said to reside within oxide structures (5). Insoluble oxides, formed during weathering, provide reactive surfaces for adsorption of trace elements, which, in turn, are occluded as precipitation continues. As one might suspect, soils that contain high amounts of Fe and Mn oxides (e.g., those derived from volcanic ash) will contain relatively high amounts of trace elements in oxide-occluded forms. Trace elements held in this manner are not believed to be readily available to plants.

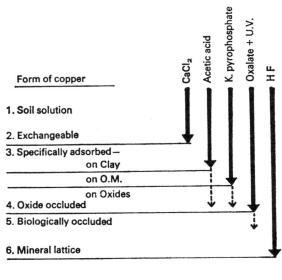

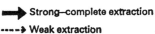

➡ Strong–complete extraction

----➔ Weak extraction

Fig. 9.4

Scheme for the fractionation of Cu in soils. (From McLaren and Crawford,[28] reproduced by permission of Oxford University Press, Oxford.)

In many soils, the bulk of the trace elements occurs either as the primary minerals or in octahedral positions of crystalline silicate clays (6). The amounts found in primary minerals are usually low in highly weathered soils of humid and subhumid regions, but they may be appreciable in unweathered soils of arid and semiarid regions. According to Viets,[29] the micronutrients contained in secondary clay minerals are even less available to plants than those of primary minerals. Some release of trace elements from primary minerals undoubtedly occurs during weathering, but except for soils that are strongly acidic, the process proceeds at an exceedingly slow rate. As will be shown later, the disintegration of minerals, with liberation of micronutrients, can be enhanced in the presence of organic chelating agents.

Fractionation Schemes

Numerous problems are involved in the determination of the forms of micronutrient cations in soil because of the low amounts present. Nevertheless, a number of fractionation schemes have been used. A typical example for Cu is shown in Fig. 9.4. In this case, *soluble plus exchangeable Cu* is determined by neutral salt extraction ($CaCl_2$), *specifically adsorbed* on clay by extraction with dilute acetic acid, *organically bound* Cu by extraction with $Na_4P_2O_7$, *oxide occluded* by treatment with oxalate under ultraviolet light, and *mineral lattice* by HF digestion of the final soil residue. Results obtained by McLaren and Crawford[28] for 24 contrasting soil types are as follows:

	% of total Cu
Soluble + exchangeable	0.1 - 0.2
Specifically adsorbed by clay	0.2 - 2.7
Organically bound	16.2 - 46.9
Oxide occluded	0.0 - 35.9
Mineral lattice	33.6 - 77.2

For the soils examined, a minimum of from one-fifth to one-half of the Cu occurred in organically bound forms. Some of the Cu identified as "soluble + exchangeable" may also have occurred in chelated forms. Oxide-occluded forms are important only in those particular soils that contain appreciable amounts of Fe or Mn oxides.

As noted earlier, the micronutrient cations in oxide occluded and mineral lattice forms are relatively unavailable to plants. In agreement with the concept of Viets,[29] McLaren and Crawford[28] concluded that the amount of Cu available to plants (exchangeable and soluble Cu) was controlled by equilibria involving specifically adsorbed forms (Cu extracted with 2.5% acetic acid) and the organically bound fraction. The suggested relationship between the three forms was as follows:

$$\text{Exchangeable and soluble Cu} \rightleftharpoons \text{Specifically adsorbed Cu} \rightleftharpoons \text{Organically bound Cu}$$

A somewhat more complex fractionation procedure has been used by Miller et al.[31] to determine the distribution of Cd, Cu, Pb, and Zn in contaminated soils. Their procedure is based on the ability of certain solvents to remove specific bound forms of the metal. The sequential fractionation scheme is as follows:

Soluble metal	H_2O (30 min)
Exchangeable metal	1 M KNO_3
Adsorbed metal	0.5 M NH_4F
Organically bound metal	0.1 M $Na_4P_2O_7$
Carbonate metal	0.1 M EDTA, pH 7
Mn oxide-bound metal	0.1 M $NH_2OH \cdot HCl$(pH 3)
Fe oxide-bound metal	Citrate–bicarbonate–dithionite solution (80°C, 15 min)
Precipitated metal	1 M HNO_3
Residual metal	Digestion with concentrated HNO_3

Significant amounts of the trace elements in some contaminated soils of industrialized northwestern Indiana were accounted for in the organically bound fraction.

Unlike Co, Cu, and Zn, both Fe and Mn can be oxidized to higher valence states, from which highly insoluble oxides and phosphates can be formed. Boron and Mo are unique in that they occur as anions ($H_2BO_3^-$ and MoO_4^{2-}) and are thus subject to losses through leaching.

Boron is present in the soil parent material as the mineral tourmaline, but its main form in soil may be in combination with organic matter, namely, as borate complexes with cis-hydroxyl groups, as follows:

$$
\begin{array}{l}
-\overset{|}{C}-OH \\
| \\
-\overset{|}{C}-OH \\
|
\end{array}
\quad + \quad H_3BO_3 \quad \rightleftharpoons \quad
\left[
\begin{array}{l}
-\overset{|}{C}-O \\
| \qquad\quad B \diagdown \quad OH \\
-\overset{|}{C}-O \diagup \qquad OH \\
|
\end{array}
\right]^{-} H^{+} \quad + \quad H_2O
$$

As organic matter is mineralized by microorganisms, B is released to available forms. The temporary appearance of B deficiency in plants during periods of drought has been attributed to reduced mineralization of organically combined B. Boron, like Mo, occurs in the soil solution as an anion ($H_2BO_3^-$) and is thus readily available to plants.

Speciation of Trace Metals in the Soil Solution

Natural waters from all sources have been found to contain trace metals in organically bound forms, including lakes, streams, estuaries, and the ocean.[32-36] Trace elements in the soil solution also occur partly in organically bound forms.[37-42] Organic ligands undoubtedly enhance the availabilities of trace elements to plants; in some instances, toxicity effects caused by ionic forms of the metal ion can be reduced or eliminated. In aqueous systems, variations in the speciation of trace elements can alter dramatically their toxicities to fish and other aquatic organisms.[35] The free (hydrated) metal ion is the most toxic; most stable complexes are nontoxic. The interaction of organic matter with Al is believed to be of considerable importance in controlling soil solution levels of Al^{3+} in acid soils.[43-45]

Reports in the scientific literature on chemical forms of trace elements in the soil solution are extremely limited and can be attributed to the following:

1   The concentrations of metal ions in the soil solution are normally very low, often of the order of 10^{-8} to 10^{-9} M, thereby creating severe analytical problems in their determination.

2   The trace element may exist in a large number of different chemical forms.

3   Extracts typical of the soil solution are not easily obtained.

4   The amounts and chemical forms of any given trace element in the soil solution will vary over time and may be affected by method of sample preparation, including drying and storage.

Several indirect approaches have been used in attempts to estimate free and complexed forms of metals in the soil solution. In one method, a complexing agent is added that forms a complex that can be removed from the system with an immiscible solvent.[38,39,43,45] Another technique has been to pass the soil extract through a cation exchange resin, in which case cationic forms are adsorbed; complexed forms pass through.[40,41]

Hodgson et al.,[38] using the immiscible displacement method, reported the following values for the percentage of three micronutrient cations in the displaced soil solutions of some New York soils that occurred in organically bound forms: Co, 0 to 69%; Cu, 76 to 99%; Zn, 5 to 90%. A subsequent study[39] with come Colorado calcareous soils gave consistently higher percentages for both Cu (98 to 99%) and Zn (28 to 99%). In both studies, organic bound forms were taken as the difference between the total concentration of metal ions in solution and the amounts recovered as the free cations. Errors can arise in estimates for free cations if the activity of the added complexing agent is affected by the soil solution, if dissolution in the organic solvent alters the natural distribution of the metal between complexed or uncomplexed form, or if the added agent disturbs the equilibria of natural complexes.[38] In view of recent work using anodic stripping voltammetry (discussed later), the estimates quoted previously for organically bound forms of the trace elements would appear to be high.

Sanders[42] applied the ion-exchange equilibrium technique to aqueous extracts of five English soils and concluded that practically all the Co, Mn, and Zn occurred as the free cations (Co^{2+}, Mn^{2+}, and Zn^{2+}). His estimates for organically bound forms (traces at most) are considerably lower than those reported by Hodgson et al.[38,39] Olomu et al.[40] found that essentially all the soluble Fe in some flooded soils was complexed with organic matter; Mn was either not complexed or weakly complexed.

Camerlynck and Kiekens[37] used a combination of cation and anion exchange resins to determine the distribution of select metals in the water-soluble fraction of a sandy soil. The forms identified included free ion species, charged complexes, neutral species, and pH-dependent amphoterics. Soluble Cu and Fe were largely present as stable complexes, Mn was largely in the free ionic form, and Zn was evenly distributed between the two. In other work, a chelating resin, adjusted to the desired pH and saturated with Ca and select trace elements to maintain metal ion activities constant in the calibrating solution ($Ca(NO_3)_2$), was used to determine free and complexed forms of Cd, Zn, and Ca in sludge-amended soils.[46] The findings indicated that some of the ligands had approximately $10^{5.5}$ times greater affinity for Cd and Zn than for Ca.

Direct determination of the concentration of free metal ions in natural waters can be done by use of ion-selective electrodes or by anodic stripping voltammetry (ASV). A major limitation of ion-selective electrodes is their rather low sensitivities; furthermore, only a few divalent cations can be measured (Cu^{2+}, Pb^{2+}, Cd^{2+}, Ca^{2+}). In both methods, electrode response is affected by pH, ionic strength, and sorption of organics on the electrode surface.[33,47] With ASV, "labile" complexes are measured along with the free cation; furthermore, unreliable results will be obtained if the ligand is reduced in the same potential region as the metal of interest.[48] Ion-selective electrodes would appear to have the greatest potential for soils polluted with heavy metals and where rather high concentrations would be expected in the soil solution.[49]

Advantages of ASV for determining the concentration of free metal cations are:[33]

1   The method is highly sensitive and is capable of measuring free metal ions at the low levels found in soil extracts and natural waters. Only small volumes of solution are required (2–10 ml).

2   Only free and labile species of the metal are measured. Some information about chemical forms may be obtained from potential shifts and analysis of the voltammetric waveshape obtained.

3   The method is nondestructive, thereby allowing response to changes in chemical and physical parameters to be monitored on the same sample.

Oxidation of lake and river water by ultraviolet irradiation in the presence of H_2O_2 generally increases the concentration of metal ions detectable by ASV, and the amount released by oxidation has been designated organically bound.[33,36] Recovery of metals is usually incomplete, and rather elaborate fractionation schemes, in conjunction with ASV, have been employed for determining combined forms, some of which may have application to soil extracts. In the scheme of Florence[34] and Batley and Gardner,[32] ASV labile and total Cu are determined in (1) the original water sample, (2) water passed through a chelating resin, (3) water subjected to ultraviolet irradiation in the presence of H_2O_2, and (4) ultraviolet irradiated water passed through a chelating resin. Seven fractions were obtained, three of which represented organically bound forms. (ASV labile and nonlabile species not removed by the resin and nonlabile organic species adsorbed by the resin.) Results obtained with river and reservoir water have shown that substantial amounts of the Cu, and lesser amounts of the Cd, Pb, and Zn, are associated with organic matter. In some cases appreciable amounts of the bound metals were accounted for as complexes with inorganic ligands. Extrapolation of these findings to the soil solution suggests that high results will be obtained for organically bound metals when taken as the difference between the total concentration in solution and the free cationic form, as done by Hodgson et al.[38,39]

The species of metal ions likely to occur in the solution phase of conventional agricultural soils and sludge-amended soils have also been predicted on the basis of analytical data for cations, anions, and organic matter, for which the multipurpose computer program GEOCHEM has been applied.[50-54] GEOCHEM is an iterative program that uses stability constant data (corrected for ionic strength) to calculate the various inorganic and ligand-bound species. As is to be expected, the approach is restricted to those systems that have been well characterized with regard to cationic (e.g., K^+, Na^+, NH_4^+, H^+, Ca^{2+}, Mg^{2+}, Cu^{2+}, Mn^{2+}, etc.) and anionic (Cl^-, HCO_3^-, NO_3^-, SO_4^{2-}, etc.) species, and, ideally, for the kinds and amounts of organic ligands. In general, these studies have indicated that those metals that form strong metal complexes (e.g., Cu^{2+}) occur mostly in organically complexed forms; those that form weak complexes (e.g., Cd^{2+}) occur mostly in free ionic forms.

Finally, it should be noted that the condition of electrochemical neutrality must be maintained in the soil solution. Accordingly, the total quantity of cations in ionic forms must equal the total anionic content. Cronan et al.[55] determined organically bound forms of trace elements in the leachates of some New Hampshire subalpine forests from the deficit between total cations and total anions.

REACTIONS OF MICRONUTRIENTS WITH ORGANIC MATTER

The great importance of organic matter in governing the reactions of trace elements in soil, and hence their availabilities to plants, requires that special attention be given to metal–organic matter complexes—their formation and role in plant nutrition and pedogenic processes.

The ability of soil organic matter to form stable complexes with metal ions has been well established. Numerous compounds are involved, including the so-called humic and fulvic acids and a myriad of individual biochemical substances. Some of the metals occurring naturally in soil, or introduced in fertilizers, are held as insoluble complexes and are unavailable to plants. On the other hand, many metals that ordinarily would convert to insoluble precipitates at the pH values found in productive agricultural soils are maintained in solution through chelation. Organic matter reactions involving metal ions have been discussed in several reviews.[3,6,8]

Significance of Complexation Reactions

The formation of metal–organic complexes would have the following effects in soil.

1   Metal ions that would ordinarily convert to insoluble precipitates at the pH values found in productive agricultural soils would be maintained in solution. Many biochemicals synthesized by microorganisms, including

amino acids and the simple aliphatic acids, form soluble complexes with metal ions. Complexes of metal ions with fulvic acids are also soluble.

2 Organic complexing agents may influence the availability of trace elements to higher plants, as well as to soil microorganisms. Organic substances play a particularly significant role in the behavior of trace elements in peat.

3 Under certain circumstances the concentration of a metal ion may be reduced to a nontoxic level through complexation. This will be particularly true when the metal–organic complex has low solubility, such as is the case with complexes of humic acid and other high-molecular-weight components of the organic matter.

4 Chelation plays a key role in the disintegration of rocks and minerals. Lichens, as well as bacteria and fungi, bring about the disintegration of rock surfaces to which they are attached through the production of organic chelating agents.

5 Complexing agents of various types may function as carriers of heavy metals in natural waters. This work has been reviewed by Reuter and Perdue.[56]

6 The interaction of Al^{3+} with organic matter may be of considerable importance in controlling soil solution levels of Al^{3+} in acid soils.[43-45]

Properties of Metal Chelate Complexes

A metal ion in aqueous solution contains attached water molecules oriented in such a way that the negative (oxygen) end of the water dipole is directed toward the positively charged metal ion. A complex arises when water molecules surrounding the metal ion are replaced by other molecules or ions, with the formation of a coordination compound. The compound that combines with the metal ion is commonly referred to as the *ligand*. The reaction between Cu and glycine, a common amino acid, is as follows.

$$Cu^{2+} \;+\; 2\overset{+}{H_3}N\text{-}CH_2\text{-}COO^- \;\rightleftharpoons$$

$$\begin{array}{c} CO\text{-}O \quad\quad NH_2\text{-}CH_2 \\ | \quad\quad \diagdown\diagup Cu \diagdown \quad | \quad\quad +\;\; 2H^+ \\ HCH\text{-}NH_2 \quad\quad O\text{-}CO \end{array}$$

I

A covalent bond consists of a pair of electrons shared by two atoms and occupying two orbitals, one of each atom. Essentially, a coordinate complex arises because the outer electron shell of the metal ion is not completely filled and can accept additional pairs of electrons from atoms that have a pair of electrons available for sharing.

Examples of groupings contained in organic compounds that have unshared pairs of electrons and that can form coordinate linkages with metal ions, are shown by structures II to V.

$$
\begin{array}{cccc}
\text{H} & \text{H} & \text{H} & \text{H} \\
| & | & | & | \\
\text{R-C=O:} & \text{R-S:} & \text{R-NH} & \text{R-O:} \\
\\
\text{II} & \text{III} & \text{IV} & \text{V}
\end{array}
$$

The order of decreasing affinity of organic groupings for metal ions is appoximately as follows:

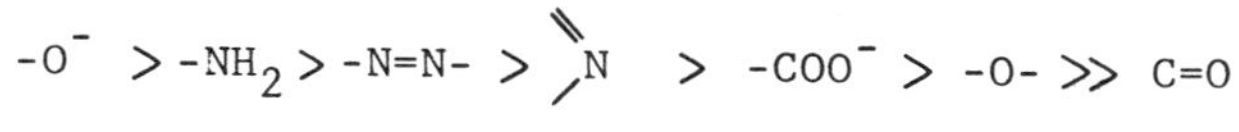

$$-O^{-} > -NH_2 > -N=N- > \diagdown_{\diagup} N > -COO^{-} > -O- \gg C=O$$

enolate   amine   azo   ring N   carboxylate   ether   carbonyl
ion

A *chelate* complex is formed when two or more coordinate positions about the metal ion are occupied by donor groups of a single ligand to form an internal ring structure. The word *chelate* is derived from the Greek *chele,* meaning a crab's claw, and refers to the pincerlike manner in which the metal is bound. If the chelating agent forms two bonds with the metal ion, it is said to be bidentate; similarly, there are terdentate, tetradentate, and pentadentate chelating agents. The formation of more than one bond between metal and organic molecule usually imparts high stability to the complex.

A typical structure of a chelate complex was shown earlier for the glycine–Cu complex (I). Other complexes are shown here for citric acid (VI) and tartaric acid (VII).

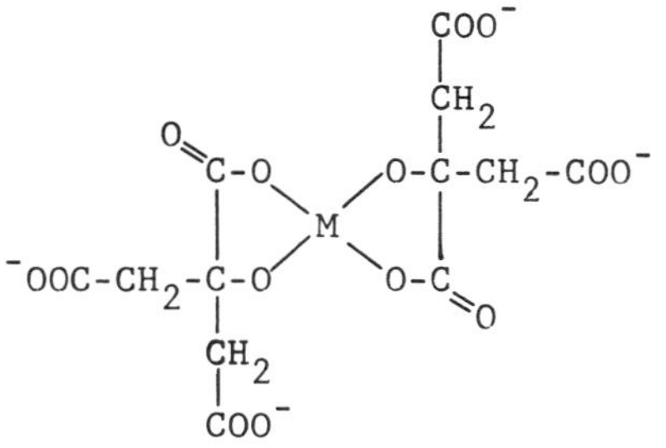

Citrate
VI

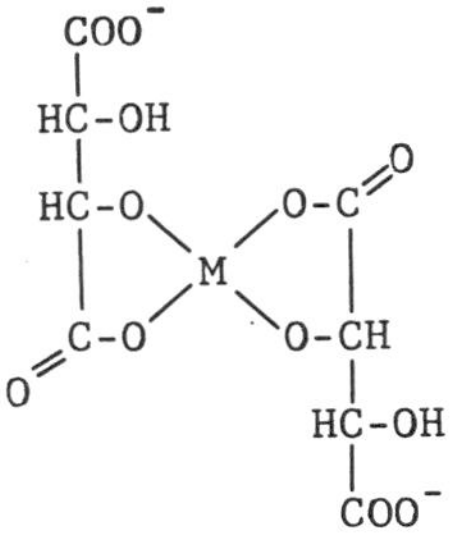

Tartrate
VII

The stability of a metal–chelate complex is determined by a variety of factors, including the number of atoms that form a bond with the metal ion, the number of rings that are formed, the nature and concentration of metal ions, and pH. The stability sequence for some select divalent cations is approximately as follows:

$$Cu^{2+} > Ni^{2+} > Co^{2+} > Zn^{2+} > Fe^{2+} > Mn^{2+}$$

This order may vary somewhat, depending on the nature of the ligand, interactions with other ions, and pH. Trivalent cations (e.g., Fe^{3+} and Al^{3+}) form stronger complexes than divalent cations.

The organic compounds in soil that form stable complexes with metal ions are of two types: (1) biochemicals of the types known to occur in living organisms, and (2) a series of complex polymers that are formed by secondary synthesis reactions and that bear no resemblance to the natural products. Included with the first group are the organic acids, polyphenols, amino acids, peptides, proteins, and polysaccharides. The second group includes the humic and fulvic acids.

The trace elements in soil that occur as insoluble complexes with organic matter are largely those that are bound to components of the humic fraction, particularly humic acids. On the other hand, the metals found in soluble complexes are mainly those associated with individual biochemical molecules, such as organic acids. As will be mentioned later, metal complexes with fulvic acids also have high water solubilities.

Biochemical Compounds as Chelating Agents

Biochemical compounds having chelating characteristics, such as simple aliphatic acids, amino acids, and polyphenols, are continuously produced in soil through the activities of microorganisms. These constituents normally have only a transitory existence. Accordingly, the amounts present in the soil solution at any one time will represent a balance between synthesis and destruction by microorganisms. Appreciable quantities of these compounds may accumulate from time to time in localized zones where biological activity is intense, such as in the rhizosphere and near decomposing plant residues. Soils amended with manures and other organic wastes may also be relatively rich in metal-binding biochemicals. Their production during the decay of plant and animal residues may enhance the availability of insoluble micronutrients to plants.

The statement has often been made that biochemical compounds are not important as natural chelators in soil because of their rapid destruction by microorganisms. However, this argument is not completely valid. The application of modern chromatographic techniques has shown that, while the concentration of any given compound or group of compounds in the soil solution may be slight, the accumulating effect of all complexing species may be appreciable. In many agricultural soils, particularly those well supplied with organic matter, the combined total of potential chelate formers in the aqueous phase may be sufficient to account for the minute quantities of metal ions in the soil solution, often in the ppm range.

The concentration of individual biochemical species in the soil solution is approximately as follows:[8]

Simple aliphatic acids	1×10^{-3} to 4×10^{-3} M
Amino acids	8×10^{-5} to 6×10^{-4} M
Aromatic acids	5×10^{-5} to 3×10^{-4} M
Hydroxamate siderophores	1×10^{-8} to 1×10^{-7} M

Typical chelate structures of metal complexes with biochemical compounds were shown earlier in structures I, VI, and VII.

Organic Acids

Simple aliphatic acids are of special interest because of their ubiquitous nature and because many of the hydroxy derivatives are effective solubilizers of mineral matter. The ones most effective in chelating metal ions are those of the di- and tricarboxylic acid types, typical examples being citric (VI) and tartaric (VII) acids.

In terms of organic matter content, the amounts of organic acids in soil are negligible. However, by expressing the quantities in terms of potential concentration in the soil solution, the likelihood that organic acids influence plant growth by functioning as vehicles for the transport of micronutrient cations to plant roots becomes more apparent. Wang et al.[57] found that the combined concentration of malonic, tartaric, and malic acids in some incubated soils frequently ranged from 20 to 80 μmoles/100 g, equivalent to from 1.0×10^{-3} M to 4.0×10^{-3} M in the soil solution at the 20% moisture level.

Organic acids produced by colonizing microorganisms are believed to act as solubilizers of mineral matter in nature. This work, reviewed elsewhere,[8] has shown that an unusually high proportion of the microorganisms associated with the "raw" soil of rock crevices and the interior of porous weathered stones produces organic acids, with decomposition of silicate minerals when tested under laboratory conditions. Fungi most active in dissolving silicates are those that produce citric or oxalic acids, both strong chelating agents.[58] A summary of data obtained by Webley et al.[59] for the incidence of silicate-dissolving bacteria, actinomycetes, and fungi among the isolates from rock sequences and weathered stones is shown in Table 9.6.

The action of organic acids in solubilizing rocks and minerals can be attributed to lowering of the pH and, more important, to the formation of metal chelates. The elements released may subsequently be taken up by plants, or adsorbed by clay minerals and humic substances.

Excretion products of roots include a variety of simple organic acids, many of which (e.g., oxalic, tartaric, and citric) are capable of forming complexes with metal ions. Differences in the susceptibilities of plant species to trace element deficiencies often have been attributed to variations in organic acid production.[3]

Table 9.6
Incidence of Silicate-Dissolving Bacteria, Actinomycetes, and Fungi among the Total
Isolates from Rock Sequences and Weathered Stones[a]

	Total Isolates	% of Total Dissolving			
		Ca Silicate	Wollastonite	Mg Silicate	Zn Silicate
Bacteria	265	83	57	65	Not tested
Actinomycetes	39	87	38	46	Not tested
Fungi	149	94	Not tested	76	96

[a] From Webley et al.[59]

The rhizosphere (soil in the immediate vicinity of the root) has been shown to be a favorable habitat for acid-producing bacteria. Rouatt and Katznelson[60] recorded higher proportions of acid-producing organisms in the rhizosphere than in root-free soil. Louw and Webley[61] concluded that acid producers were preferentially stimulated in the rhizosphere.

Sugar Acids

Sugar acids also may be important as solubilizers of mineral matter. Gluconic (VIII), glucuronic (IX), and galacturonic (X) acids are all common metabolites of soil microorganisms. Habitats rich in organic matter have been found to contain large numbers of microorganisms that produce 2-ketogluconic acid (XI).[58] This compound has also been observed in the rhizosphere of crop plants.[62] Its ability to solubilize insoluble Ca phosphates in the pH range 2.4 to 6.4 may be due primarily to its acidic nature rather than the ability to chelate Ca.[63]

$$
\begin{array}{cccc}
\text{COOH} & \text{CHO} & \text{CHO} & \text{COOH} \\
| & | & | & | \\
\text{HCOH} & \text{HCOH} & \text{HCOH} & \text{C=O} \\
| & | & | & | \\
\text{HOCH} & \text{HOCH} & \text{HOCH} & \text{HOCH} \\
| & | & | & | \\
\text{HCOH} & \text{HCOH} & \text{HOCH} & \text{HCOH} \\
| & | & | & | \\
\text{HCOH} & \text{HCOH} & \text{HCOH} & \text{HCOH} \\
| & | & | & | \\
\text{CH}_2\text{OH} & \text{COOH} & \text{COOH} & \text{CH}_2\text{OH} \\
\textbf{VIII} & \textbf{IX} & \textbf{X} & \textbf{XI}
\end{array}
$$

Amino Acids

Amino acids may also play a role in mineral nutrition and pedogenic processes through their ability to form soluble chelate complexes with metal ions. Several reviews on the subject show that a wide array of free amino acids has been reported in soils. The free amino acid content of the soil is influenced by climatic conditions, moisture status, plant type and stage of growth, residue additions, and cultural conditions. Levels seldom exceed 2 µg/g or 4.5 kg/ha plow depth, but they may be seven times higher in "rhizosphere" soil. Their concentration in the soil solution at the 20% moisture level will be of the order of 8×10^{-5} to 6×10^{-4} M.

A priori, it would appear that amino acids play a role secondary to organic acids as solubilizers of mineral matter in soil. It should be mentioned, however, that while the contribution of any single compound or group of compounds may be slight, the accumulating effect of all complexing species (including those mentioned in the next section) may be appreciable.

Hydroxamate Siderophores

Considerable attention has recently been given to the role of hydroxamate siderophores in enhancing the availability of Fe in soils.[64-67] These sub-stances, which contain the reactive group

$$-\overset{\overset{\displaystyle O}{\displaystyle \|}}{C}-NOH-,$$

represent a group of microbially produced Fe^{3+} transport ligands with stability constants as high as 10^{32}. Greater amounts are produced when the organism is under Fe stress. The overall reaction leading to chelation of Fe is as follows:[66]

Because hydroxamate siderophores have such a high affinity for ferric Fe^{3+}, it is believed that they act as scavenging molecules in environments where the insoluble ferric Fe^{3+} ion predominates, such as in neutral or alkaline soils where mobility is controlled by the low solubility of ferric oxides.

Biologically significant levels of hydroxamate siderophores have been observed in soils.[65] The amounts contained in the rhizosphere of plants appear to be 10 to 50% higher than those in the bulk soil.[66] Hydroxamate siderophores also have been shown to be produced by soil fungi, including ectomycorrhizal fungi that live in intimate association with plant roots.[67]

Lichen Acids

Lichens bring about the dissolution of mineral substances from rocks and minerals through the production of a special class of organic acids represented by structures (XII) and (XIII).

XII   XIII

Geographically, the lichens are distributed widely over land masses of the earth, including the Antarctic continent and the desert regions of Australia and southwestern USA. Small but detectable amounts of lichen acids undoubtedly appear sporadically in these natural habitats. Their occurrence in productive agricultural soils is doubtful.

Other Compounds

Other chelating compounds occurring naturally in soil, albeit in minute amounts, include organic phosphates, phytic acid, chlorophyll and chlorophyll-degradation products, simple sugars (formation of borate complexes), porphyrins, phenolic compounds, and auxins. The significance of these constituents in complexing metals in soil is unknown.

Trace Metal Interactions with Humic and Fulvic Acids

The importance of humic and fulvic acids in modifying the chemical properties of trace elements has been well established. As noted in Chapter 1, these constituents consist of a series of acidic, yellow- to black-colored polyelectrolytes having molecular weights that range from a few hundred to several hundred thousand. In classical terminology, humic acid is the material extracted from soils with alkaline solutions and precipitated upon acidification; fulvic acid is the alkali-soluble material that remains in solution (see Fig. 1.5). Fulvic acids have considerably lower molecular weights than humic acids, and for this reason, they are believed to play a particularly important role in maintaining micronutrient cations in the soil solution.

The work of Geering and Hodgson[68] suggests that both individual biochemical compounds (aliphatic acids, amino acids, etc.) and fulvic acid-type constituents are involved in the formation of mobile complexes in soils, with the fulvic acids being the most effective in complexing metals. Results obtained with lake and river waters have shown that over 80% of the soluble organic matter consists of polymeric substances resembling fulvic acids.[56,69]

The ability of humic and fulvic acids to form stable complexes with metal ions can be attributed to their high content of oxygen-containing functional groups, including COOH, phenolic-, alcoholic-, and enolic-OH, and C=O.

Structures commonly considered to be present in humic substances, and that have the potential for binding with metal ions, include the following:

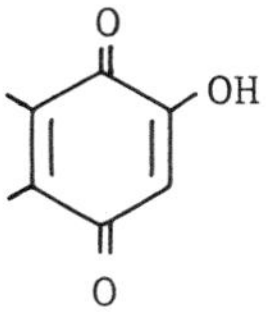

XV 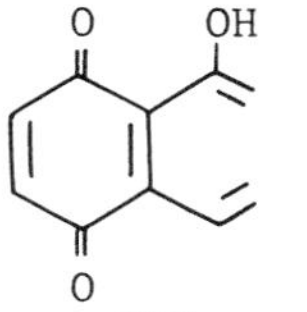 XVI 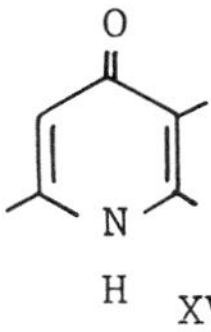 XVII 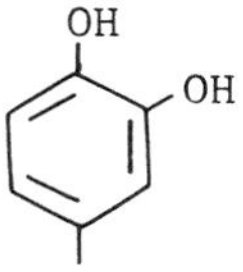 XVIII

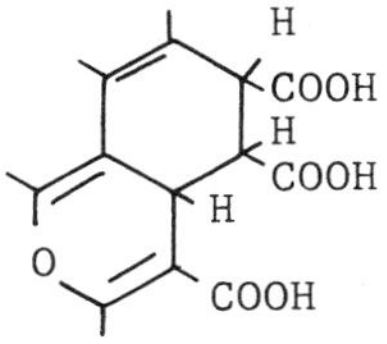

XIX 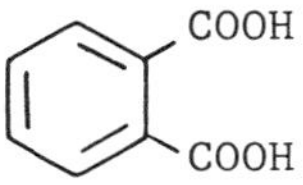 XX 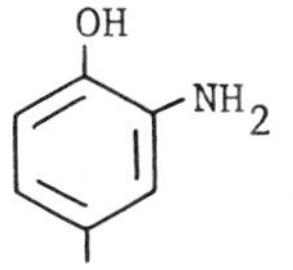 XXI 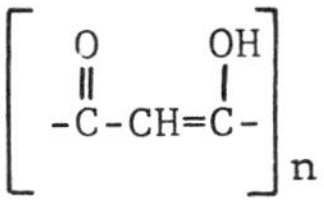 XXII

Schnitzer[70] and Gamble et al.[71] concluded that two types of reactions are involved in metal–fulvic acid interactions, the most important one involving both phenolic OH and COOH groups. A reaction of lesser importance involved COOH groups only. The reactions are:

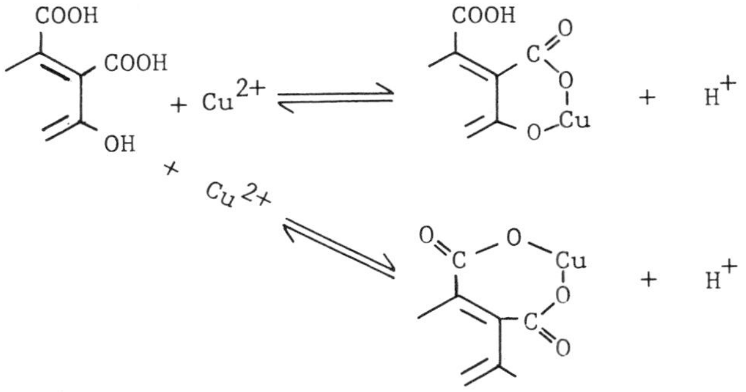

The formation of phthalate-type complexes (bottom reaction) is likely because humic acids have been shown to contain COOH groups that are located on adjacent positions of aromatic rings. Positive proof for the formation of salicylate-like ring structures (top reaction) has yet to be achieved.

Results of infrared spectroscopy studies have confirmed that COOH groups, or more precisely carboxylate (COO^-), play a prominent role in the complexing of metal ions by humic and fulvic acids. Some evidence indicates that OH, C=O, and NH groups may also be involved.[72–74] The suggestion has been made that, in addition to the preceding, complexes may be formed with conjugated ketonic structures, according to the following reaction:[73]

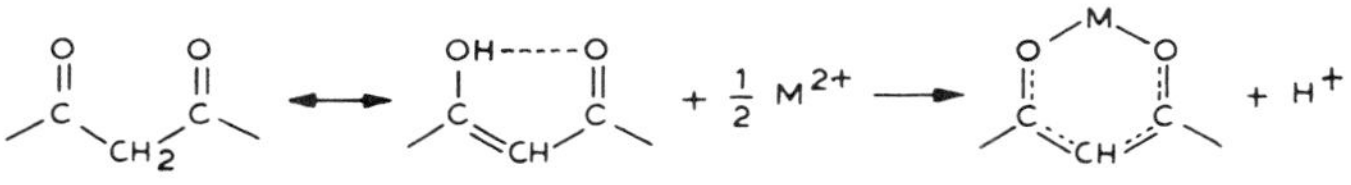

Considerable controversy exists as to the extent to which COO^- linkages

are covalent or ionic. The asymmetric stretching vibration of COO^- in ionic bonds occurs in the 1630 to 1575 cm^{-1} region; when coordinate linkages are formed, the frequency shifts to between 1650 and 1620 cm^{-1}. Frequency shifts with metal–humate complexes have been variable and slight, a result that may be due to the formation of mixed complexes. Interpretations in the 1620 cm^{-1} region are complicated further because of interference from covalent bonding with other groups.[73]

Results of electron spin resonance spectroscopy (ESR) studies have also been inconclusive. Lakatos et al.[75] reported that Cu^{2+} was bound to humic acid by a N donor atom and two carboxylates (COO^-). On the other hand, McBride[76] concluded that only oxygen donors (i.e., COO^-) were involved; furthermore, a single bond was observed. Goodman and Cheshire[77] and Cheshire et al.[78] obtained evidence suggesting that Cu retained by a peat humic acid after acid washing was coordinated to porphyrin groups, from which they concluded that a small fraction of the Cu in peat was strongly fixed in the form of porphyrin-type complexes.

Metal Ion-Binding Capacity

Approaches used to determine the binding capacities of humic substances for metal ions include proton release,[79,80] metal ion retention as determined by competition with a cation-exchange resin,[81] dialysis,[82] anodic stripping voltammetry,[83,84] and ion-selective electrode measurements.[85,86] In general, the maximum amount of any given metal ion that can be bound is approximately equal to the content of acidic functional groups, primarily COOH. Exchange acidities of humic substances vary greatly, but they generally fall within the range of 1.5 to 5.0 meq/g. This corresponds to a retention of from 48 to 160 mg of Cu^{2+} per g of humic acid. Assuming a C content of 56% for humic acids, one Cu^{2+} atom would be bound per 20 to 60 C atoms in the saturated complex.

Factors influencing the quantity of metal ions bound by humic substances include pH, ionic strength, molecular weight, and functional group content. For any given pH and ionic strength, trivalent cations are bound in greater amounts than divalent cations; for the latter, those forming strong coordination complexes (e.g., Cu^{2+}) will be bound to a greater extent than weakly coordinated ones (e.g., Ca^{2+} and Mg^{2+}).

Solubility Characteristics

Humic and fulvic acids form both soluble and insoluble complexes with polyvalent cations, depending on degree of saturation. Because of their lower molecular weights and higher contents of acidic functional groups, metal complexes of fulvic acids are more soluble than those of humic acids.

A number of processes affect the solubility characteristics of metal–humate and metal–fulvate complexes in soils, as well as in natural waters. A major factor is the extent to which the complex is saturated with metal ions. Other factors affecting solubility include pH, adsorption of the complex to mineral matter (e.g., clay), and biodegradation. Under proper pH conditions, trivalent cations, and to some extent divalent cations, are effective in precipitating humic substances from very dilute solutions; monovalent cations are generally effective only at relatively high particle concentrations.

Flocculation of humic substances in natural waters can result from changes in water chemistry. Thus cation-induced coagulation of humic colloids is of importance in the removal of bound Fe and other elements from river water during mixing with seawater in coastal estuaries. The concentrations of trace metals in the ocean are extremely low, a result that has been attributed by Turekian[87] to the role of particles (organic and inorganic) in sequestering metals during every step of the transfer from the continent to the ocean floor.

Immobilization of micronutrients by interaction with humic substances can occur either through the formation of insoluble complexes or through solid-phase complexation to humate present as a coating on clay surfaces. Sorption is possible through direct exchange at the clay-organic interface or through the formation of soluble complexes that subsequently become associated with mineral surfaces through adsorption. Some cations link humic complexes to clay surfaces; others occupy peripheral sites and are available for exchange with ligands of the soil solution.

Unlike simple biochemical compounds and the low-molecular-weight fulvic acids, humic acids tend to form complexes with metal ions that are insoluble and thus relatively unavailable to plants. However, a portion of the bound metal may be "labile," that is, in dynamic equilibrium with specific adsorbed and exchangeable forms (see section on *Fractionation Schemes*). Zunino and Martin[10] developed a unified concept for the role of organic substances on the translocation and biological availability of trace elements in soil (Fig. 9.5). The first stage is characterized by the solubilization of trace elements through chelation by simple organic compounds (e.g., organic acids). These were termed *type I complexes*. With time the metals become sequestered by humic substances to form *type II complexes,* which were believed to be more stable and less biologically available than type I complexes. As chelating sites become saturated, complexes less stable than type II are formed, the so-called *type III complexes*. Biochemical chelating agents compete successfully for the metal ions of type III complexes to form complexes of type I, thereby facilitating uptake by the plant. Other research (reviewed elsewhere[8]) has shown that trace metals (notably Cu) become increasingly available to plants as combining sites become saturated. Also, it has long been suspected that low levels of micronutrients in highly organic soils (Histosols) are so tightly complexed that they cannot be taken up by crop plants.

As noted earlier, Al toxicity is a major problem in many acid soils (<pH 5.5). However, acid soils rich in native organic matter, or amended with large quantities of organic residues, give low Al concentrations in the soil

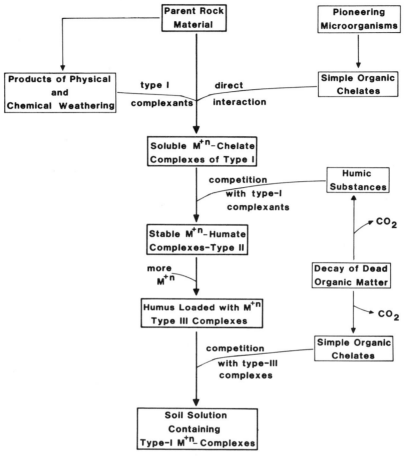

Fig. 9.5

Role of organic substances on the translocation and biological availability of trace elements. (Adapted from a drawing by Zunino and Martin.[10])

solution and permit good growth of crops under conditions where toxicities would otherwise occur.[43–45]

Reduction Properties

Humic substances have the ability to reduce Fe^{3+} to Fe^{2+}, anionic MoO_4^{2-} to MO^{5+}, VO_3^- to VO^{2+}, and Hg^{2+} to Hg^0.[88–92] Reduction of ionic species is of considerable importance in soil and water systems because the solubility characteristics of the metal ions (and hence mobilities) are modified. Evidence for reduction of V by humic substances has been provided by electron spin resonance (ESR) spectroscopy.[88,89] The ESR approach also has been used in conjunction with Mössbauer spectroscopy to obtain information on oxidation states and site symmetrics of Fe bound by humic and fulvic acids.[93,94]

A red sludgelike deposit, consisting primarily of a mixture of Fe oxides and insoluble organic matter (bacterial cell bodies and waste products), is often formed in tile drainage systems, such as in Florida citrus groves.[95] The deposit prevents the tile from functioning properly by blocking the passageway and is believed to be formed by Fe bacteria that oxidize and precipitate reduced Fe and other chemicals in the drainage waters.[95]

Role in Translocation of Metal Ions

In addition to their effects on the solubilization and disintegration of soil-forming rocks and minerals (discussed earlier), organic substances serve as agents for the translocation of metal ions to lower soil horizons. The process has been referred to as *cheluviation,* although it has not been established with certainty that all the complexes are of the type that form chelate rings. Complexation results in differential leaching of metal ions according to their ability to form coordination complexes with organic ligands: Fe, Al, and other strongly complexed elements being eluted to a greater extent than weakly complexed ones.

Eluviation of metal ions as soluble organic complexes occurs to some extent in all soils, but the process is most pronounced in the Spodosols (formerly called Podzols). These soils have developed under climatic and biologic conditions that have resulted in the mobilization and transport of considerable quantities of sesquioxides into the subsoil. The organic surface layer of the soil, A or O horizon, consists largely of acid decomposition products of forest litter, which is underlain by a light-colored eluvial horizon, E, which has lost substantially more sesquioxides than silica. The E horizon is, in turn, underlain by a dark-colored illuvial horizon, B, in which the major accumulation products are sesquioxides and organic matter.

Stobbe and Wright[96] reviewed the major processes involved in the formation of Spodosols and reported that the prevailing concept at the time was that polyphenols, organic acids, and possibly other complexing substances in water percolating from the surface litter bring about the solution of sesquioxides and the formation of soluble metal–organic complexes. Bloomfield[97,98] observed that aqueous extracts of pine needles, leaves, and bark were able to dissolve Fe and Al oxides, with reduction of ferric Fe^{3+} and the formation of stable organic complexes with ferrous Fe^{2+}. The solubilization and reduction of ferric Fe^{3+} iron was attributed to the joint action of carboxylic acids and polyphenols, but especially the latter. The participation of organic acids is also to be expected because these constituents are known to occur in the raw humus layers of forest soils, as well as in soil leachates.

One theory that has received wide attention by soil scientists and others

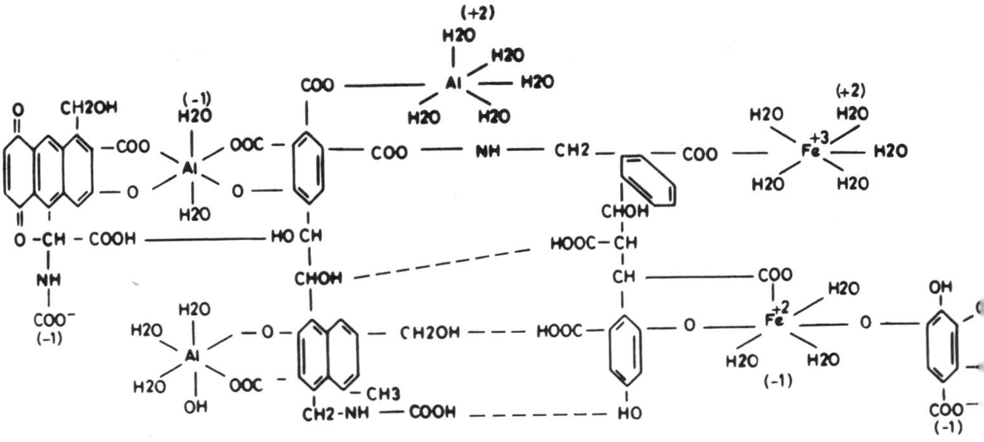

Fig. 9.6

Hypothetical polymerized structure for the binding of Al^{3+}, Fe^{3+}, and Fe^{2+} by humic matter. Several different forms of bonding are represented. (From F. De Coninck, "Formation of Spodic Horizons," *Geoderma* **24**, 101; reproduced by permission of Elsevier Scientific Publishing Company.)

is that the mobilization and transport of sesquioxides in the Spodosol is due, at least in part, to polymeric phenols (i.e., fulvic acids) formed in the overlying leaf litter through decay by microorganisms. According to this concept, mobile organic colloids percolating downward through the soil profile form complexes with Fe and Al until a critical metal content is reached, following which precipitation occurs.[99] Partial decay of organic matter in the B horizon by microorganisms would further saturate the complex. Once started, the accumulation process would be self-perpetuating, since the free oxides thus formed would cause further precipitation of the sesquioxide–humus complex. On periodic drying, the organic matter compex may harden, thereby restricting movement below the accumulation zone.

De Coninck[100] concluded that, in the mobile state, metal–humate complexes are highly hydrated (i.e., they are hydrophilic); transition to the solid state involves loss of hydration water and the complexes become hydrophobic. A hypothetical structure for the unhydrated complex is shown in Fig. 9.6.

The concept that organic substances are responsible for the translocation of Fe and Al in the Spodosol has not been universally accepted. For example, Anderson et al.[101] concluded that the insoluble organic matter–metal complexes of Spodosol B horizons are formed in situ by the interaction between humic substances in leaching waters and previously precipitated inorganic sesquioxides. Farmer et al.[102] attributed the formation of the B horizon in Hydromorphic Humus Podzols to coprecipitation arising from mixing of organic-rich surface waters with Al from groundwater.

ENVIRONMENTAL ASPECTS OF THE TRACE ELEMENT CYCLE

From an environmental standpoint, the trace elements can be categorized as (1) those derived from the soil parent material but that exist in such abnormally high amounts, or are so readily available, that they constitute a threat to the health of plants and animals, and (2) those that are introduced into the soil as toxic pollutants. Toxicities due to natural accumulations have been discussed in previous sections and will not be considered further. Also, the subject of heavy metals applied to soils in sewage sludges will be excluded; this topic was covered in Chapter 3.

Specific Trace Elements

A discussion of environmental problems involving specific trace metals (Cd, Cu, Pb, Hg, and arsenic) follows. Although covered individually, pollution of soil by one trace element is frequently accompanied by additions of others, a typical example being the occurrence of Cd, Cu, Ni, and Pb in aerosols originating from mining and smelting operations. Evidence has recently been obtained for the accumulation of trace elements (Cd, Cu, Pb, Ni, Zn, etc.) in the forest floor (i.e., surface organic horizon) of New England forest soils, apparently due to anthropogenic deposits from the atmosphere.[103-105]

An extensive literature has been developed on toxic metals in the environment. The material that follows represents a synopsis from the reviews of Allaway,[15] Lagerwerff,[17] Lisk,[18] and Loneragan et al.,[5] from which specific reference can be obtained. The effect of heavy metals on the activities of soil microorganisms, including C and N transformations, has been reviewed by Tyler.[20]

Cadmium

Concern about the effect of Cd stems from its tendency to accumulate in plants and animals. Low but prolonged intake of Cd leads to hypertension and other disorders in animals and man, in part because of replacement of Zn in certain enzymes. Cadmium is more mobile in soil, and more easily adsorbed by plants, than other heavy metals, notably Pb and Cu. Accordingly, the potential for moving from soil to plant to man is greater.

Cadmium reaches the soil from a variety of sources, including phosphorus fertilizers (notably rock phosphate and superphosphate), Cd-containing tire dust, and combustion products of coal, wood, and urban organic trash. Emissions of various types are likely to contain Cd because of its relatively high volatility.

The amount of Cd discharged annually in the USA by car tire wear has been estimated at 6,000 kg, much of which accumulates in roadside soils. High levels of Cd have been observed in vegetable crops grown on roadside soils, as well as in the milk of dairy cows grazing in contaminated areas.

The entry of Cd into the food cycle can be partially controlled by lime applications to soil, which lead to precipitation of Cd as the highly insoluble carbonate, sulfate, or phosphate.

Copper

Contamination of soils by Cu-containing materials arises from many sources but most notably from mining and smelting operations. Elevated concentrations of Cu typically occur downwind from the point source of pollution. The Cu content of soils around smelters decreases exponentially with distance and depends on the kind of ore and the process used.

High levels of Cu, often exceeding 1,000 µg/g, have been recorded in soils from continued use of Cu-containing fungicides for controlling diseases of citrus, grapes, hops, and certain vegetables.

Copper is frequently added to the diet of pigs and poultry to improve rates of food conversion and growth. Spreading of the Cu-enriched manure on land can lead to toxicity problems, depending on soil type, climate, topography, plant species, and rate and manner of manure application.

Lead

Exogenous sources of Pb in soil include fossil fuels (e.g., aerial Pb originating from leaded gasoline), mining and smelting operations, and fertilizer impurities, among others. Furthermore, Pb is sometimes added to soil as a pesticide, such as Pb arsenate to control fungal diseases of certain crops.

As is the case with Cu and other trace element contaminants, the concentration of Pb in soil decreases with distance from the point source of pollution. The dependence of Pb concentration in roadside soils to the distance from the density of traffic, and to the direction of prevailing winds, has been well documented.[17] A typical result is given in Fig. 9.7.

Mercury

Mercury in soil may result from additions of mercurial fungicides, fallout from the combustion of fossil fuels, and other sources. Worldwide production of Hg amounts to more than 7×10^6 kg annually, of which between 25 and 50% is believed to be discharged by intentional or unintentional dumping. About 3,000 metric tons of Hg are released annually to the environment through the combustion of coal.

Mercury-containing fungicides, such as phenylmercuric acetate and various methylmercury compounds, constitute a major source of Hg residues in some soils. These substances have been used for seed treatments and in orchard sprays, such as for apples.

The fate of Hg residues in soil is unknown. Inactivation of both organic and inorganic compounds can occur through fixation by organic matter and clay. Application of lime to raise the pH to 6.5 or above can minimize its uptake by plants. Under certain circumstances, inorganic Hg compounds can be reduced to metallic Hg, the vapor of which is toxic to plants and microorganisms.

Inorganic and organic Hg compounds in industrial wastes find their way

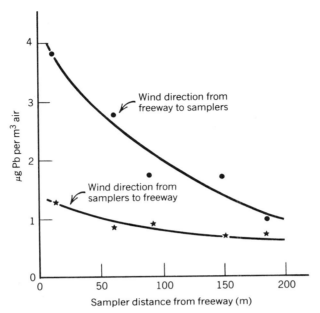

Fig. 9.7

Relationship between Pb concentration in particulate air samples as a function of distance from the freeway. Wind movement was in the direction of the test site during the day (upper curve) and away from the test site during late evening and early morning (lower curve). Soil Pb levels follow a similar trend. (From Page et al.[106])

into the bottom muds of lakes, bays, and other water bodies, where highly toxic methylmercury compounds can be formed (see Chapter 3).

Arsenic

Major sources of arsenic pollution from human activities are metal smelting, combustion of fossil fuels, and use of arsenic-containing pesticides. Organic arsenicals are used in poultry feeds and thereby represent a source of pollution when the poultry litter is applied to soil. Arsenic toxicities to plants have been observed on fields heavily contaminated with arsenic pesticides, notably old orchard soils.

Factors affecting the toxicity of arsenic to plants include soil characteristics (i.e., pH, organic matter content, texture, and available Ca, Fe, and Al), plant variety, and nature of the arsenic compound. Additions of lime and organic matter may aid in overcoming arsenic phytotoxicity. In contrast, phosphate fertilization may accentuate phytotoxicity by competing for fixation sites.

Soil and Plant Factors Affecting Cycling of Toxic Elements

A number of cultural practices can be used to control the environmental cycling of trace elements in soil. They include management practices directed toward decreasing the availability of trace elements, such as additions of lime to increase pH and of plant residues to provide chelating sites, deep tillage to lower the concentration of the toxic element in the surface soil, and selection of plants based on rooting characteristics (deep-rooted versus shallow-rooted ones) and the ability to assimilate and accumulate toxic elements in the harvested portion of the crop. Crop selection is of major importance, as plant species growing on the same soil are known to contain markedly different concentrations of the trace elements.

The topic of abatement methods for reducing toxicities of trace elements have been discussed in detail by Allaway,[15] Lagerwerff,[17] and Lisk.[18]

SUMMARY

Considerable variation exists in the content of any given micronutrient in soil, depending primarily on the nature of the parent material from which the soil was formed. Both deficiencies and toxicities are common. The availability of micronutrients to plants is affected by chemical reactions involving the formation of insoluble precipitates as well as transformations carried out by microorganisms. The return of plant residues to the soil leads to recycling of micronutrients, a factor of considerable importance in micronutrient-deficient soils.

Plants vary greatly in their requirements and ability to assimilate micronutrients. Abnormalities in the concentration of trace elements in plants (e.g., Mo and Se) can cause nutritional disorders in animals.

Organic substances of various types play a prominent role in the binding of micronutrients in soil. Both soluble and insoluble complexes are formed. Biochemical compounds produced by microorganisms (e.g., organic acids, hydroxamate siderophores) enhance the availability of micronutrients to plants, especially Fe in deficient soils.

Humic substances generally form insoluble complexes with transition metal cations. These naturally occurring polyelectrolytes contain high amounts of oxygen-containing functional groups (COOH; phenolic-, alcoholic-, and enolic-OH; C=O structures), which acocunts for their unusual ability to combine with trace metals.

Aluminium toxicity is a major problem in many acid soils. However, acid soils rich in native organic matter, or amended with large quantities of organic residues, give low Al^{3+} concentrations in the soil solution and permit good growth of crops under conditions where toxicities would otherwise occur.

REFERENCES

1 K. B. Krauskopf, "Geochemistry of Micronutrients," in J. J. Mortvedt, P. M. Giordano, and W. L. Lindsay, Eds., *Micronutrients in Agriculture*, Soil Science Society of America, Madison, Wis., 1972, pp. 7–40.

2   J. Kubota and W. H. Allaway, "Geographic Distribution of Trace Element Problems," in J. J. Mortvedt, P. M. Giordano, and W. L. Lindsay, Eds., *Micronutrients in Agriculture,* Soil Science Society of America, Madison, Wis., 1972, pp. 525–554.

3   J. F. Hodgson, *Adv. Agron.* **15,** 119 (1963).

4   B. E. Davies, Ed., *Applied Soil Trace Elements,* Wiley, New York, 1980.

5   J. F. Loneragan, A. D. Robson, and R. D. Graham, Eds., *Copper in Soils and Plants,* Academic Press, New York, 1981.

6   G. W. Leeper, *Six Trace Elements in Soils,* Melbourne University Press, Melbourne, Australia, 1970.

7   R. L. Mitchell, "Trace elements," in F. E. Bear, Ed., *Chemistry of the Soil,* Reinhold, New York, 1955, pp. 253–285.

8   F. J. Stevenson and M. S. Ardakani, "Organic Matter Reactions Involving Micronutrients in Soil," in J. J. Mortvedt, P. M. Giordano, and W. L. Lindsay, Eds., *Micronutrients in Agriculture,* Soil Science Society of America, Madison, Wis., 1972, pp. 79–114.

9   V. Sauchelli, *Trace Elements in Agriculture,* Van Nostrand Reinhold, New York, 1969.

10   H. Zunino and J. P. Martin, *Soil Sci.* **123,** 65 (1977).

11   D. A. Robb and W. S. Pierpoint, *Metals and Micronutrients,* Academic Press, New York, 1983.

12   W. Mertz, *Science* **213,** 1332 (1981).

13   S. R. Olsen, "Micronutrient Interactions," in J. J. Mortvedt, P. M. Giordano, and W. L. Lindsay, Eds., *Micronutrients in Agriculture,* Soil Science Society of America, Madison, Wis., 1972, pp. 243–264.

14   C. A. Price, H. E. Clark, and E. A. Funkhouser, "Functions of Micronutrients in Plants," in J. J. Mortvedt, P. M. Giordano, and W. L. Lindsay, Eds., *Micronutrients in Agriculture,* Soil Science Society of America, Madison, Wis., 1972, pp. 231–242.

15   W. H. Allaway, *Adv. Agron.* **20,** 235 (1968).

16   R. L. Chaney and P. M. Giordano, "Micronutrients as Related to Plant Deficiencies and Toxicities," in L. F. Elliott and F. J. Stevenson, Eds., *Soils for Management of Organic Wastes and Waste Waters,* Soil Science Society of America, Madison, Wis., 1977, pp. 235–279.

17   J. V. Lagerwerff, "Lead, Mercury and Cadminum as Environmental Contaminants," in J. J. Mortvedt, P. M. Giordano, and W. L. Lindsay, Eds., *Micronutrients in Agriculture,* Soil Science Society of America, Madison, Wis., 1972, pp. 593–636.

18   D. J. Lisk, *Adv. Agron.* **24,** 267 (1972).

19   D. Purves, *Environ. Pollution* **3,** 17 (1972).

20   G. Tyler, "Heavy Metals in Soil Biology and Biochemistry," in E. A. Paul and J. N. Ladd, Eds., *Soil Biochemistry,* Vol. 5, Dekker, New York, 1981, pp. 371–414.

21   G. W. Leeper, *Managing the Heavy Metals on the Land,* Sheaffer and Roland, Chicago, 1978.

22   R. E. Lucas and B. D. Knezek, "Climatic and Soil Conditions Promoting Micronutrient Deficiencies in Plants," in J. J. Mortvedt, P. M. Giordano, and W. L. Lindsay, Eds., *Micronutrients in Agriculture,* Soil Science Society of America, Madison, Wis., 1972, pp. 265–288.

23   J. B. Jones, Jr., "Plant Tissue Analysis for Micronutrients," in J. J. Mortvedt, P. M. Giordano, and W. L. Lindsay, Eds., *Micronutrients in Agriculture,* Soil Science Society of America, Madison, Wis., 1972, pp. 319–346.

24   M. Alexander, *Introduction to Soil Microbiology, 2nd ed.,* Wiley, New York, 1977.

25   H. J. Atkinson, G. R. Giles, and J. G. Desjardins, *Can. J. Agric. Sci.* **34,** 76 (1954).

26   D. C. Whitehead, "Nutrient Minerals in Grassland Herbage," in *Commonwealth Bureau of Pastures and Field Crops Mimeo.* Publ. No. I, 1966, pp. 1–83.

27 D. C. Fraser, *Econ. Geol.* **56**, 1063 (1961).

28 R. G. McLaren and D. V. Crawford, *J. Soil Sci.* **24**, 172 (1973).

29 F. G. Viets, Jr., *J. Agr. Food. Chem.* **10**, 174 (1962).

30 M. B. McBride, "Forms and Distribution of Copper in Solid and Solution Phases of Soil," in J. F. Loneragan, A. D. Robson, and R. D. Graham, Eds., *Copper in Soils and Plants,* Academic Press, New York, 1981, pp. 24–45.

31 W. P. Miller, W. W. McFee, and J. M. Kelly, *J. Environ. Qual.* **12**, 579 (1983).

32 G. E. Batley and D. Gardner, *Estuarine Coastal Marine Sci.* **7**, 59 (1978).

33 H. Blutstein and J. D. Smith, *Water Res.* **12**, 119 (1978).

34 T. M. Florence, *Water Res.* **11**, 681 (1977).

35 T. M. Florence, *Trends Anal. Chem.* **2**, 162 (1983).

36 J. Gardiner and M. J. Stiff, *Water Res.* **9**, 517 (1975).

37 R. Camerlynck and L. Kiekens, *Plant and Soil* **68**, 331 (1982).

38 J. F. Hodgson, H. R. Geering, and W. A. Norvell, *Soil Sci. Soc. Amer. Proc.* **29**, 665 (1965).

39 J. F. Hodgson, W. L. Lindsay, and J. F. Trierweiler, *Soil Sci. Soc. Amer. Proc.* **30**, 723 (1966).

40 M. O. Olomu, G. J. Racz, and C. M. Cho, *Soil Sci. Soc. Amer. Proc.* **37**, 220 (1973).

41 J. R. Sanders, *J. Soil Sci.* **33**, 679 (1982).

42 J. R. Sanders, *J. Soil Sci.* **34**, 315 (1983).

43 P. R. Bloom, M. B. McBride, and R. M. Weaver, *Soil Sci. Soc. Amer. J.* **43**, 488 (1979).

44 W. L. Hargrove and G. W. Thomas, *American Society of Agronomy Special Publ.* **40**, 151 (1981).

45 B. R. James, C. J. Clark, and S. J. Riha, *Soil Sci. Soc. Amer. J.,* **47**, 893 (1983).

46 L. L. Hendrickson and R. B. Corey, *Soil Sci. Soc. Amer. J.* **47**, 467 (1983).

47 F-L. Greter, J. Buffle, and W. Haerdi, *J. Electroanal. Chem.* **101**, 211 (1979).

48 J. R. Tuschall, Jr., and P. L. Brezonik, *Anal. Chem.* **53**, 1986 (1981).

49 N. Cavallaro and M. B. McBride, *Soil Sci. Soc. Amer. J.* **44**, 729 (1980).

50 D. Behel Jr., D. W. Nelson, and L. E. Sommers, *J. Environ. Qual.* **12**, 181 (1983).

51 W. E. Emmerich, L. J. Lund, A. L. Page, and A. C. Chang, *J. Environ. Qual.* **11**, 182 (1982).

52 B. Lighthart, J. Baham, and V. V. Volk, *J. Environ. Qual.* **12**, 543 (1983).

53 R. T. Mahler, F. T. Bingham, G. Sposito, and A. L. Page, *J. Environ. Qual.* **9**, 359 (1980).

54 G. Sposito, F. T. Bingham, S. S. Yadav, and C. A. Inouye, *Soil Sci. Soc. Amer. J.* **46**, 51 (1982).

55 S. C. Cronan, W. A. Reiners, R. C. Reynold, Jr., and G. E. Lang, *Science* **200**, 309 (1978).

56 J. H. Reuter and E. M. Perdue, *Geochim. Cosmochim. Acta* **41**, 325 (1977).

57 T. S. C. Wang, S-Y. Cheng, and H. Tung, *Soil Sci.* **104**, 138 (1967).

58 D. M. Webley and R. B. Duff, *Nature* **194**, 364 (1962).

59 D. M. Webley, M. E. K. Henderson, and I. F. Taylor, *J. Soil Sci.* **14**, 102 (1963).

60 J. W. Rouatt and H. Katznelson, *J. Appl. Bacteriol.* **24**, 164 (1961).

61 H. A. Louw and D. M. Webley, *J. Appl. Bacteriol.* **22**, 216 (1959).

62 A. Moghimi, M. E. Tate, and J. M. Oades, *Soil Biol. Biochem.* **10**, 283 (1978).

63 A. Moghimi and M. E. Tate, *Soil Biol. Biochem.* **10**, 289 (1978).

64 H. A. Akers, *Soil Sci.* **135**, 156 (1983).

65 G. R. Cline, P. E. Powell, P. J. Szaniszlo, and C. P. P. Reid, *Soil Sci.* **136**, 145 (1983).

66 P. E. Powell, P. J. Szaniszlo, G. R. Cline, and C. P. P. Reid, *J. Plant Nutr.* **5**, 653 (1982).

67 P. J. Szaniszlo, P. E. Powell, C. P. P. Reid, and G. R. Cline, *Mycologia* **73**, 1158 (1981).

68 H. R. Geering and J. F. Hodgson, *Soil Sci. Soc. Amer. Proc.* **33**, 54 (1969).

69 T. A. Jackson, *Soil Sci.* **119**, 56 (1975).

70 M. Schnitzer, *Soil Sci. Soc. Amer. Proc.* **33**, 75 (1969).

71 D. S. Gamble, M. Schnitzer, and I. Hoffman, *Can. J. Chem.* **48**, 3197 (1970).

72 S. A. Boyd, L. E. Sommers, and D. W. Nelson, *Soil Sci. Soc. Amer. J.* **43**, 893 (1979).

73 A. Piccolo and F. J. Stevenson, *Geoderma* **27**, 195 (1981).

74 P. Vinkler, B. Lakatos, and J. Meisel, *Geoderma* **15**, 231 (1976).

75 B. Lakatos, T. Tibai, and J. Meisel, *Geoderma* **19**, 319 (1977).

76 M. B. McBride, *Soil Sci.* **126**, 200 (1978).

77 B. A. Goodman and M. V. Cheshire, *J. Soil Sci.* **27**, 337 (1976).

78 M. V. Cheshire, M. I. Berrow, B. A. Goodman, and C. M. Mundie, *Geochim. Cosmochim. Acta* **41**, 1131 (1977).

79 F. J. Stevenson, *Soil Sci.* **123**, 10 (1977).

80 H. van Dijk, *Geoderma* **5**, 53 (1971).

81 M. L. Crosser and H. E. Allen, *Soil Sci.* **123**, 176 (1977).

82 H. Zunino and J. P. Martin, *Soil Sci.* **123**, 188 (1977).

83 R. D. Guy and C. L. Chakrabarti, *Chem. Can.* **28**, 26 (1976).

84 T. A. O'Shea and K. H. Mancy, *Anal. Chem.* **48**, 1603 (1976).

85 W. T. Bresnahan, C. L. Grant, and J. H. Weber, *Anal. Chem.* **50**, 1675 (1978).

86 J. Buffle, F-L. Greter, and W. Haerdi, *Anal. Chem.* **49**, 216 (1977).

87 K. K. Turekian, *Geochim. Cosmochim. Acta* **41**, 1139 (1977).

88 B. A. Goodman and M. V. Cheshire, *Geochim. Cosmochim Acta* **39**, 1711 (1975).

89 B. A. Goodman and M. V. Cheshire, *Nature* **299**, 618 (1982).

90 B. Lakatos, L. Korecz, and J. Meisel, *Geoderma* **19**, 149 (1977).

91 R. K. Skogerboe and S. A. Wilson, *Anal. Chem.* **53**, 228 (1981).

92 M. Szilágyi, *Soil Sci.* **111**, 233 (1971).

93 S. M. Griffith, J. Silver, and M. Schnitzer, *Geoderma* **23**, 299 (1980).

94 M. Senesi, S. M. Griffith, and M. Schnitzer, *Geochim. Cosmochim. Acta* **41**, 969 (1977).

95 W. F. Spencer, R. Patrick, and H. W. Ford, *Soil Sci. Soc. Amer. Proc.* **27**, 134 (1963).

96 P. C. Stobbe and J. R. Wright, *Soil Sci. Soc. Amer. Proc.* **23**, 161 (1959).

97 C. Bloomfield, *J. Soil Sci.* **4**, 5 (1953).

98 C. Bloomfield, *J. Soc. Food Agr.* **8**, 389 (1957).

99 B. C. Deb, *J. Soil Sci.* **1**, 112 (1949).

100 F. De Coninck, *Geoderma* **24**, 101 (1980).

101 H. A. Anderson, M. L. Berrow, V. C. Farmer, A. Hepburn, J. D. Russell, and A. D. Walker, *J. Soil Sci.* **33**, 125 (1982).

102 V. C. Farmer, J. O. Skjemstad, and C. H. Thompson, *Nature (London)* **304**, 342 (1983).

103 A. J. Friedland, A. H. Johnson, T. G. Siccama, and D. L. Mader, *Soil Sci. Soc. Amer. J.* **48**, 422 (1984).

104 A. J. Friedland, A. H. Johnson, and T. G. Siccama, *Water, Air and Soil Pollution* **21**, 161 (1984).

105 A. H. Johnson, T. G. Siccama, and A. J. Friedland, *J. Environ. Qual.* **11**, 577 (1982).

106 A. L. Page, T. J. Tanji, and M. S. Joshi, *Hilgardia* **41**, 1 (1970).

INDEX

Acetylene:
    inhibition of nitrous oxide reduction,
        135
    nitrogen fixation measurements, 118
Acid rain, 217, 227–228, 303
Acid sulfate soils, 312
Actinomycetes:
    decomposition of organic matter by, 58
    nitrification, 163–164
    nitrogen fixation by, 120, 128
    numbers in soil, 7, 9
Adenosine triphosphate (ATP):
    in biochemical nitrogen transformations,
        114, 119, 162
    biomass estimates from, 9–10
    structure of, 243
Agricultural wastes, 78–95
Agrobacterium, denitrification by, 134
Air pollution:
    carbon dioxide, 78, 102
    odors, 81, 89
    ozone layer, 224–227
Alcaligines, denitrification by, 134
Algae, *see* Blue-green algae
Alnus, as source of fixed nitrogen, 117,
        129–130
Aluminum toxicity, 69, 228, 357–358
Anabaena, nitrogen fixation by, 120–121
Amines:
    formation of nitrosamines from, 97, 228–229
    volatilization from plants, 146
    in wet sediments, 37
Amino acids:
    ammonia formation from, 157
    as chelating agents, 348, 351–
        353
    reaction with quinones, 173–175,
        299–300
    in soils, 175–178, 188–189, 192, 298–
        299, 351
    sulfur-containing, 298–299

Amino sugars:
    ammonia formation from, 159–160
    in soils, 175–178, 188–189, 192
Ammonia:
    absorption by soil, 132
    enzymatic formation of, *see* Ammonifi-
        cation
    of primitive atmosphere, 109
    in rainwater, 106, 131
    reactions with organic matter, *see* Ammonia
        fixation
    in soil hydrolysates, 174–177, 188–189,
        192
    sorption by soil, 132
    volatilization of:
        from animal wastes, 87
        effect of pH on, 144
        field measurements of, 142–143
        optimum conditions for, 144–145
        from plants, 146
        from rice paddy fields, 143
        from sewage sludge, 89
        from soil, 142–145, 168
    see also Ammonium
Ammonia fixation:
    agricultural significance of, 202
    availability of fixed nitrogen, 205
    factors affecting, 203–204
    by humic substances, 202–205
    by lignin, 202
Ammonia volatilization, *see* Ammonia
Ammonification:
    of amino acids, 157
    of amino sugars, 159–160
    definition of, 106, 156–157
    of nucleic acids, 158–159
    of urea, 160
Ammonium:
    in animal manures, 85, 87
    in aquatic systems, 109, 157
    exchangeable, 143–144